Materials for Solar Cell Technologies I

Edited by

**Inamuddin[1], Tauseef Ahmad Rangreez[2], Mohd Imran Ahamed[3]
and Rajender Boddula[4]**

[1]Department of Applied Chemistry, Zakir Husain College of Engineering and Technology, Faculty of
Engineering and Technology, Aligarh Muslim University, Aligarh-202 002, India

[2]Department of Chemistry, National Institute of Technology, Srinagar, Jammu and Kashmir 190006,
India

[3]Department of Chemistry, Faculty of Science, Aligarh Muslim University, Aligarh 202 002, India

[4]CAS Key Laboratory of Nanosystem and Hierarchical Fabrication, National Center for Nanoscience and
Technology, Beijing 100190, PR China

Published by **Materials Research Forum LLC**
Millersville, PA 17551, USA

Published as part of the book series
Materials Research Foundations
Volume 88 (2021)
ISSN 2471-8890 (Print)
ISSN 2471-8904 (Online)

Print ISBN 978-1-64490-108-3
eBook ISBN 978-1-64490-109-0

Distributed worldwide by

Materials Research Forum LLC
105 Springdale Lane
Millersville, PA 17551
USA
https://www.mrforum.com

Manufactured in the United States of America
10 9 8 7 6 5 4 3 2 1

Table of Contents

Preface

Material Challenges in Next Generation Solar Cells
Aamir Ahmed, Sandeep Arya 1

Graphene Materials for Third Generation Solar Cell Technologies
Onoriode P. Avbenake 29

**Carbon Nanomaterials Beyond Graphene for Solar Cell and
Electrochemical Sensing**
Fethi Achi, Abdellah Henni, Sabah Menaa, Amira Bensana 62

**New Generation Transparent Conducting Electrode Materials for
Solar Cell Technologies**
Sandeep Pandey, Manoj Karakoti, Amit Kumar, Sunil Dhali, Aniket Rana,
Kuldeep K. Garg, Rajiv K Singh, Nanda Gopal Sahoo 86

Hollow Nanostructures for Application in Solar Cells
Peetam Mandal, Abha bhargava and Mitali Saha 129

Monocrystalline Silicon Solar Cells
M. Rizwan, Waheed S. Khan, S. Aleena 148

Low Band-Gap Materials for Solar Cells
Yadavalli Venkata Durga Nageswar, Vaidya Jayathirtha Rao 176

Absorber Materials for Solar Cells
Pallavi Jain, Palak Pant, Sapna Raghav, Dinesh Kumar 236

Keyword Index
About the Editors

Preface

Solar cells are the ideal choice of energy-harvesting photovoltaic technology that can potentially meet the increasing global energy demands. Solar cells composed of varied semiconductor materials are arising all over the world to convert solar radiation into electricity by utilization of sunlight with zero greenhouse gas emissions. Expansion of materials discoveries and production processes has been a crucial part in solar cell technologies that has attracted global attention and extensive research in the photovoltaic and energy conversion field. However, there are many challenges to overcome before photovoltaics can provide clean, abundant, and cheap energy on a larger scale. Thus, the challenge of developing efficient and stable materials for emerging solar cell technologies has stimulated the investigation of materials including conductive polymers, semiconductors, transition-metal compounds, alloys, and carbon materials, etc. This book gives a comprehensive and unified summary of current research progress and development trends of materials for solar cell technologies. Key topics include fabrication methods for solar energy materials and their utilization in various sorts of solar cell design and their merits and demerits, theoretical insights, and versatile applications in the present market. This book chronicles vital strides in solar cell device development research and benefits material scientists, professionals, faculty, postgraduate and graduate students operating in chemistry, physics, semiconductor technology, photochemistry, materials science, light science, and nanotechnology. Based on thematic topics, the book edition contains the following 8 chapters.

Chapter 1 is a discussion about the challenges in the development, energy efficiency, and commercialization of next-generation materials for the solar cell technology. The enhancement is necessary for these materials. And at the same time, research for new and more efficient materials must continue as well.

Chapter 2 elucidates the application of pure graphene in third-generation solar cells. The discourse was in light of the utilization of the unique crystal and electronic properties of graphene in dye-sensitized solar cells, perovskite solar cells, organic photovoltaic cells, and Schottky junction towards improved power conversion efficiency.

Chapter 3 discusses the techniques for the synthesis of solar cells based on graphene and carbon nanomaterials. The chapter reports recent functionalization strategies used to construct sensing platforms with graphene and carbon nanomaterial. The performance of electrochemical biosensors and the effect of carbon-based materials on the analytical parameters were also discussed.

Chapter 4 discusses the newly explored transparent conducting electrode (TCE) materials, an alternative to ITO based TCEs for solar cell applications. Additionally, the chapter describes the criteria for the selection of the TCE, cost-benefit analysis, and sustainability of these alternative materials for futuristic cost-effective solar cell technologies.

Chapter 5 summarized the developments of hollow nanostructured photoelectrodes required in solar cells. The rise in efficiency from the first generation hollow nanostructured solar cell up to date and the advantages of carbonaceous hollow nanostructures as electrode materials over the metallic hollow nanostructures has also been discussed in this chapter.

Chapter 6 discusses monocrystalline silicon solar cells (MSSC) from the perspective of the global energy crisis and the importance of renewable energy. The typical MSSC structure, recent advancements such as PERL and HIT solar cells, enrichment methods for power conversion efficiency, and optimization methods via control of optical losses are deliberated in detail.

Chapter 7 discusses the importance of BHJ organic solar cells and the requirement of low bandgap materials, taking recently reported syntheses of materials useful for organic solar cells. Information on fullerene acceptors, non-fullerene acceptors, polymer donors, and small organic molecule donors applied for organic solar cell device preparation is collated.

Chapter 8 discusses the advancement in solar cells by using up-conversion materials and different metal-based absorber materials. The chapter focusses on the graphene, sulfides, and metal nanoparticle-based absorber materials. Finally, at the end of the chapter, material setbacks for the 21st century are discussed.

Editors

Inamuddin, Tauseef Ahmad Rangreez, Mohd Imran Ahamed and Rajender Boddula

Materials for Solar Cell Technologies I   Materials Research Forum LLC
Materials Research Foundations **88** (2021) 1-28   https://doi.org/10.21741/9781644901090-1

Chapter 1

Material Challenges in Next Generation Solar Cells

Aamir Ahmed, Sandeep Arya

Department of Physics, University of Jammu, Jammu, Jammu and Kashmir-180006

snp09arya@gmail.com

Abstract

Solar cells have emerged as a substitute for fuels, generating energy which is both renewable and pollution-free at reasonable prices. On the commercial scale, the silicon-based solar cells are still being used despite their efficiency decreasing over time. With the advancement in technology, efforts are being made to develop new materials for solar cells with higher efficiency and stability. The development of materials such as multijunctions, ultrathin films, quantum dots, dye sensitized materials, and perovskites has opened a new dimension to the solar cell technology. These are often referred to as next-generation materials for solar cell technology. In this chapter, an effort has been made to address the various issues these new generation solar technologies face and why there is a need to search for various new materials in order to improve and make these technologies commercially viable.

Keywords

Solar Cells, Photovoltaics (PVs), Perovskites, Device, Quantum Dots, Multijunction, Power Conversion Efficiency (PCE), Nanocrystalline, Efficiency, Shockley-Queisser (SQ) Limit, Quantum Dot Sensitized Solar Cell (QDSSC)

Contents

Material Challenges in Next Generation Solar Cells..1

1.   Introduction..2

2.   Factors influencing the performance of solar cells...............................4

     2.1   Cost of production 4

     2.2   Efficiency..4

Materials for Solar Cell Technologies I   Materials Research Forum LLC
Materials Research Foundations **88** (2021) 1-28   https://doi.org/10.21741/9781644901090-1

3.   **Materials for future generation solar cells** 6

    3.1   Multijunction solar cells 6

    3.2   Ultrathin films 9

    3.3   Dye-sensitized solar cells (DSCs) 10

    3.4   Quantum dot solar cells 11

    3.4.1   Colloidal quantum dots (CQDs) 11

    3.4.2   Carbon quantum dot (CQD) 12

    3.4.3   Quantum dot sensitized solar cells (QDSSC) 13

    3.5   Organic solar cells 14

    3.6   Perovskite solar cells (PSCs) 16

Conclusion 17

References 18

1.   Introduction

In 1839, Becquerel discovered the photovoltaic effect and since then, the research in this field has achieved many new heights [1]. A large-scale application of material research was first mentioned when the photosensitivity of selenium (first semiconductor) was discovered and wafers were fabricated with the conversion efficiency of about 1% [2]. But it was the year 1954, when the first photocell made from silicon with an efficiency of about 6% was fabricated [3] and this event triggered the development of photovoltaic technology.

In the last 60 years, increasing pressure from global warming and the depletion of oil deposits have been major alarming issues in the world. In the meantime, solar cells have emerged as a substitute for fuels, generating energy that is both renewable and pollution-free at reasonable prices. In 1990, the contribution of the photovoltaic systems to the net production of electricity in the Organization for Economic Co-operation and Development was below 0.01% (almost negligible) which has increased up to 1.7% in recent times [4]. As per the statistics of the International Energy Agency, the photovoltaics (PVs) will surpass the global installation capacity of about 300GW in the year 2017. In the US alone, the size of power generated by solar photovoltaics installed is around 60GW which is predicted to double in the coming 5 years [5]. This rapid growth has been possible due to the dedicated fundamental research converting the laboratory results into commercial systems with the highest efficiency and performances. Before addressing the challenges these photovoltaic systems face, it is important to know about the types of these solar cells in brief for a better understanding of the subject.

The main types of solar cells are:

1. *Silicon solar cells*

    This was the first area that emerged in the PVs [6], covering about 80% of the PVs installed in the world [7,8] and having a market representation of 90% [9]. Both mono and polycrystalline forms of silicon are used [10]. These are also known as "First Generation Solar Cells". An image of the silicon solar panel is given in the Figure 1.

2. *Thin film solar cells*

    These are second-generation solar cells developed as an alternative to amorphous silicon cells and were based on copper indium gallium selenide (CIGS) and CdTe. Thin film solar cells have better mechanical properties, flexibility and low cost of production, but possess lower efficiency. With the development of thin film solar cells, a new field of electrochemistry came into being.

3. *3rd-generation solar cells*

    The 3rd generation solar cells include perovskite solar cells, dye-sensitized solar cells, organic solar cells and the new concepts that are developing for the better attainment of efficiency in solar cell technology. These are also debated as 'emerging concepts' because of their low market saturation.

Figure 1   Silicon solar panels from Spain Paneles solares y repetidor al fondo
(Creative Commons Attribution (CC BY 2.0)).

In this chapter, the new generation solar cells or precisely the third generation solar cells having an extensive application from integration to the space applications are discussed. It is very important to maintain the balance between the investment required to produce a solar cell and the energy produced by it, which is directly dependent upon the conversion efficiency and lifetime of a solar cell. This can be made easy to understand with an example of a silicon solar cell whose efficiency starts to decrease with an increase in temperature or decrease in intensity of illumination when compared to other technologies in its competition. The research going on in this field has led to many remarkable results and advances that have brought down the cost of electricity generated by solar cells and also made them economical in various parts of the world. First, the factors that greatly influence the solar cell technology are discussed and later on the new materials that are being developed in this technology along with the challenges they face are elaborated.

2.   Factors influencing the performance of solar cells

2.1   Cost of production

The drive to cut production costs is one reason why research in the field of solar cell technology has long focused on alternative materials. Over the last few decades, silicon PV cells have reduced the production expenses and emerged as the largest used product in solar cell technology. In comparison to silicon-based solar cells, the fabrication of thin films is less expensive and silver connections are also not required. Moreover, the energy payback time of thin films is comparable to silicon base solar cells [11]. With more advent in technology, it becomes less expensive to generate energy from solar cells. In 1975, the production cost for one watt of peak power was 100 USD, 100 MJ and it is estimated to reach figures of 0.5 USD and 15 MJ in the coming years. Similar is the case for perovskites and other materials which are discussed separately in the material section.

2.2   Efficiency

Conversion efficiency is the most important indicator which is used to compare and assess the solar cell technology. It is actually the ratio of solar energy input to the electrical energy output of a solar cell. The efficiency combines various parameters of the solar cell technology such as fill factor; short-circuit current and open-circuit voltage. All these parameters are dependent upon the type of material used and the defects arising during the manufacturing process. In addition to these factors, further defects are induced while the solar cells are mounted into a solar module. Moreover, a famous limit for efficiency known as Shockley-Queisser (SQ) limit [12] has been set and any loss i.e. conversion, electrical or optical will result in an efficiency lower than the SQ limit. A comparison between the best module and the best cell of each major technology has been shown in Figure 2 [13]. From the figure 2, the difference between the best cell and the best cell module indicates how much could be gained by improving the laboratory prototypes into integrated systems and the discrepancy between the best cell efficiency and the upper SQ limit gives a picture of the room for improvement. Over the last five years, most of the mature technologies have shown little evolution despite approaching the SQ limit [13,14].

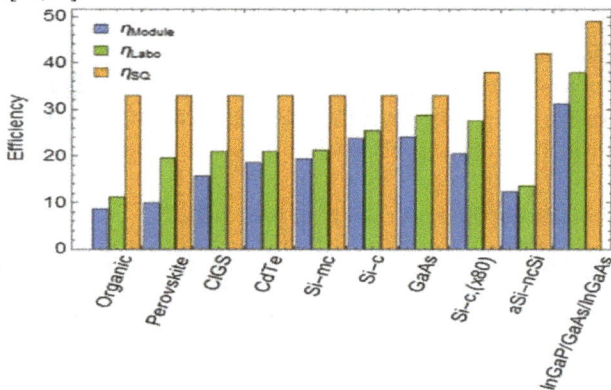

Figure 2   Best module (blue), best cell (green), and Shockley–Queisser (SQ) efficiency (yellow)of major technologies. Si-mc,Si-c, Si-c(x80), and aSi-ncSi stand for multi crystalline, crystalline, crystalline under 80 suns illumination and amorphous/ nanocrystalline silicon, respectively. Data from [13] (Creative Commons Attribution (CC BY 4.0)).

It must be noted that the efficiency does not provide any information whether the material used for making a solar cell can be utilized for surfaces with a larger area. In order to expand an efficient material to larger surfaces, the homogeneity must be controlled during the growth process. And besides efficiency, another method can be used for the comparison of output power supplied by the cell units i.e. the product of the efficiency of cell unit with its surface area. In Figure 3 [13], the graph between mature technologies up-scaled to large surface and the emergent technologies limited to smaller surfaces shows that the up-scaled mature technologies have been able to deliver more than 100 W.

These factors will play a key role in deciding the type of material that will be used for making better solar cells and these terms will be repeated number of times as we move to the main part of this chapter.

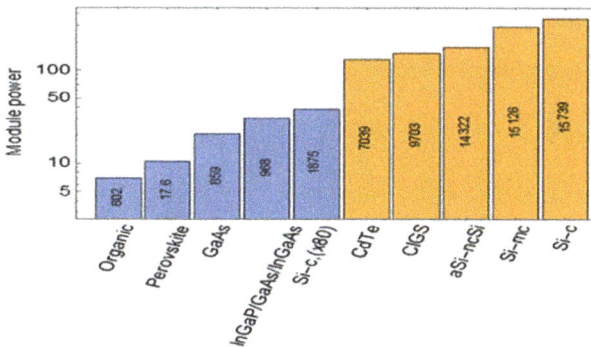

Figure 3   Output power of the best module for each technology. For each technology, the surface of the best module is indicated in cm². Mature technologies (1st and 2nd generations) are depicted in yellow, 3rd generation in blue. Data from [13] (Creative Commons Attribution (CC BY 4.0)).

3.   Materials for future generation solar cells

To counter the challenges in the energy conversion by solar cells, the use of next-generation materials is being predicted by the researchers. The material that is not in use at present or the studies are going on for its use in manufacturing solar cells may be included in the next-generation solar cell material. The term also denotes the application of dyes and organic materials for the fabrication of solar cells as most of the solar cells are made up of inorganic semiconductors (mostly silicon) which are not as efficient as the organic semiconductors. It also includes the use of nanomaterials for making solar cells using various manufacturing techniques.

3.1   Multijunction solar cells

Solar cells with an efficiency of about 40%, far in excess to the single junction cell; the multijunction solar cells are the most efficient solar cells developed so far. Multijunction solar cells are theoretically as well as practically much more efficient than the single junction cells [12,15]. An image of a multijunction solar cell is shown in Figure 4 [16]. The advantage of multijunction cells in terms of efficiency is demonstrated in Table 1 [17], in which the efficiencies of various cells are compared.

Materials Research Forum LLC
https://doi.org/10.21741/9781644901090-1

Table 1   *Efficiencies for leading single and multijunction solar cell technologies [17] Reprinted with permission (Copyright (2010) Elsevier).*

Devices	No. of junctions	Efficiency (%)	Suns
Si	1	25.0 ± 0.5	1
GaAs	1	26.4 ± 0.8	1
CIGS	1	19.4 ± 0.6	1
CdTe	1	16.7 ± 0.5	1
GaInP/GaAs/GaInAs	3	35.8 ± 1.5	1
Si	1	27.6 ± 1.0	92
GaAs	1	29.1 ± 1.3	117
GaInP/GaInAs/Ge	3	41.6 ± 2.5	364

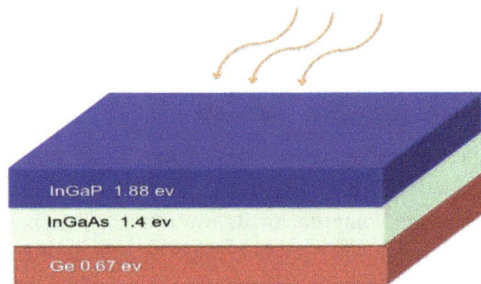

Figure 4   *Image of a multijunction solar cell [16] copyright @ Stanford University (2012).*

Multijunctions possess more applications at the commercial scale than their application at the laboratory. The "concentrator photovoltaics" make use of lenses and mirrors for collecting and focusing light onto a smaller multijunction solar cell, reducing cost and achieving high efficiency [18]. Increasing concentration of light results in an increased efficiency, which is due to an increase in the population of minority charge carriers thus increasing the voltage produced by the solar cell. In practice, the efficiency of most of the solar cells is limited due to a voltage drop across the internal series resistance of a cell. And because of this voltage drop, all cells do not find an application in concentration operation. But multijunction solar cells have exhibited an efficiency of about 40% with concentrator operation [19]. Multijunction solar cells are also preferred for space applications due to their higher efficiency. They are used in solar panels of telecommunication satellites; for example, in Mars rovers. At present, the multijunction solar cell is the only concept that has been able to overcome the Shockley-Queisser limit and find its applications in the industrial world.

The idea of multijunction solar cells is established around the concept that a solar cell attains its optimum efficiency when a photon of light having wavelength equivalent to the band gap is incident upon the cell. In these cells, various wavelengths of light are absorbed due to the presence of different layers. Each layer absorbs a wavelength corresponding to its bandgap, thus making these cells more efficient in converting solar energy into electrical energy. The material with the highest bandgap is always placed on the top of layers. In Figure 5 [13], the solar spectrum is divided into various ranges of wavelength and each wavelength is converted with a cell of the suitable bandgap.

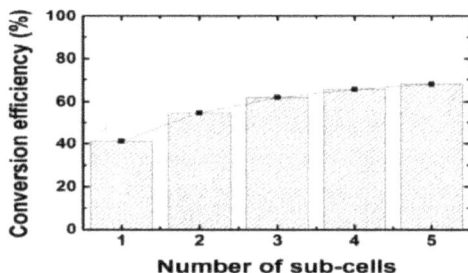

Figure 5   Output power provided by multijunction cells in the radiative limit, as a function of the number of sub cells, under aM1.5d for a 1000 sun concentration. data from [13] (Creative Commons Attribution (CC BY 4.0)).

There are two configurations that are mostly used in multijunction solar cells i.e. 2 and 4-wire configurations. The cells are connected in series in 2-wire configuration and maximum performance is attained when the current in various subcells is equal. The series connection between different junctions is achieved by using highly re-combinative layers such as tunnel junctions. While in the 4-wire configuration, all cells are connected individually; this reduces the compulsion of matching current. As the current is extracted individually form each cell and in order to allow lateral current between layers, contact grid and low resistivity layers (transparent) are required in between layers. Also, there arises a complexity due to the similarity between the materials used in the different layers of a device in both configurations. The materials belonging to III-IV groups of the periodic table are mostly used in multijunction cells as they provide a large range of spectrum of available band gaps. The material combination which best suits to fulfill the needs of a multijunction cell is Ge/GaAs/InGaP, reaching the efficiency of about 41.6% under 364 suns [20]. But at present, still no combination of materials is available that can be regarded as perfect, but due to various emerging techniques such as wafer bonding [21-23], metamorphic growth [19], inverted stacks [24-26], dilute nitride [27] and multi-quantum wells [28-30], such devices can be made available at the commercial scale. Despite having the potential to be much more efficient than the traditional solar cells, high production costs and continuous development and research means that these are not currently available commercially. The production of multijunction solar cells is a difficult and complicated process in which expensive materials are used. Since the price of solar cells has fallen over the years, a similar pattern is expected to be followed in case of multi-junction cells for the future coming years and there is a hope of achieving 50% efficiency with these cells.

3.2   Ultrathin films

The thickness of a solar cell is one of the limiting factors in conventional solar cells, as the absorptivity of the cell is associated with its thickness. And the thickness of a solar cell needs to be sufficient over a range of bandwidth so that the most of the photons incident upon the cell are absorbed. Thus, there is a need for the development of ultrathin solar cells. Because a thin solar cell can help in the attainment of higher efficiency and also reduces the cost of production for expensive solar cells by reducing their thickness. An image of an ultrathin film is shown in Figure 6 [31]. Moreover, the optical intensity of cell is increased due to absorption over small thickness which enhances the open-circuit voltage (V_{oc}) of a solar cell. The volume recombination of carriers generated in the cells is also limited by reducing the thickness of the cell.

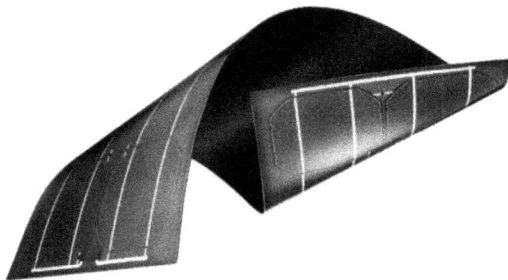

Figure 6   *Image of an ultrathin film [31] (Creative Commons Attribution (CC BY 2.0)).*

Ultrathin films of copper indium gallium selenide (CIGS), amorphous thin films of silicon and CdTe are commercially used in solar cell technology. The thickness of ultrathin films varies from a few nanometers to micrometers. In the laboratory, an efficiency of about 21% has been attained for CdTe and CIGS solar cells. Recently in 2019, researchers at the Centre de Nanosciences et de nanotechnologies (C2N) in collaboration with German Fraunhofer ISE have developed an ultrathin film of thickness 205 nm made of GaAs with an efficiency of 20%. Previously made ultrathin films with 20% efficiency were of thickness in micrometers, but due to the development of this nanometer thick ultrathin film, the use of rarely available materials can be reduced.

The development of making ultrathin solar cells very absorbing and cost-effective is one of the goals of III-IV solar cell technology. Their development will lead to the better usage of material, better collection of charge carriers, high open-circuit voltage which will ultimately lead to the attainment of increased efficiency. But the usage of these ultrathin films has never reached more than 20% on the market scale that is all due to the dependence of their efficiency on the semiconductor chosen and the technology used for their growth. Moreover, the production of low charge carriers per incident photon is also an issue. Solar Frontier, the world's largest CIS solar energy provider has developed an ultrathin film of area 0.5 cm^2 with an efficiency of about 22.3%.

3.3   Dye-sensitized solar cells (DSCs)

Dye-sensitized solar cells have been investigated extensively due to their high theoretical efficiency, easy fabrication, low cost, and the concept seemed to be the most promising among the photovoltaic devices. In these solar cells, a molecular dye is used for the efficient absorption of the solar spectrum. The molecules of the dye are bound to TiO_2 and then either by a conducting polymer or an electrolyte, a porous network is infiltrated into it. An exciton (generated in the dye) is detached due to the quick dose of an electron into the oxide. In the last 20 years, major improvements in the performance of DSCs have been made due to continuous research efforts [32]. Record efficiency of 13% has been achieved in the DSC devices using liquid electrolytes and zinc-porphyrin based co-sensitized systems [33]. And by using an organic hole transporting material (HTM) such as the most widely used spiro- OMeTAD(2,2',7,7'-tetrakis-(N, N-di-p-methoxyphenylamine)-9,9'-spirobiflourine) in place of liquid electrolyte, many advantages have been gained. The sensitization of metal oxide (semiconductor) with organic or metallo-organic dyes is the main concept behind the use of these organic-inorganic cells. In the DSC technology, there are many benefits in the fabrication process such as the availability of cheap, nontoxic materials in abundance which result in the high power conversion efficiencies(PCEs) of about 14% and 7.5% for liquid and solid devices [34,35]. Over the past few years, growth in the usage of dyes free from ruthenium (Ru) has led to the achievement of about 13% PCEs for devices with a smaller area, thus making these dyes a potential competition for silicon devices [33,36]. In order to further increase the PCEs, the co-sensitization and fine-tuning of electronic properties of semiconducting oxide with its complementary dye seem to be a very efficient method [34,36,37]. Also the development of D-π-[M]-π-A including ruthenium-di-acetylide has led to the introduction of a new group of dyes with optical properties suitable for DSCs operation and displaying a color range of red, blue, violet and green. The electronic properties and DFT studies of green or blue color displaying dyes have revealed that these may prove to be very promising in the near future of DSCs technology [38]. The use of liquid electrolytes has although led to the development of efficient DSCs, but the issue of their commercialization has forced the researchers to focus their work on solid-state DSCs. In the first step, indoline donor-based dyes having fluorine as a substituent group were used for making DSCs [39]. With an introduction of alkylene chains to the fluorine substituent group, the performance of DSCs was further improved, as these chains suppress the recombination of electrons between electrolyte and conduction band of TiO_2. Despite their low efficiency as compared to Ru-based dyes, these dyes can be used for DSCs as they have low toxicity and are highly stable. The low efficiency of DSC's is due to low electron mobility, defects in material and poor contact to electrodes. In addition to this, a high concentration of defects and reliability are other main problems. So there is a need for inventing new DSC materials for this PV technology to pave the way into the future.

3.4   Quantum dot solar cells

Quantum dots are the zero-dimensional particles having a size range of less than 10 nanometers. Due to their small sizes, the quantum confinement effect takes place and materials show different optoelectronic properties than the bulk of the material. And various solar cells have been proposed due to these unique optoelectronic properties of quantum dots. The band structure of quantum dot depends upon its size as the smaller the size, the larger is the bandgap. So, varying the size of quantum dot according to the bandgap required for a solar cell can be very helpful in achieving higher efficiency. Quantum Dot solar cells can be of great importance in the field of PV technology, where solar cells are quantum dots based multi-junctions. Some of the quantum dots that have been proposed and are mostly discussed are explained briefly here.

3.4.1 Colloidal quantum dots (CQDs)

Colloidal quantum dots (CQDs) are the materials when dispersed in solution, can be used for making LEDs [40], solar cells [41], biosensors and biomarkers [42]. CQDs are also well-suited with methods such as dip-coating and spin-coating [43], spray-coating, and microstamping [44]. Much of interest has been laid on the use of Pb based heterojunction quantum dots, which are fabricated by the deposition of PbS layer (quantum dot layer) on top of the ZnO layer (in Figure 7) [45]. The solar cell fabricated on these materials has shown a conversion efficiency of about 11% [46,47]. For synthesizing quantum dots, electrodeposition has emerged as one of the fast-growing techniques. In this technique, an electrical potential is applied across the counter electrode and conducting area of substrate causing redox reaction in the liquid which results in the formation of quantum dots on the surface. And by changing the temperature of the bath, the potential applied across the bath and pH value of the electrolyte, the bandgap and lattice constant of the quantum dots formed can be varied. The possibility of varying the bandgap of the semiconductors using the electrodeposition method has made this technique very useful in solar cell technology [48,49]. But this process is only applicable to the materials with electrical conduction.

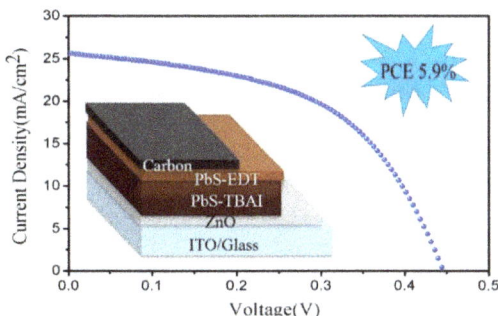

Figure 7   Image of CQD with PbS layer on top of ZnO [45]. Reprinted with permission (Copyright (2017) Elsevier).

3.4.2 Carbon quantum dot (CQD)

Carbon quantum dot (CQD) is a type of carbon nanomaterial with a size of less than 10 nm consisting of graphene and carbon dots, showing properties such as good solubility and strong luminescence. Their property of luminescence is of main importance in the field of solar cell technology. These quantum dots can be prepared by using top-down and bottom-up approach. One of the methods used for the preparation of CQD is the pyrolysis of an organic precursor. The diagrammatic representation of the process is shown in Figure 8 [50]. An approach to increase the optical properties of photoluminescent materials is the doping and the doping of N, S, P has brought significant changes in the optical properties of CQD. The variable fluorescence, good exception, and donation of an electron makes CQD very efficient in photoluminescence quenching [51].

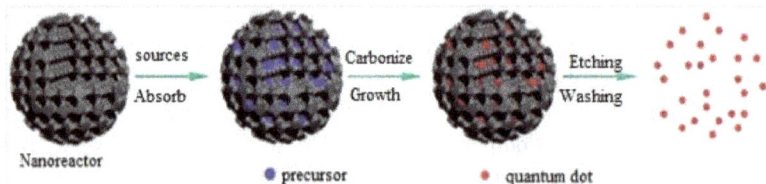

Figure 8   Schematic illustration of the preparation of CQDs via confined pyrolysis of an organic precursor in nanoreactors [50] (Creative Commons Attribution (CC BY 3.0)

3.4.3 Quantum dot sensitized solar cells (QDSSC)

In quantum dot sensitized solar cells, quantum dots are substituted for dye molecules. The quantum dots have the advantage of tunable optical properties with size and better formation of heterojunction with conductors. The QDSSCs are based on the working principle of sensitizing a semiconductor of wide bandgap with a semiconductor having bandgap in near or visible IR region [52]. TiO_2 is mostly used as a wide bandgap semiconductor, but ZnO [53] and SnO_2 [54-56] have also been reported. Whereas quantum dots of semiconductors such as CdS, CdSe, PbS, and InAs are used as narrow bandgap semiconductors due to their flexible electrical and optical characteristics [57-60]. In a quantum dot sensitized PV cell, molecules of a dye are absorbed onto the surface of TiO_2 particles of size 10-3-nm which itself has been sintered into a nanocrystalline TiO_2 layer. The photo-excitation of dye molecules produces electrons that are transmitted from an excited state of dye into the conduction band of TiO_2 resulting in charge separation and production of photovoltaic effect. An image of a solar cell based on QD is shown in Figure 9 [61].

From the characterization, it can be derived that the colloidal quantum dots of PbS are capable of producing photovoltaic effects for both single-junction and multijunction solar cells. In addition to QDs based on Pb and Cd, there are many options for the fabrication of CQDs. The quantum dot solar cells are not commercially available yet, but many companies are beginning to show interest in their manufacture. And various companies have recognized that quantum dot PVs are the future of solar cell technology [62]. But the main problem with the QD photovoltaics is the low yield. Moreover, the quantum dots of heavy metals such as lead or cadmium chalcogenides, PbSe and CdSe are poisonous when exposed to and must be preserved in a stable polymer shell [63]. And to make environmentally safe QD solar cells, the application of $CuInS_2$ and $AgSiS_2$ is a good option [64]. Many material challenges are ahead of the researchers to produce quantum dot-based materials with high yield and no issues of toxicity for applications in solar cell technology.

Figure 9   Image of a dye sensitized quantum dot solar cell [61] (Creative Commons Attribution (CC BY 2.0)).

3.5   Organic solar cells

Part of the third generation photovoltaic systems, organic solar cells are made of organic or polymer materials (organic molecules or organic polymers). The development of organic solar cells is an inexpensive and emerging PV technology. The success in recent years has led to improvements and achievement of efficiency of about 4 to 5% for commercial systems and 6 to 8% for laboratory systems (OrgaPVnet, 2209). And there are reports of PCE of about 13% for single junction solar cells [65]. Despite their low efficiency, organic solar cells find many advantages than other PV technologies which make them interesting for particular applications. They can be used for nomadic applications because of their flexibility and lightweight, and the probability of tuning their shape, color and transparency opens up a path to many new applications [66]. Moreover, these have a low carbon print [67], low energy payback time [68] and are low intensity efficient [69]. Fullerenes and their derivatives are the most frequently used materials as an acceptor in organic photovoltaic devices. The derivative of fullerene commonly known as $PC_{60}BM$ i.e. [6,6]-phenyl-C61-butyric acid methyl ester has been the key material used as an acceptor for OPV technology [70]. And in order to increase open-circuit voltage and the absorption in visible region, derivatives like $PC_{70}BM$ i.e. [6,6]-phenyl-C71-butyric acid methyl ester [71,72] and Indene-C60 bisadduct ($IC_{60}BA$)

[73] are used and efficiency of about 11.5% has been recorded with such acceptors [74]. But these materials also come with shortcomings such as the poor absorption in the visible region [70], loss of energy [75] and instability [76]. But due to recent advances in the research, alternative materials to be used as acceptors in OPV devices have been studied. To overcome the limitations of PCBM, the research group of Holliday et al. [77,78] developed IDTBR, reducing the energy losses and achieving the energy efficiency of about 10% [79]. ITIC, an n-type semiconductor in combination with PBDB-T showed an efficiency of about 11% [80]. In both IDTBR and PBDB-T, the improvement in efficiency is due to an additional generation of charge carriers. In order to replace PCBM, perylene dimide and naphthalene dimide (N2200) have been developed and between these two, N2200 has been used to achieve an efficiency of 8.27%. Recent reports published have also shown a record conversion efficiency of 13% using non-fullerene acceptors [65]. Image of an organic solar cell is shown in Figure 10 [81].

Figure 10   Image of an organic solar cell [81] (Creative Commons Attribution (CC BY 3.0)).

The problem with OPVs is that their performance does not remain stable over time. With the aging of these solar cells, efficiency starts to decrease rapidly (also known as burn-in) which may result into the failure of these solar cells [82]. The nature of organic semiconductor used and the method used to develop these solar cells also incur multiple losses in these devices. In 2006, to develop OPVs with long life, a new method of inverted architecture was developed which has proved to be an important step in the development of this technology [83-85]. The development of non-fullerenes has proved to be a solution to many degradation problems of fullerene OPVs [86,87]. But the search for other organic derivatives with better achievement of efficiency is still a concern for researchers and continuous efforts are being made to search for new materials.

3.6   Perovskite solar cells (PSCs)

Recently a great deal of attention was gained by Perovskite solar cells (PSCs) due to their low production cost, diversity of structure and excellent optoelectronic properties (high diffusion length of carriers and strong absorption) of perovskite materials such as methyl ammonium lead iodide (MAPbI$_3$) [88-92]. In a short span of time, the PCE of PSCs has shown a great improvement reaching a value of about 20% [93] and PSCs were found to be more radiant than amorphous silicon and CdTe solar cells. One of the methods for the production of PSCs is the use of evaporated perovskite layer, but PSCs with a solution-processed perovskite absorber has been the most efficient configuration. The first use of perovskite material (based on lead halide) was as an absorber in dye-sensitized solar cells (DSCs) and since then many methods have been applied to increase their optoelectronic properties [89]. The use of solid hole transport material spiro-OMeTAD was one of the approaches that led to the attainment of high efficiency in PSCs [94]. Also, the introduction of a step involving anti-solvent in the spin coating has led to a smooth perovskite layer with larger particle size [95], which eventually led to the better performance [96,97] and efficiency over 18%. From this, it became quite clear that the grain size of perovskite has an important impact over the performance of PSCs. And it was also recently expressed by two independently working groups that perovskite solar cells follow the Rau's relation, which implies that the photo voltage can be predicted by measuring external quantum efficiency and photocurrent generation efficiency. A measure problem faced in the PSCs is the hysteresis obtained in the current density and voltage curve (J-V) [98] as shown in Figure 11 [13]. The hysteresis obtained is assumed to be from the ferroelectric polarization, accumulation of charge at the interface of cell and migration of ions in the perovskite [99-102]. A successful approach of using a reversed structure with PTAA and NiOx as hole and electron transport layer has led to reduction of hysteresis and achievement of 18% efficiency [103]. This also suggests that by experimenting with the structure of perovskite, the hysteresis can be reduced.

Another issue with PSCs is the degradation of the perovskite layer with moisture [104]. The degradation of perovskite leads to the formation of PbI$_2$ (in the case of MAPbI$_3$), which is toxic as well as water soluble. The inclusion of this compound into the water sources and fields can lead to the contamination of the environment and cause an eco-toxicological problem. The thermal stability of PSCs is also not very good as during the solar cell operation, the temperature can increase strongly which will eventually lead to the degradation of its crystal structure [105] and any changes in the crystal structure may lead to hysteresis. To address the issue of stability in PSCs, the introduction of Cs, Br, and formamidinium in MAPbI₃ may lead to the achievement of highly stable PSCs [96]. Also, the replacement of OMeTAD because of its low stability, can lead to the attainment of better stability in PSCs [106]. TiO_2 because of its photocatalytic activity is also affecting the stability of PSCs [104] and materials like SnO_2 and $BaSnO_3$ are likely the better replacements having already achieved an efficiency of above 20% [107,108]. Introduction of Al_2O_3 into the structure of PSCs in order to prevent unwanted chemical reactions has been used by many research groups to enhance the stability of PSCs. Some of these PSCs have maintained an efficiency of 90% despite being exposed to air for 24 days [109,110].

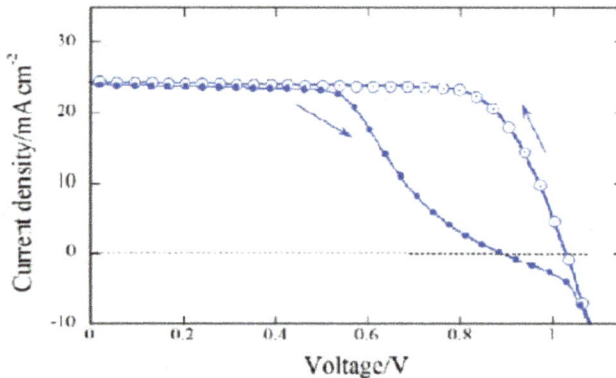

Figure 11   J-V curve of perovskite solar cells [13] (Creative Commons Attribution (CC BY 4.0)).

Low cost and high efficiency are the main reason why PSCs are studied extensively. These are considered to be a credible alternative to the silicon solar cell technology and researches have revealed their potential for being a replacement. But the presence of lead leading to eco-toxicity, hysteresis in J-V curves and stability issues have thrown some challenges in their commercialization. But there is hope in the near future for the commercialization of these perovskite layer based solar cells [92,96,111,112].

Conclusion

Optoelectronics have seen much advancement due to constant progress in the manufacturing methods of bulk and thin film solar cells. These advances have eventually led to a reduction in cost, improvement in reliability and making solar cells emerge as a potential option for power generation. Solar cells provide an alternative to the energy crisis by providing a cost and energy-efficient source of energy. The search for a material with higher efficiency is still going on and a great deal of work has already been done on the theoretical efficiency of solar cells. Now there is a need to search for materials and techniques to develop such efficient solar cells. Although so many materials have been discussed above and all of them are facing some issues, some being efficient but costly while others being cheaper but less efficient. On the commercial scale, still, the silicon-based solar cells are being used despite their efficiency getting decreased with time. In order to harness the solar energy in a superior way, plenty of research needs to be done regarding the examining of more efficient material with better efficiency and better life cycle at much cheaper costs. It is obvious from the theoretical and experimental results that the current techniques are not enough to build such solar cells. Hence, there is a need to develop new techniques which can provide better control on the dimension. But the ownership cost of new equipment required in these new techniques must be lower than the ones that are currently being used in the fabrication of solar cells.

References

[1]   E. Bequerel, Recherches sur les effets de la radiation chimique de la lumière solaire, au moyen des courants électriques, CR Acad. Sci. 9 (1839) 145-149

[2]   W.G. Adams, R.E. Day, The action of light on selenium, Philos. Trans. R. Soc. Lond. 167 (1877) 313-349. https://doi.org/10.1098/rstl.1877.0009

[3]   D.M. Chapin, C.S. Fuller, G.L. Pearson, A new silicon *p-n* junction photocell for converting solar radiation into electrical power, J. Appl. Phys. 25 (1954) 676–677. https://doi.org/10.1063/1.1721711

[4]   D.J. Feldman, R.M. Margolis, Q2/Q3 2018 Solar industry update. National renewable energy lab. (NREL), Golden, CO (United States), (2018). https://doi.org/10.2172/1485577

[5]   International Energy Agency (IEA). Electricity Information 2016. Paris, France: IEA. Available at: https://www.eia.gov/outlooks/ieo/pdf/0484(2016).pdf

[6]   V. Petrova-Koch, R. Hezel, A. Goetzberger, High efficient low-cost photovoltaics: recent developments, Berlin Heidelberg: Springer 2008. https://doi.org/10.1007/978-3-030-22864-4

[7]   U. Gangopadhyay, S. Jana, S. Das, State of art of solar photovoltaic technology. InConference papers in science Hindawi 2013 (2013). https://doi.org/10.1155/2013/764132.

[8]   Energy SP. Technology roadmap. 2014 [cited 2017 May11]. Available from: http://www.bpva.org.uk/media/215436/technologyroadmapsolarphotovoltaicenergy_2 014edition.pdf

[9]   M. Schmela., Global market outlook for solar power/2016-2020

[10]   S. Battersby, News feature: The solar cell of the future, Proc. Natl. Acad. Sci. 116 (2019) 7-10. https://doi.org/10.1073/pnas.1820406116

[11]   K. P. Bhandari, J.M. Collier, R.J. Ellingson, D.S. Apul, Energy payback time (EPBT) and energy return on energy invested (EROI) of solar photovoltaic systems: A systematic review and meta-analysis, Renew. Sustain. Energy Rev. 47 (2015) 133-141. https://doi.org/10.1016/j.rser.2015.02.057

[12]   W. Shockley, H.J. Queisser, Detailed balance limit of efficiency of p-n junction solar cells, J. Appl. Phys. 32 (1961) 510–519. https://doi.org/10.1063/1.1736034

[13]   S. Almosni, A. Delamarre, Z. Jehl, D. Suchet, L. Cojocaru, M. Giteau, B. Behaghel, A. Julian, C. Ibrahim, L. Tatry, H. Wang, Material challenges for solar cells in the twenty-first century: directions in emerging technologies, Sci. Technol. Adv. Mater. 19 (2018) 336-69. https://doi.org/10.1080/14686996.2018.1433439

[14]   Photovoltaic Research | NREL [Internet]. [cited 2017 May 12]. Available from: https://www.nrel.gov/pv/

[15]   C.H. Henry, Limiting efficiencies of ideal single and multiple energy gap terrestrial solar cells, J. Appl. Phys. 51 (2008), https://doi.org/doi.org/10.1063/1.328272

[16]   Rahim Esfandyarpour. multijunction solar cells, December 12, 2012. Available from: http://large.standford.edu/courses/2012/ph240/esfandyarpour-r2/

[17]   D.J. Friedman, Progress and challenges for next-generation high-efficiency multijunction solar cells, Curr. Opin. Solid State Mater. Sci. 14 (2010) 131-138. https://doi.org/10.1016/j.cossms.2010.07.001

[18]   R.M. Swanson, The promise of concentrators, Prog. Photovolt. Res. Appl. 8 (2000) 93–111

[19]   W. Guter, J. Schone, S.P. Phillips, M. Steiner, F. Siefer, A. Wekkeli, E. Welser, E. Oliva, A.W. Bett, F. Dimroth, Current matched triple junction solar cell reaching 41.1% conversion efficiency under concentrated sunlight, Appl. Phys. Lett. 94 (2009) 223504. https://doi.org/10.1063/1.3148341

[20]   R.R. King, A. Boca, W. Hong, X.Q. Liu, D. Bhusari, D. Larrabee, K.M. Edmondson, D.C. Law, C.M. Fetzer, S. Mesropian, N.H. Karam, Band-gap-engineered architectures for high-efficiency multi-junction concentrator solar cells, 24th EPSEC. (2009) 55–61. https://doi.org/10.4229/24thEUPVSEC2009-1AO.5.2

[21]   P.T. Chiu, D.C. Law, R.L. Woo, S.B. Singer, D. Bhusari, W.D. Hong, A. Zakaria, J. Boisvert, S. Mesropian, R. King, N.H. Karam, Direct semiconductor bonded 5 J cell for space and terrestrial applications, IEEE J. Photovolt. 4 (2014) 493–497. https://doi.org/10.1109/JPHOTOV.2013.2279336

[22]   R. Cariou, J. Benick, P. Beutel, N. Razek, C. Flotgen, M. Hermle, D. Lackner, S.W. Glunz, A.W. Bett, M. Wimplinger, F. Dimroth, Monolithic two- terminal III-V//Si triple-junction solar cells with 30.2% efficiency under 1-Sun AM1.5 g, IEEE J. Photovolt. 7 (2017) 367–373. https://doi.org/10.1109/JPHOTOV.2016.2629840

[23]   Press release-New world record for solar cell efficiency at 46%-Fraunhofer ISE [Internet], Fraunhofer institute for solar energy systems ISE. (cited 2017). Available from: http://www.ise.fraunhofer. de/en/press-media/press-releases/2014/new-world-record-for-solar-cell-efficiency-at-46-percent.html

[24]   Press release-sharp develops concentrator solar cell with world's highest conversion efficiency of 43.5% | Press Releases | Sharp Global [Internet]. (cited 2017). Available from: http://www.sharp-world.com/ corporate/news/120531.html

[25]   P. Patel, D. Aiken, A. Boca, B. Cho, D. Chumney, M.B. Clevenger, A. Cornfeld, N. Fatemi, Y. Lin, J. McCarty, F. Newman, P. Sharps, J. Spann, M. Stan, J. Steinfeldt, C. Strautin, T. Varghese, Experimental results from performance improvement and radiation hardening of inverted metamorphic multi-junction solar cells, IEEE J. Photovolt. 2 (2012) 377–381. https://doi.org/10.1109/JPHOTOV.2012.2198048

[26]   S. Wojtczuk, P. Chiu, X. Zhang, D. Pulver, C. Harris, M. Timmons, Bi-facial growth InGaP/GaAs/InGaAs concentrator solar cells, IEEE J. Photovolt. 2 (2012) 371-376. https://doi.org/10.1109/JPHOTOV.2012.2189369

[27]   V. Sabnis, H. Yuen, M. Wiemer, High-efficiency multi-junction solar cells employing dilute nitrides, AIP Conf Proc. Toledo (2012). https://doi.org/10.1063/1.4753823

[28]   K.W.J. Barnham, G. Duggan, A new approach to high-efficiency multi-bandgap solar cells, J. Appl. Phys. 67 (1990) 3490–3493. https://doi.org/10.1063/1.345339

[29]   H. Fujii, K. Toprasertpong, Y. Wang, K. Watanabe, M. Sugiyama, Y. Nakano, 100-period, 1.23-eV bandgap InGaAs/GaAsP quantum wells for high-efficiency GaAs solar cells: toward current- matched Ge-based tandem cells, Prog. Photovolt. Res. Appl. 22 (2013) 784–795. https://doi.org/10.1002/pip.2454

[30]   K. Toprasertpong, H. Fujii, T. Thomas, M. Führer, D.A. Álvarez, D.J. Farrell, K. Watanabe, Y. Okada, N. J. Daukes, M. Sugiyama, Y. Nakano, Absorption threshold extended to 1.15 eV using InGaAs/GaAsP quantum wells for over-50%-efficient

lattice-matched quad-junction solar cells, Prog. Photovolt. Res. Appl. 24 (2016) 533–542. https://doi.org/10.1002/pip.2585

[31]   Available from: https://www.sunflaresolar.com/our-technology

[32]   M.K. Nazeeruddin, A. Kay, I. Rodicio, R. Humpbry-Baker, E. Miiller, P. Liska, N. Vlachopoulos, M. Gratzel, Conversion of light to electricity by cis-X2bis (2,2′-bipyridyl- 4,4′-dicarboxylate) ruthenium(II) charge-transfer sensitizers (X = Cl-, Br, I-, CN-, and SCN-) on nanocrystalline titanium dioxide electrodes, J. Am. Chem. Soc. 115 (1993) 6382–6390. https://doi.org/10.1021/ja00067a063

[33]   S. Mathew, A. Yella, P. Gao, R.H. Baker, B.F.E. Curchod, N.A. Astani, I. Tavernelli, U. Rothlisberger, M.K. Nazeeruddin, M. Gratzel, Dye-sensitized solar cells with 13% efficiency achieved through the molecular engineering of porphyrin sensitizers, Nat. Chem. 6 (2014) 242–247. https://doi.org/10.1038/nchem.1861

[34]   K. Kakiage, Y. Aoyama, T. Yano, K. Oya, J. Fujisawab, M. Hanaya, Highly efficient dye-sensitized solar cells with collaborative sensitization by silyl-anchor and carboxy-anchor dyes, Chem. Commun. 51 (2015) 15894–15897. https://doi.org/10.1039/C5CC06759F

[35]   X. Zhang, Y. Xu, F. Giordano, M. R. Schreier, N. Pellet, Y. Hu, C. Yi, N. Robertson, J. Hua, S.M. Zakeeruddin, H. Tian, M. Gratzel, Molecular engineering of potent sensitizers for very efficient light harvesting in thin-film solid-state dye-sensitized solar cells, J. Am. Chem. Soc. 138 (2016) 10742–10745. https://doi.org/10.1021/jacs.6b05281

[36]   A. Yella, C.L. Mai, S.M. Zakeeruddin, S.N. Chang, C.H. Hsieh, C.Y. Yeh, M. Gratzel, Molecular engineering of push-pull porphyrin dyes for highly efficient dye-sensitized solar cells: the role of benzene spacers, Angew Chem. Int. Ed. 53 (2014) 2973–2977. https://doi.org/10.1002/anie.201309343

[37]   C. Lee, R.Y. Lin, L. Lin, C. Li, T. Chu, S. Sun, J.T. Lin, K. Ho, Recent progress in organic sensitizers for dye-sensitized solar cells, RSC Adv. 5 (2015) 23810–23825. https://doi.org/10.1039/C4RA16493H

[38]   R. Sivakumar, R. Recabarren, S. Ramkumar, A. Manivel, J. A. Morales, D. Contreras, M. Paulraj, Ruthenium (II) complexes incorporating carbazole-diazafluorenebased bipolar ligands for dye sensitized solar cell applications, New J. Chem. 41 (2017) 5605-5612. https://doi.org/10.1039/C7NJ01019B

[39]   T. Horiuchi, T. Yashiro, R. Kawamura, S. Uchida, H. Segawa, Indoline dyes with benzothiazole unit for dye-sensitized solar cells, Chem. Lett. 45 (2016) 517-529. https://doi.org/10.1246/cl.160084

[40]   J.M. Caruge, J.E. Halpert, V. Wood, V. Bulovic, M.G. Bawendi, Colloidal quantum-dot light-emitting diodes with metal-oxide charge transport layers, Nat. phot. 2 (2008) 247-250. https://doi.org/10.1038/nphoton.2008.34

[41]   W.U. Huynh, J.J. Dittmer, A.P. Alivisatos, Hybrid nanorod polymer solar cells, Science. 295 (2002) 2425– 2427. https://doi.org/10.1126/science.1069156

[42]   M. Bruchez, M. Moronne, P. Gin, S. Weiss, A.P. Alivisatos, Semiconductor nanocrystals as fluorescent biological labels, Science. 281 (1998) 2013-2016. https://doi.org/10.1126/science.281.5385.2013

[43]   J.M. Luther, M. Law, Q. Song, C.L. Perkins, M.C. Beard, A.J. Nozik, Structural, optical, and electrical properties of self-assembled films of PbSe nanocrystals treated with 1, 2-ethanedithiol, ACS nano. 2 (2008) 271-280. https://doi.org/10.1021/nn7003348

[44]   G.H. Carey, A.L. Abdelhady, Z. Ning, S.M. Thon, O.M. Bakr, E.H. Sargent, Colloidal quantum dot solar cells, Chem. Rev. 115 (2015) 12732-12763. https://doi.org/10.1021/acs.chemrev.5b00063

[45]   J. An, X. Yang, W. Wang, J. Li, H. Wang, Z. Yu, C. Gong, X. Wang, L. Sun, Stable and efficient PbS colloidal quantum dot solar cells incorporating low-temperature processed carbon paste counter electrodes, Sol. Energy. 158 (2017) 28-33. https://doi.org/10.1016/j.solener.2017.07.074

[46]   X. Lan, S. Masala, E.H. Sargent, Charge-extraction strategies for colloidal quantum dot photovoltaics, Nat. Mater. 13 (2014) 233–240. https://doi.org/10.1038/nmat3816

[47]   M. Liu, O. Voznyy, R. Sabatini, F.P. Arquer, R. Munir, A.H. Balawi, X. Lan, F. Fan, G. Walters, A.R. Kirmani, S. Hoogland, Hybrid organic–inorganic inks flatten the energy landscape in colloidal quantum dot solids, Nat. Mater. 16 (2017) 258-263. https://doi.org/10.1038/nmat4800

[48]   X.Y. Yu, J.Y. Liao, K.Q. Qiu, D.B. Kuang, C.Y. Su, Dynamic study of highly efficient CdS/CdSe quantum dot-sensitized solar cells fabricated by electrodeposition, ACS Nano. 5 (2011) 9494-9500. https://doi.org/10.1021/nn203375g

[49]   R.S. Mane, C.D. Lokhande, Chemical deposition method for metal chalcogenide thin films, Mater. Chem. Phys. 65 (2000) 1-31. https://doi.org/10.1016/S0254-0584(00)00217-0

[50]   Y. Wang, A. Hu, Carbon quantum dots: synthesis, properties and applications, J. Mater. Chem. C. 2 (2014) 6921-6939. https://doi.org/10.1039/c4tc00988f

[51] R. Wang, K.Q. Lu, Z.R. Tang, Y.J. Xu, Recent progress in carbon quantum dots: synthesis, properties and applications in photocatalysis, J. Mater. Chem. A. 5 (2017) 3717-3734. https://doi.org/10.1039/C6TA08660H

[52] M.A. Green, Third generation photovoltaics: solar cells for 2020 and beyond. Low-dimensional systems and nanostructures, Phys. E. 14 (2002) 65-70. https://doi.org/10.1016/S1386-9477(02)00361-2

[53] G. Hodes, Comparison of dye and semiconductor-sensitized porous nanocrystalline liquid junction solar cells, J. Phys. Chem. C, 112 (2008) 17778–17787. https://doi.org/10.1021/jp803310s

[54] I. Robel, V. Subramanian, M. Kuno, P.V. Kamat, Quantum dot solar cells, Harvesting light energy with CdSe nanocrystals molecularly linked to mesoscopic TiO_2 films, J. Am. Chem. Soc. 128 (2006) 2385-2393. https://doi.org/10.1021/ja056494n

[55] A.M. Zaban, O.I. Mićić, B.A. Gregg, A.J. Nozik, Photosensitization of nanoporous TiO_2 electrodes with InP quantum dots, Langmuir. 14 (1998) 3153-3156. https://doi.org/10.1021/la9713863

[56] I. Mora-Sero, S. Gimenez, F. Fabregat-Santiago, R. Gomez, Q. Shen, T. Toyoda, J. Bisquert, Recombination in quantum dot sensitized solar cells, Acc. Chem. Res. 42 (2009) 1848-1857. https://doi.org/10.1021/ar900134d

[57] J.H. Bang, P.V. Kamat, Solar cells by design: photoelectrochemistry of TiO_2 nanorod arrays decorated with CdSe, Adv. Funct. Mater. 20 (2010) 1970-1976. https://doi.org/10.1002/adfm.200902234

[58] V.G. Pedro, X. Xu, I.M. Sero, J. Bisquert, Modeling high-efficiency quantum dot sensitized solar cells, ACS Nano. 4 (2010) 5783-5790. https://doi.org/10.1021/nn101534y

[59] X.Y. Yu, J.Y. Liao, K.Q. Qiu, D.B. Kuang, C.Y. Su, Dynamic study of highly efficient CdS/CdSe quantum dot-sensitized solar cells fabricated by electrodeposition, ACS Nano. 5 (2011) 9494-9500. https://doi.org/10.1021/nn203375g

[60] C. Cheng, S.K. Karuturi, L. Liu, J. Liu, H. Li, L.T. Su, A.I. Tok, H.J. Fan, Quantum-dot-sensitized TiO_2 inverse opals for photoelectrochemical hydrogen generation, Small. 8 (2012) 37-42. https://doi.org/10.1002/smll.201101660

[61] Available from: https://www.aurelautomation.com/new-technologies/dssc-dye-sensitized- solar-cell/attachment/dssc/

[62] M. Chatsko, 3 Wild solar power technologies that could secure the industry's future. (2018), Retrieved from https://www.fool.com/investing/2018/07/19/3-wild-solar-power-technologies-that-could-secure.aspx

[63]   M. Bernechea, N.C. Miller, G. Xercavins, D. So, A. Stavrinadis, G. Konstantatos, Solution-processed solar cells based on environmentally friendly $AgBiS_2$ nanocrystals, Nat. Phot. 10 (2016) 521. https://doi.org/10.1038/nphoton.2016.108

[64]   G. Wang, H. Wei, J. Shi, Y. Xu, H. Wu, Y. Luo, D. Li, Q. Meng, Significantly enhanced energy conversion efficiency of $CuInS_2$ quantum dot sensitized solar cells by controlling surface defects, Nano Energy. 35 (2017) 17-25. https://doi.org/10.1016/j.nanoen.2017.03.008

[65]   W. Zhao, S. Li, H. Yao, S. Zhang, Y. Zhang, B. Yang, J. Hou, Molecular optimization enables over 13% efficiency in organic solar cells, J. Am. Chem. Soc. 139 (2017) 7148–7151. https://doi.org/10.1021/jacs.7b02677

[66]   S. Berny, N. Blouin, A. Distler, H.J. Egelhaaf, M. Krompiec, A. Lohr, O.R. Lozman, G.E. Morse, L. Nanson, A. Pron, T. Sauermann, Solar trees: first large-scale demonstration of fully solution coated, semitransparent, flexible organic photovoltaic modules, Adv. Sci. 3 (2016) 1500342. https://doi.org/10.1002/advs.201500342

[67]   N. Espinosa, R. Garcia-Valverde, A. Urbina, F.C. Krebs, A life cycle analysis of polymer solar cell modules prepared using roll-to-roll methods under ambient conditions, Sol. Energy Mater. Sol. Cells. 95 (2011) 1293-1302. https://doi.org/10.1016/j.solmat.2010.08.020

[68]   S. Lizin, S.V. Passel, E.D. Schepper, W. Maes, L. Lutsen, J. Manca, D. Vanderzande, Life cycle analyses of organic photovoltaics: a review, Energy Environ. Sci. 6 (2013) 3136-3149. https://doi.org/10.1039/C3EE42653J

[69]   H.K. Lee, Z. Li, J.R. Durrant, W.C. Tsoi, Is organic photovoltaics promising for indoor applications?, Appl. Phys. Lett. 108 (2016) 253301. https://doi.org/10.1063/1.4954268

[70]   S.E. Shaheen, C.J. Brabec, N.S. Sariciftci, F. Padinger, T. Fromherz, J.C. Hummelen, 2.5% efficient organic plastic solar cells, Appl. Phys. Lett. 78 (2001) 841-843. https://doi.org/10.1063/1.1345834@apl.2019.APLCLASS2019.issue-1

[71]   M.M. Wienk, J.M. Kroon, W.J. Verhees, J. Knol, J.C. Hummelen, P.A. van Hal, R.A. Janssen, Efficient methano [70] fullerene/MDMO-PPV bulk heterojunction photovoltaic cells, Angew Chem. Int. Ed. 42 (2003) 3371-3375. https://doi.org/10.1002/anie.200351647

[72]   J. Subbiah, P.M. Beaujuge, K.R. Choudhury, S. Ellinger, J.R. Reynolds, F. So, Combined effects of MoO_3 interlayer and $PC_{70}BM$ on polymer photovoltaic device performance, Org. Electron. 11 (2010) 955-958. https://doi.org/10.1016/j.orgel.2010.02.006

[73]   G.J. Zhao, Y.J. He, Y. Li, 6.5% efficiency of polymer solar cells based on poly(3-hexylthiophene) and Indene-C60 bisadduct by device optimization, Adv. Mater. 22 (2010) 4355–4358. https://doi.org/10.1002/adma.201001339

[74]   J. Zhao, Y. Li, G. Yang, K. Jiang, H. Lin, H. Ade, W. Ma, H. Yan, Efficient organic solar cells processed from hydrocarbon solvents, Nat. Energy. 1 (2016) 1-7. https://doi.org/10.1038/nenergy201527

[75]   M.C. Scharber, On the efficiency limit of conjugated polymer: fullerene-based bulk hetero-junction solar cells, Adv. Mater. 28 (2016) 1994–2001. https://doi.org/10.1002/adma.201504914

[76]   S. Bertho, G. Janssen, T.J. Cleij, B. Conings, W. Moons, A. Gadisa, J. D'Haen, E. Goovaerts, L. Lutsen, J. Manca, D. Vanderzande, Effect of temperature on the morphological and photovoltaic stability of bulk heterojunction polymer: fullerene solar cells, Sol. Energ. Mater. Sol. Cell. 92 (2008) 753-760. https://doi.org/10.1016/j.solmat.2008.01.006

[77]   S. Holliday, R.S. Ashraf, C.B. Nielsen, M. Kirkus, J.A. Röhr, C.H. Tan, E. Collado-Fregoso, A.C. Knall, J.R. Durrant, J. Nelson, I. McCulloch, A rhodanine flanked nonfullerene acceptor for solution-processed organic photovoltaics, J. Am. Chem. Soc. 137 (2015) 898-904. https://doi.org/10.1021/ja5110602

[78]   S. Holliday, R.S. Ashraf, A. Wadsworth, D. Baran, S.A. Yousaf, C.B. Nielsen, C.H. Tan, S.D. Dimitrov, Z. Shang, N. Gasparini, M. Alamoudi, High-efficiency and air-stable P3HT-based polymer solar cells with a new non-fullerene acceptor, Nat. Commun. 7 (2016) 11585. https://doi.org/10.1038/ncomms11585

[79]   D. Baran, T. Kirchartz, S. Wheeler, S. Dimitrov, M. Abdelsamie, J. Gorman, R.S. Ashraf, S. Holliday, A. Wadsworth, N. Gasparini, P. Kaienburg, Reduced voltage losses yield 10% efficient fullerene free organic solar cells with> 1 V open circuit voltages, Energy Env. Sci. 9 (2016) 3783-3793. https://doi.org/10.1039/C6EE02598F

[80]   W. Zhao, D. Qian, S. Zhang, S. Li, O. Inganas, F. Gao, J. Hou, Fullerene-free polymer solar cells with over 11% efficiency and excellent thermal stability, Adv. Mater. 28 (2016) 4734-4739. https://doi.org/10.1002/adma.201600281

[81]   A.S. Gertsen, M.F. Castro, R.R. Sondergaard, J.W. Andreasen, Scalable fabrication of organic solar cells based on non-fullerene acceptors, Flex. Print. Electron. 5 (2020) 014004. https://doi.org/10.1088/2058-8585/ab5f57

[82]   S.A. Gevorgyan, N. Espinosa, L. Ciammaruchi, B. Roth, F. Livi, S. Tsopanidis, S. Zufle, S. Queiros, A. Gregori, G.A. Benatto, M. Corazza, Baselines for lifetime of organic solar cells, Adv. Energy Mater. 6 (2016) 1600910. https://doi.org/10.1002/aenm.201600910

[83] M.S. White, D.C. Olson, S.E. Shaheen, N. Kopidakis, D.S. Ginley, Inverted bulk-heterojunction organic photovoltaic device using a solution-derived ZnO under layer, Appl. Phys. Lett. 89 (2006) 143517. https://doi.org/10.1063/1.2359579

[84] G. Li, C.W. Chu, V. Shrotriya, J. Huang, Y. Yang, Efficient inverted polymer solar cells, Appl. Phys. Lett. 88 (2006) 253503. https://doi.org/10.1063/1.2212270

[85] S.K. Hau, H.L. Yip, N.S. Baek, J. Zou, K. O'Malley, A.K. Jen, Air-stable inverted flexible polymer solar cells using zinc oxide nanoparticles as an electron selective layer, Appl. Phys. Lett. 92 (2008) 225. https://doi.org/10.1063/1.2945281

[86] H. Cha, J. Wu, A. Wadsworth, J. Nagitta, S. Limbu, S. Pont, Z. Li, J. Searle, M.F. Wyatt, D. Baran, J.S. Kim, An efficient, "burn in" free organic solar cell employing a nonfullerene electron acceptor, Adv. Mater. 29 (2017) 1701156. https://doi.org/10.1002/adma.201701156

[87] N. Gasparini, M. Salvador, S. Strohm, T. Heumueller, I. Levchuk, A. Wadsworth, J.H. Bannock, J.C. de Mello, H.J. Egelhaaf, D. Baran, I. McCulloch, C.J. Brabec, Burn-in free nonfullerene-based organic solar cells, Adv. Energy Mater. 7 (2017) 1700770. https://doi.org/10.1002/aenm.201700770

[88] M.M. Lee, J. Teuscher, T. Miyasaka, T.N. Murakami, H.J. Snaith, Efficient hybrid solar cells based on meso-superstructured organometal halide perovskites, Science. 338 (2012) 643-647. https://doi.org/10.1126/science.1228604

[89] A. Kojima, K. Teshima, Y. Shirai, T. Miyasaka, Organometal halide perovskites as visible-light sensitizers for photovoltaic cells, J. Am. Chem. Soc. 131 (2009) 6050-6051. https://doi.org/10.1021/ja809598r

[90] M. Gratzel, The light and shade of perovskite solar cells, Nat. Mater. 13 (2014) 838–842. https://doi.org/10.1038/nmat4065

[91] N-G. Park, Perovskite solar cells: an emerging photovoltaic technology, Mat. Today. 18 (2015) 65–72. https://doi.org/10.1016/j.mattod.2014.07.007

[92] N.J. Jeon, J.H. Noh, W.S. Yang, Y.C. Kim, S. Ryu, J. Seo, S.I. Seok, Compositional engineering of perovskite materials for high-performance solar cells, Nature. 517 (2015) 476-480. https://doi.org/10.1038/nature14133

[93] M.A. Green, K. Emery, Y. Hishikawa, W. Warta, E.D. Dunlop, Solar cell efficiency tables (Version 45), Prog. Photovolt. Res. Appl. 23 (2015) 1-9. https://doi.org/10.1002/pip.2573

[94] N.G. Park, Organometal perovskite light absorbers toward a 20% efficiency low-cost solid-state mesoscopic solar cell, J. Phys. Chem. Lett. 4 (2013) 2423– 2429. https://doi.org/10.1021/jz400892a

Materials Research Forum LLC
https://doi.org/10.21741/9781644901090-1

[95] N.J. Jeon, J.H. Noh, Y.C. Kim, W.S. Yang, S. Ryu, S.I. Seok, Solvent engineering for high-performance inorganic–organic hybrid perovskite solar cells, Nat. Mater. 13 (2014) 897-903. https://doi.org/10.1038/nmat4014

[96] M. Saliba, T. Matsui, J.Y. Seo, K. Domanski, J.P. Correa-Baena, M.K. Nazeeruddin, S.M. Zakeeruddin, W. Tress, A. Abate, A. Hagfeldt, M. Gratzel, Cesium-containing triple cation perovskite solar cells: improved stability, reproducibility and high efficiency, Energy Environ. Sci. 9 (2016) 1989-97. https://doi.org/10.1039/C5EE03874J

[97] T. Salim, S. Sun, Y. Abe, A. Krishna, A.C. Grimsdale, Y.M. Lam, Perovskite-based solar cells: impact of morphology and device architecture on device performance, J. Mater. Chem. A. 3 (2015) 8943-8969. https://doi.org/10.1039/C4TA05226A

[98] H.J. Snaith, A. Abate, J.M. Ball, G.E. Eperon, T. Leijtens, N.K. Noel, S.D. J.T. Stranks, Wang, K. Wojciechowski, W. Zhang, Anomalous hysteresis in perovskite solar cells, J. Phys. Chem. Lett. 5 (2014) 1511-1515. https://doi.org/10.1021/jz500113x

[99] L. Cojocaru, S. Uchida, P.V. Jayaweera, S. Kaneko, J. Nakazaki, T. Kubo, H. Segawa, Origin of the hysteresis in I-V curves for planar structure perovskite solar cells rationalized with a surface boundary-induced capacitance model, Chem. Lett. 44 (2015) 1750-1752. https://doi.org/10.1246/cl.150933

[100] S. van Reenen, M. Kemerink, H.J. Snaith, Modeling anomalous hysteresis in perovskite solar cells, J. Phys. Chem. Lett. 6 (2015) 3808-3814. https://doi.org/10.1021/acs.jpclett.5b01645

[101] M.T. Neukom, S. Zufle, E. Knapp, M. Makha, R. Hany, B. Ruhstaller, Why perovskite solar cells with high efficiency show small IV-curve hysteresis, Sol. Energ. Mater. Sol. Cell. 169 (2017) 159-166. https://doi.org/10.1016/j.solmat.2017.05.021

[102] P. Calado, A.M. Telford, D. Bryant, X. Li, J. Nelson, B.C. O'Regan, P.R. Barnes, Evidence for ion migration in hybrid perovskite solar cells with minimal hysteresis, Nat. Commun. 7 (2016) 1-10. https://doi.org/10.1038/ncomms13831

[103] J.H. Heo, H.J. Han, D. Kim, T.K. Ahn, S.H. Im, Hysteresis-less inverted $CH_3NH_3PbI_3$ planar perovskite hybrid solar cells with 18.1% power conversion efficiency, Energy Environ. Sci. 8 (2015) 1602-1608. https://doi.org/10.1039/C5EE00120J

[104] F. Li, M. Liu, Recent efficient strategies for improving the moisture stability of perovskite solar cells, J. Mater. Chem. A. 5 (2017) 15447-15459. https://doi.org/10.1039/C7TA01325F

[105] N.G. Park, T. Miyasaka, M. Gratzel, Organic-inorganic halide perovskite photovoltaics, Cham, Switzerland: Springer. (2016). https://doi.org/10.1007/978-3-319-35114-8

[106] L. Calió, S. Kazim, M. Gratzel, S. Ahmad, Hole-transport materials for perovskite solar cells, Angew Chem. Int. Ed. 55 (2016) 14522–14545. https://doi.org/10.1002/anie.201601757

[107] E.H. Anaraki, A. Kermanpur, L. Steier, K. Domanski, T. Matsui, W. Tress, M. Saliba, A. Abate, M. Gratzel, A. Hagfeldt, J.P. Correa-Baena, Highly efficient and stable planar perovskite solar cells by solution-processed tin oxide, Energy Environ. Sci. 9 (2016) 3128-3134. https://doi.org/10.1039/C6EE02390H

[108] S.S. Shin, E.J. Yeom, W.S. Yang, S. Hur, M.G. Kim, J. Im, J. Seo, J.H. Noh, S.I. Seok, Colloidally prepared La-doped $BaSnO_3$ electrodes for efficient, photostable perovskite solar cells, Science. 356 (2017) 167-171. https://doi.org/10.1126/science.aam6620

[109] X. Dong, X. Fang, M. Lv, B. Lin, S. Zhang, J. Ding, N. Yuan, Improvement of the humidity stability of organic–inorganic perovskite solar cells using ultrathin Al_2O_3 layers prepared by atomic layer deposition, J. Mater. Chem. A. 3 (2015) 5360-5367. https://doi.org/10.1039/C4TA06128D

[110] G. Niu, W. Li, F. Meng, L. Wang, H. Dong, Y. Qiu, Study on the stability of $CH_3NH_3PbI_3$ films and the effect of post-modification by aluminum oxide in all-solid-state hybrid solar cells. J. Mater. Chem. A. 2 (2014) 705–710. https://doi.org/10.1039/c3ta13606j

[111] F. Giordano, A. Abate, J.P. Baena, M. Saliba, T. Matsui, S.H. Im, S.M. Zakeeruddin, M.K. Nazeeruddin, A. Hagfeldt, M. Graetzel, Enhanced electronic properties in mesoporous TiO_2 via lithium doping for high-efficiency perovskite solar cells. Nat. Commun. 7 (2016) 10379. https://doi.org/10.1038/ncomms10379

[112] H. Yoon, S.M. Kang, J.K. Lee, M. Choi, Hysteresis-free low-temperature-processed planar perovskite solar cells with 19.1% efficiency, Energy Environ. Science. 9 (2016) 2262-2266. https://doi.org/10.1039/C6EE01037G

Materials Research Forum LLC
https://doi.org/10.21741/9781644901090-2

Chapter 2

Graphene Materials for Third Generation Solar Cell Technologies

Onoriode P. Avbenake

Chemical and Petroleum Engineering Department, Bayero University, Kano

School of Science and Technology, Pan-Atlantic University, Ibeju-Lekki, Lagos, Nigeria

paulavbenake@gmail.com

Abstract

Photovoltaic technology is the most sustainable source of renewable energy because sunlight radiation is free and readily available. Therefore, the materials required accessing this energy source, cost and the efficiency of conversion from solar to electricity is the topic of interest in continued research. Graphene as a sp^2-hybridized 2-dimensional carbon with unique crystal and electronic properties comprising high charge carrier mobility, optical transparency, inexpensive, excellent mechanical strength and flexibility with chemical stability and inertness among others is a suitable material for application in various units of the different architectures in third generation solar cells. It can be applied as a semiconductor layer, electrolyte and counter-electrode in dye-sensitized solar cells; electrode, perovskite, electron and hole transporting layers in perovskite solar cells; and electrode, hole transporting layer and electron acceptor and donor in organic solar cells; in addition to graphene/silicon Schottky junction. Following the application of graphene in various units of the third generation architecture, the power conversion efficiency has increased from 1.9% to over 22%, with ongoing research expected to develop a more stable design with longevity comparable to commercially available silicon-based p-n junction.

Keywords

Solar Cells, Graphene, Dye-Sensitized Solar Cells (DSSCs), Perovskite Solar Cells (PSCs), Organic Solar Cells (OPV), Schottky Junction

Materials for Solar Cell Technologies I   Materials Research Forum LLC
Materials Research Foundations **88** (2021) 29-61   https://doi.org/10.21741/9781644901090-2

Contents

Graphene Materials for Third Generation Solar Cell Technologies...........**29**

1.   Introduction...**30**

2.   Dye-sensitized solar cells (DSSCs)..**32**

    2.1   Semiconductor layer 34

    2.2   Electrolyte...36

    2.3   Counter electrode (CE)..37

3.   Perovskite solar cells (PSCs)..**38**

    3.1   Electrode 39

    3.2   Perovskite 40

    3.3   Electron transport layer (ETL) 41

    3.4   Hole transport layer (HTL)..42

4.   Organic solar cells (OPV) **42**

    4.1   Electrode 44

    4.2   Hole transport layer (HTL)..45

    4.3   Electron acceptor and donor 45

5.   Graphene/silicon Schottky junction..**46**

Conclusion..**47**

References..**48**

1.   Introduction

Fossil fuel accounts for about 86% of the total world energy consumption [1] with reported decline in conventional resources and a switch to unconventional resources to augment supply [2] which is projected to cease in 50 years [3]. Coupled with the environmental footprint, renewable resources champions the various sustainable sources of energy globally. Among the lots, the photovoltaic technology is a relatively simple process where incident photon from sunlight is absorbed on a photoactive material to excite the electron (e^-) in the valence band generating electron and hole (h^+) pairs which are transported to their distinct electrodes to produce electricity.

Moreover, the efficiency of a solar cell can be optimized by maintaining the valency of the counter electrode, the concentration of the electrolyte, the conductivity of the electron transporting material and, stability within the perovskite crystal due to prolonged working condition among others. In the wake of novel device design, light absorbing materials, materials combinations and nanotechnology; photovoltaic technology has enjoyed tremendous improvement in power conversion efficiency (PCE) since its discovery in 1839 [4]. From the first-generation of photovoltaic cells which are silicon-based inspired by abundance silicon materials as a good semiconductor based on a p-n junction technology first reported in 1954 [5] to cadmium thin-film as a supportive semiconductor [6] and currently to the third-generation of photovoltaic cells based on pure organic or organic-inorganic blends [7–10].

Furthermore, cost remains the means of propulsion for this generational evolution of photovoltaic cells since, high-purity silicon are required for the first generation [11] and cadmium is toxic with low efficiency [12]. Meanwhile, carbon is a distinctive element and is readily available, existing in various forms and shapes with interesting optoelectronic properties. It exists in various forms as; porous carbon, carbon black, carbon nanotubes (nanowires, fibres), diamond, fullrenes, graphite, and graphene with metallic, semimetallic, conducting, semiconducting and insulating properties which underpins their potential application in energy storage and conversion [13]. Consequently, there is no gainsaying that carbon is the basis of the third-generation photovoltaic cells with diverse designs like; dye-sensitized solar cells (DSSCs), perovskite solar cells (PSCs), organic solar cells or photovoltaic cells (OPV), quantum dot solar cells, and Schottky junction.

Graphene was discovered 15 years ago by physicists Geim and Novoselov when they employed a sticky tape to peel the surface of graphite [14]. It is a sp^2-hybridized 2-dimensional carbon with unique crystal and electronic qualities different from bulk graphite [15]. Other members of the graphene family are the 0- dimensional fullerenes, 1-dimensional nanotubes and 3- dimensional graphite (figure 1) [14]. Furthermore, graphene distinguishes itself from other carbon forms with high charge carrier mobility, low charge carrier density, optical transparency, excellent mechanical strength and flexibility, chemical stability and inertness, zero energy bandgap, high surface area, high transmittance in the visible range, outstanding thermal and electrical conductivity, and cheap large-scale production among others, which places graphene as a one-stop material for application in various units of the different architectures of third-generation photovoltaic cells.

Although, there are comprehensive review articles [16-29] and book chapters [30-34] on application of graphene and graphene-based materials like; graphene oxide (GO) [35-39] and reduced graphene oxide (rGO) [40-44] in photovoltaic cells. They did not reflect the progress attained in recent times and are mainly constrained to specific third-generation photovoltaic architecture. In this chapter, an attempt to review the applications of pure graphene in various units of dye-sensitized solar cells (DSSCs), perovskite solar cells (PSCs), organic photovoltaic cells (OPV), and Schottky junction has been made.

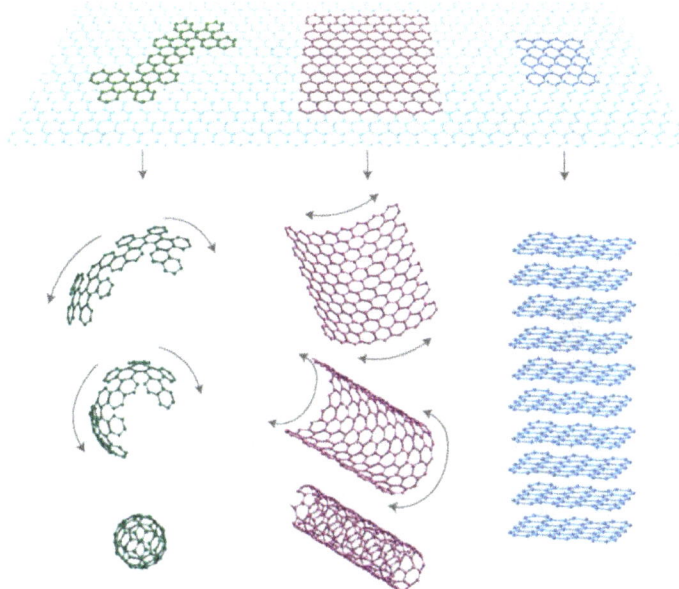

Figure 1 Mother of all graphitic forms. Graphene is a 2D building material for carbon materials of all other dimensionalities. It can be wrapped up into 0D buckyballs, rolled into 1D nanotubes or stacked into 3D graphite. [14] Reprinted with permission of Springer Nature.

2. Dye-sensitized solar cells (DSSCs)

DSSCs were developed by O'Regan and M. Grätzel in 1991 [45]. Above all they are light-harvesting device based on a photosensitizer. The photosensitizer is a liquid dye with the ability to split excitons into charges and holes, with a metal oxide semiconductor employed as the electron acceptor. The main constituents of a typical DSSC are; the photoanode, the photosensitizer, the electrolyte and the counter electrode. Graphene has successfully been employed in the various units except the photosensitizer which is a liquid dye molecule.

Materials for Solar Cell Technologies I   Materials Research Forum LLC
Materials Research Foundations **88** (2021) 29-61   https://doi.org/10.21741/9781644901090-2

Basically, the processes that occur during solar to power conversion in DSSCs are; excitation of the dye due to photo radiation, injection of electron into the conduction band of the solar cell, subsequent transportation of the electron to the counter electrode, and regeneration and restoration of the dye to the ground state via iodide/tri-iodide (I^-/I_3^-) (reaction 1) couple electrochemical reactions in the electrolyte (figure 2).

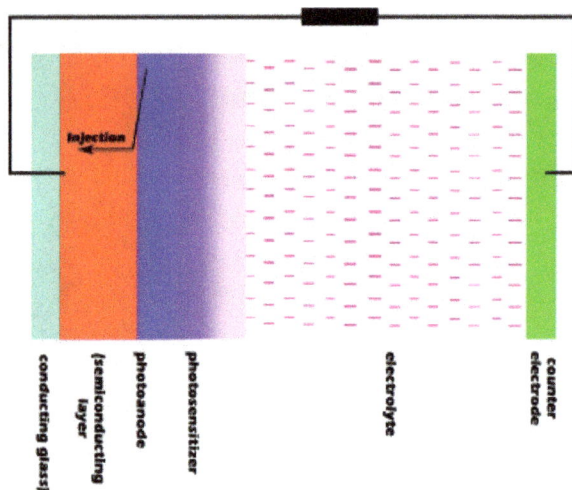

Figure 2 Schematic illustration of the device architecture of dye-sensitized solar cells (DSSCs).

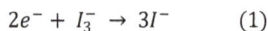

$$2e^- + I_3^- \rightarrow 3I^- \qquad (1)$$

Generally, platinum is the catalyst of choice for electron transport and reduction of I_3^- to I^- at the counter electrode [46,47], liquid based electrolytes promotes high contacts at electrolyte-electrode interface [48] and TiO_2 is a relatively cheap and readily available semiconductor for photoanodes [49]. However, slow electron transport through TiO_2 due to traps favours the recombination with oxidized dye or electrolyte over movement to the counter electrode, leakages and poor charge transfer properties in liquid electrolytes limits the restoration of the dye, and high cost of platinum coupled with corrosion by liquid electrolyte increases the cost of traditional DSSCs. These drawbacks encouraged the integration of graphene in the semiconductor layer, electrolyte and counter electrode units of DSSCs.

2.1 Semiconductor layer

Dark current in a solar cell is as a result of recombination of injected electrons to restore the dye. Rather than the electrons be transported through the semiconducting layer and via the external circuit to the counter electrode (figure 2), they move backward to regenerate the oxidized dye. Several factors are responsible for this phenomenon, chief of which is the presence of traps in the semiconductor. Mehmood et al. [50], attempted to reduce this recombination by employing nanocomposite photoanodes. They acknowledged various carbon allotropes (single/multi walled carbon nanotubes SWCNTs,/MWCNTs, and graphene) that have been blended with TiO_2 and greatly improved the PCE of DSSCs from 4.9% to 7.79% [51-54]. However, in their case they improved both the photosensitizer and semiconductor. An efficient Ruthenium based photosensitizer was blended with a wide-range light absorption material for the former and graphene with TiO_2 for the latter to achieve a 9.45% PCE due to 3D conductive networks introduced by graphene which renders an active pathway for electron transfer. Meanwhile, Nemade et al. [55] fabricated polyaniline (PANi) – graphene composite and further introduced TiO_2 as impurity to enhance the electronic property of the DSSCs semiconducting layer. They recorded a 6.48% PCE with PANi loaded graphene composite and 8.63% with TiO_2 nanoparticles loaded PANi-graphene composites as a result of the homogeneous composite between PANi and graphene, their lower band gap (1.92 eV) and the high electrical conductive property of graphene for the PANi loaded graphene composite and increased photocurrent density as a result of TiO_2 with reduction in charge recombination ascribed to the presence of graphene sheets in the TiO_2 nanoparticles loaded PANi-graphene composites design.

Besides the presence of traps in the semiconducting layer of DSSCs, the thickness of the photoanode (semiconducting layer + conducting glass) is another obstacle for efficient electron transport. As a consequence, the thickness is typically limited to 8 – 15 μm [56,57]. Lo and Leung [58] developed an eco-friendly method to synthesize TiO_2/graphene nanofibres defects-free for efficient conductivity in thicker photoanodes. Pure TiO_2 nanofibre with photoanode thickness of 13.1 μm gave a PCE of 7.28% compared with TiO_2/graphene nanofibre with photoanode thickness of 23.2 μm and PCE of 8.9%. In a bid to prevent heighten regeneration of the oxidized dye, the graphene were insulated from the electrolyte by the TiO_2. Accordingly, in their scheme the N719 dye were anchored to the TiO_2 only. The photogenerated electrons flow from the dye to the

TiO_2, then the graphene, and finally towards the fluorine-doped tin oxide (FTO) electrode. The oxidized dyes were then regenerated by the electrolyte (Scheme 1).

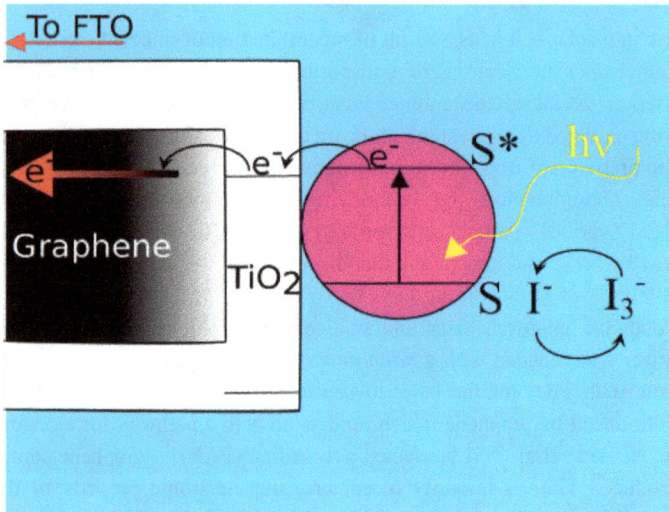

Scheme 1 Electron flow of TiO₂/graphene nanofibres photoanode. [58] Reproduced with permission from Elsevier.

The fabrication method of blending TiO_2 and graphene is highly influential in the final property of the composite. As demonstrated by Lo and Leung [58] in their shear exfoliation synthesis of TiO_2/graphene nanofibres. Some selected fabrication methods employed in DSSCs are; simultaneous reduction-hydrolysis technique with 7.1% PCE [59], solvothermal technique with 7.25% PCE [60], and electrospun TiO_2-graphene composite with 7.6% PCE [61]. In a seemingly corroborating context, Sadikin et al. [62] studied the effect of spin coating number of graphene on TiO_2 on the performance as photoanode in a DSSC. Their result shows that the photoanode blend with 1 spin-coating of graphene has 19.73% more PCE than pure TiO_2, with the efficiency fluctuating with higher coating layers. Kumar et al. [63] synthesized graphene quantum dots (GQDs) by electrochemical cyclic voltammetry technique for binder free GQDs-TiO_2 photoanode application in DSSCs to report 4.43% PCE.

2.2   Electrolyte

The 200,000 $cm^2V^{-1}s^{-1}$ carrier mobility of graphene was exploited by Tsai et al. [64] to top the electric conductivity of the liquid electrolyte in DSSCs thereby increasing reduction of oxidized dye molecules and improving the efficiency of the solar cells. 20 mg graphene nanosheet increases the PCE to 9.26% from 7.71% for pure electrolyte containing a mixture of 798 mg of 1-buty-3-methylimidazolium iodide (BMII), 38 mg of I_2, 0.36 mL of 4-tert-butyloyridine (4-TBP), 59 mg of guanidine thiocyanate (GuSCN), 4.25 mL of acetonitrile, and 0.75 mL of valeronitrile.

As previously stated, liquid-based electrolyte suffers from leakages and poor charge transfer which are unobtainable in solid electrolyte but are comparatively less efficient due to low contact in the electrolyte-electrode interface. Therefore, gel-based electrolyte, a state between solid and liquid capitalize on high ionic conductivity and long-term stability. Rehman et al. [65], synthesized polyvinyl acetate (PVAc)/graphene nanocomposite for use in liquid electrolyte in DSSCs (plate 1). Due to the high solubility of PVAc in organic solvents, electrolyte leakage and evaporation was not observed and charge transport was improved because of the presence of graphene to give a PCE higher than pure PVAc and comparable to pure liquid electrolyte which is established to decrease with time due to leakages and evaporation. In the same vein, Liu et al. [66] absorbed polyaniline-graphene/PtNi (PANi-G/PtNi) and liquid electrolyte having redox iodide/triiodide couples into poly(acrylic acid)-poly(ethylene glycol) (PAA-PEG) matrix to create a high energy level (HEV) at highly concentrated (PANi-G/PtNi) region and low energy level (LEV) at low concentrated region. In a bid to control charge transport they contacted the HEV region with the photoanode and LEV region with the counter electrode to achieve 8.64% PCE as against 5.92% with pure PAA-PEG electrolyte.

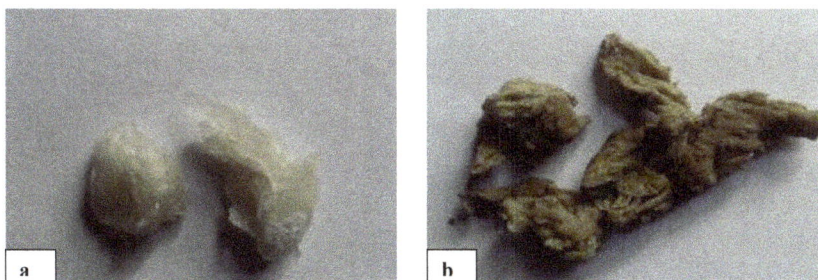

Plate 1   The prepared (a) PVAc and (b) PVAc/graphene nanocomposite images [65]. Reproduced with permission from Elsevier.

Materials for Solar Cell Technologies I
Materials Research Foundations **88** (2021) 29-61

Materials Research Forum LLC
https://doi.org/10.21741/9781644901090-2

2.3 Counter electrode (CE)

The CE in DSSCs is one of the important units that determines the efficiency of the device because of its dual functions of extracting electrons from the conduction band and providing electrons to the electrolyte for I_3^- conversion to I^- (reaction 1). Pt is a noble metal with high electrocatalytic activity for efficient electrolyte reduction and excellent electrical conductivity for fast charge transport. It was initially suitable for use as CE in DSSCs prior to the discovery that it reacts with the electrolyte to form PtI_4 [67]. Coupled with its high cost and scarcity, graphene and various blends have been utilized as suitable alternatives. Siwach et al. [68] employed a combination of pure graphene as CE and ZnO-graphene nanocomposites as photoanodes to fully exploit the charge separation and electron transport properties of graphene. The set-up with graphene as CE gave a PCE of 2.26%, with multi-walled carbon nanotubes (MWCNTs) having 2.04% and 3.17% with pure Pt. Robinson et al. [69] experimented with SnO_2/ZnO as photoanode to arrive at a PCE of 5.50% with graphene and 7.17% with Pt. Also, Seo et al. [70] recorded a PCE of 4.95% for multi-layer graphene film against 6.66% for pure Pt .

It could be observed from the aforementioned literatures that pristine graphene exhibited relatively poor catalytic activity towards the electrolyte due to the location of the active sites. This irregularity could be corrected by inserting heteroatoms into the basal plane of graphene. Tseng et al. [71] attempted to rectify the edge effect by increasing the accessible active sites from doping graphene hollow nanoballs with nitrogen directly on carbon cloth. They achieved a PCE of 7.53% with the nitrogen doped graphene (NG) which is comparable to 7.70% with sputtered Pt. The PCE of graphene subsequently outstrip Pt when doped with Selenium a chalcogen group element [72] or the NG was doped with Ni or Cu a transition metal element [73]. The success of the former was attached to the crafting method applied to dope selenium on graphene with abundant edge sites and fully activated basal planes. Ball-milling was employed to generate enriched edges and high-temperature annealing to activate the basal planes and tailor the defect density which include; voids, edges, and extrinsic Se species to induce potential active sites for the electrolyte regeneration. As a result, 8.42% PCE was achieved with Se-doped graphene compared to 7.88% with Pt. For the latter, Ni-doped MoS_2-NG gave 2.85% PCE, Cu-doped MoS_2-NG gave 2.62% PCE while Pt was 2.41%. Kim et al. [74] equally noted that the neutral polarity of graphene restrict its efficient charge transfer at the graphene/electrolyte interface. They doped graphene nanoparticles (GnPs) with metalloids (Se, Te and Sb) and comparatively investigated their performance in both ruthenium dye (N719) and organic dye (SGT-130). Their results shows that the photovoltaic performance of graphene doped CE is higher than Pt with PCE of 10.14% for SeGnPs and 10.25% for TeGnPs as against 9.98% for Pt with N719 photosensitizer and 8.22% for SeGnPs and 8.21% for TeGnPs as against 8.19% for Pt with SGT-130 photosensitizer. Similarly, Murugadoss et al. [75] designed $Co_{0.5}Ni_{0.5}Se$/graphene nanohybrid and tuned the mass ratio between $Co_{0.5}Ni_{0.5}Se$ and graphene to optimize the physical properties and electrochemical performance for application as CE in DSSCs. They reported a PCE of 9.42% for the nanohybrid doped graphene, higher than 7.68% for standard Pt. Likewise, Te-doped graphene was used in a DSSC employing *chlorophyll* as natural photosensitizer extracted from *Calotropis gigantea* plant to give a PCE of 0.271% compared to 0.196% for Pt [76].

Beside doping, direct blend of graphene with n-type semiconductors as CE has been shown to exhibit PCE higher than Pt. Zhou et al. [77] applied a blend of In_2S_3 nanoflakes with graphene as CE to obtain a PCE of 7.32% compared to 6.48% for pure Pt. Areerob et al. [78] fabricated blends of graphene $La_2W_6O_{15}$-nickel selenide/cobalt selenide (GLW-NiCoSe) CE gave a PCE of 8.14% as against 7.2% for pure Pt. Furthermore, sulfides with characteristics diverse morphology, simple preparation and low price are suitable materials for graphene blends as CE. CoS grafting graphene nanosheets showed high PCE of 7.28% [79] and $CoFe_2O_4$ nanoparticles on graphene nanosheets gave 9.04% PCE compared to 6.54% for standard Pt [80].

3.   Perovskite solar cells (PSCs)

The invention of perovskite materials in 2009 spurred research into perovskite solar cells (PSCs) [81]. Perovskite materials were introduced to replace the dye in DSSCs which rapidly surpassed 22% PCE [82-84] but the stability under prolong working condition is still in question. PSCs are based on n-i-p (direct) or p-i-n (inverted) designs depending on the direction of the carrier transport. In both configurations the photon is incidence on the transparent conducting oxide (TCO), and the difference is the perovskite active later is sandwiched between the electron transporting layer (ETL) and hole transporting layer (HTL). The ETL is directly behind the TCO for the n-i-p configuration while the HTL is just before the metal cathode and the reverse is the case for the p-i-n design (figure 3).

The HTL is the acceptor layer which is in contact with the cathode and the ETL is the donor layer in contact with the anode (TCO). The perovskite material is responsible for light absorption and charge generation, while the semiconductor (donor material) generates electron–hole pairs.

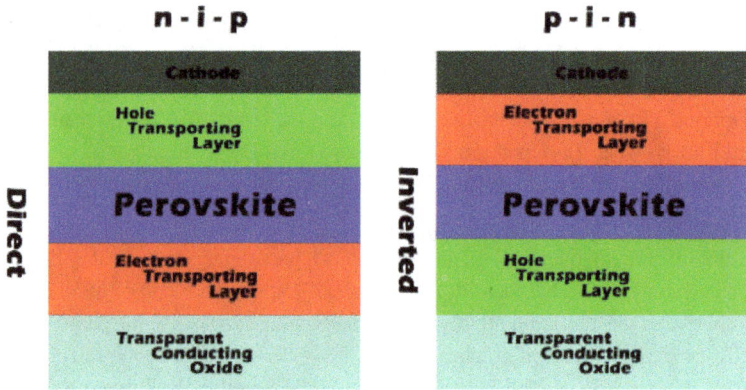

Figure 3 Schematic illustration of the device architecture of perovskite solar cells (PSCs).

For practical purposes, flexible solar cells are preferred to rigid and brittle ones in the case of cloaking and wrapping, therefore the ETL has to be compatible with both the TCO in terms of flexibility on one hand and the perovskite in terms of charge carrier accumulation and recombination on the other. Also the perovskite crystals decompose in the presence of moisture/water and atmospheric oxygen. Generally, cost is the driving force for continued research in photovoltaic cells. Consequently, the unique band structure of graphene encourages doping to control the carrier density and type (electron-like or hole-like) [14]. In addition, their flexibility, higher carrier transport performance and elimination of moisture diffusion have facilitated their application in PSCs.

3.1   Electrode

Indium-tin oxides (ITO) or fluorine-doped tin oxides (FTO) as transparent conducting oxides (TCO) in PSCs have produced remarkable PCE of over 22% [83]. Although, their long-term stability is infamously poor, another set-back in the use of ITO or FTO as TCOs is their non-flexibility. Brittle TCOs cause crack during bending subsequently damaging the device. Jang et al. [85] fabricated bis (trifluoromethanesulfonyl)-amide (TFSA) doped graphene/ triethylenetetramine (TETA) doped graphene as anode/cathode incorporating TCO function to achieve a PCE of 10.56/10.73 % from either surface illumination with excellent bending stabilities.

Materials for Solar Cell Technologies I   Materials Research Forum LLC
Materials Research Foundations **88** (2021) 29-61   https://doi.org/10.21741/9781644901090-2

In cases of severe bending, the graphene electrode could be stacked on aminopropyl triethoxysilane (APTES) or polyethylene terephthalate (PET) to ensure reinforcement. Kim et al. [86] designed multiple layer graphene on PET as anode without TCO to improve the flexibility of the PSCs. They reported the highest PCE at 2 layers graphene sheets with 13.35% and 13.94% for the forward and reverse scans. Although, ITO set-up recorded 16.20 and 16.50%, the PCE dropped to ~30% the initial value after 1000 bending cycles compared to ~90% the initial value for multilayer graphene-PET composites. The same group extended their work to Ag nanowires-doped graphene on PET [87]. The silver was incorporated to shift the Fermi level of graphene above the Dirac point, and the layer was placed between the graphene and perovskite active layer to enjoy protection from graphene against oxidation. The set-up recorded a PCE of 13.45% and 13.55% for the forward and reverse scans while maintaining ~94% of the initial PCE after 1000 bending cycles.

Indeed, the combination of APTES and PET as substrates could exponentially increase the flexibility of the solar device. Shin et al. [88] synthesized graphene/APTES transparent conductive electrode on flexible PET and evaluated the bending stability after 3000 bending cycles to be ~80% the initial value of 15.03% and 14.60%.

3.2 Perovskite

As previously observed with semiconductor layer of DSSCs, traps are also evident in the perovskite material of PSCs. The low thermal stability of the ABX_3 perovskite crystals where; $A - Cs^+$, methylammonium (MA^+), formamidinium (FA^+), $B^{2+} - Pb^{2+}$, Sn^{2+}, and $X^- - Cl^-$, Br^-, I^-, presence of 'B^{2+}' at the surface and grain boundary, and under-coordinated 'A' vacancy are the main cause of traps in PSCs. Therefore, Gan et al. [89] synthesized GN (graphite-N)- graphene quantum dots (GQDs) as a Lewis base to hybridize with the perovskite solution and interact with the Pb^{2+} and also shift the Fermi energy level of the perovskite film. These two phenomena eradicated traps and promoted charge transport to the external circuit. They reported a PCE of 19.8% with the GN-GQDs/perovskite composite as compared to 17.1% with the pristine perovskite layer. Another method of reducing grain boundaries is to increase perovskite crystallization. Li and Leung [90] introduced 1D graphene nanofibers to the perovskite framework to promote nucleation and subsequent growth of the perovskite crystal which surpassed the grain boundaries and reduce sites of electron-hole recombination. 19.83% PCE of the graphene suspended perovskite layered device was achieved against 17.51% for pristine perovskite.

Materials for Solar Cell Technologies I   Materials Research Forum LLC
Materials Research Foundations **88** (2021) 29-61   https://doi.org/10.21741/9781644901090-2

3.3   Electron transport layer (ETL)

The conventional mesoscopic perovskite device utilizes mesoporous TiO_2 as the electron transport layer (ETL) and 2,2',7,7'-tetrakis-(N,N-di-p-methoxyphenyl amine)-9,9'-spiro bifluorene (spiro-MeOTAD) as the hole transport layer (HTL) [91–93]. However, the poor charge transfer in TiO_2 and SnO_2 inducing hysteresis and low stability, the high temperature sintering process of TiO_2 and large energy barrier between TiO_2 and the work function of FTO limits their direct application in mesoscopic PSCs. Shin et al. [88] activated the tunable band gap property of GQDs by blending it with phenyl C61 butyric acid methyl ester (PCBM) as ETL synthesized at temperatures less than 100°C to achieve a maximum PCE of 16.41%. Xia et al. [94] substituted TiO_2 for SnO_2 to reduce the high temperature sintering process from 500°C to 150°C but encountered numerous trap states in the latter originating from oxygen vacancies. They also observed that the 0.57 eV energy barrier between SnO_2/perovskite interface promotes charge carrier accumulation and recombination. Subsequently, ultrathin GQDs layer was deposited on the SnO_2 nano-crystalline film to optimize the interface contact and match the energy level of SnO_2 and $CH_3NH_3PbI_3$ perovskite material (figure 4) [94]. As a result they obtained a high PCE of 16.5% compared to 13.61% without GQDs [95].

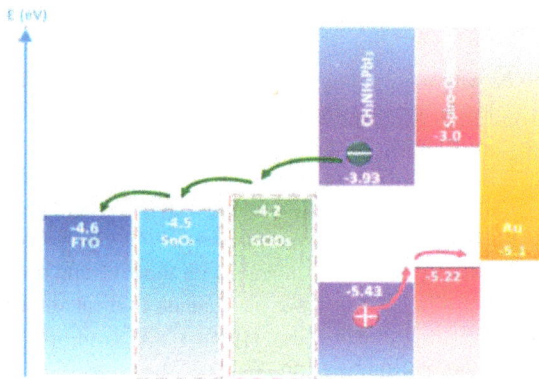

Figure 4 Schematic illustration of the energy diagram and electron transfer process for GQDs modified ETLs. [94] Reproduced with permission from Elsevier.

TiO_2 has also been blended with edge-enriched graphene nanoribbons (GNRs) [96] and graphene nanoplatelets (GnPs) [97] to complement the work function of FTO, TiO_2 ($W_{s,GNRs} = -4.3$ eV, $W_{s,FTO} = -4.7$ eV, CB $= -4.0$ eV) and balance the number of charge carriers per unit interface area per unit time with the hole transporting layer respectively. As such the interfacial charge transfer cascaded from TiO_2 to GNRs to FTO in the mesoscopic PSCs to achieve a PCE of 17.69% compared to 15.87% for pure TiO_2 ETL for the former and an increase of PCE from ~15% to ~19% for the latter. PCE as high as 20.45% was equally reported with mesoporous TiO_2 film decorated with GQDs as ETL [98]. The composite was synthesized by annealing in air at 100°C. Prior to annealing, the GQDs ethanol solution was spin-coated on the mesoporous TiO_2 film. The GQDs serve as a medium to facilitate electron transportation from the perovskite active layer to the external circuit by-passing the TiO_2 consequently reducing series resistance. This case was substantiated by lower PCE of 18.57% recorded in the set-up without GQDs.

3.4   Hole transport layer (HTL)

The two major units that determine the efficiency of PSCs are; electron/hole (exciton) separation and charge transport/transfer. Consequently, less research is dedicated to optimizing the HTL. The only work executed in the past two years incorporating pure graphene is density functional theory (DFT) calculation of the electronic property of graphene/$CH_3NH_3PbI_3$ [99]. They discovered that holes were ejected into graphene leading to a p-type doping due to the work function difference between graphene and PbI_2- (CH_3NH_3I-) surface.

4.   Organic solar cells (OPV)

Following DSSCs development in 1991, the research group of Heeger [100] demonstrated the use of organic materials as the photoactive layer in 1992. The process of energy conversion from solar to electricity is similar to PSCs with the organic layer replaced with perovskite material. Generally, OPV is a p–n type junction between an electron donor (cathode) with hole transporting organic semiconductor and electron acceptor (transparent anode) with electron transporting organic semiconductor. The roles of the electrodes and transporting layers are similar to those of DSSCs and PSCs. However, additional layer like buffers could be infused to maximize performance.

OPV also boasts of standard and inverted architecture, but unlike the PSCs, in standard OPV the HTL is directly behind the TCO (indium-doped tin oxides) and the ETL is just before the metal cathode, while the reverse is the case for inverted OPV (figure 5).

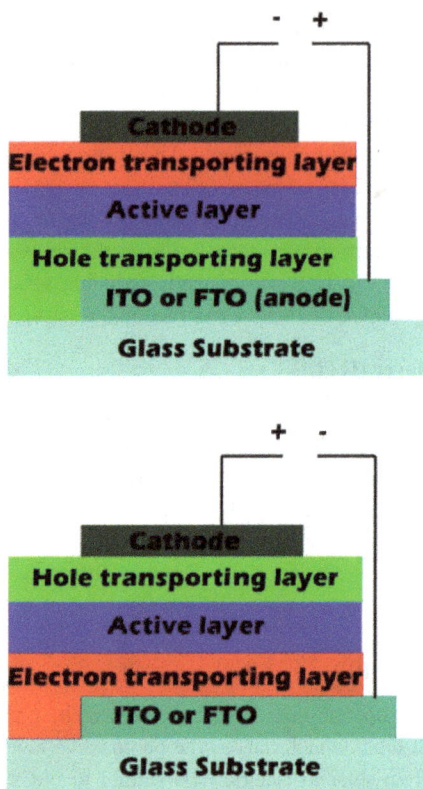

Figure 5 Schematic illustration of the device architecture of organic photovoltaic (OPV).

Similar to PSCs, the expensive and less durable indium-doped tin oxides (ITO) can be blended with graphene to improve the visible light transmittance and charge transport thereby reducing the cost of the solar cell by employing a single layer with dual purpose of TCO and ETL. Graphene can as well improve the charge carrier mobility when used as hole transporting layer by reducing the energy barrier between the electrode and active layer of the OPV. Also the edge effect and quantum confinement states of graphene poised it for efficient use as electron acceptor and donor when infused in the active layer of OPV.

Materials for Solar Cell Technologies I   Materials Research Forum LLC
Materials Research Foundations **88** (2021) 29-61   https://doi.org/10.21741/9781644901090-2

4.1   Electrode

Coupled with the brittle, scarce and chemically unstable properties of ITO which discouraged their substitution in PSCs, a 100 nm thick ITO transmission value is 85% [101] and they have very low sheet resistance ~15 Ω/square which additionally makes them unsuitable for use in OPV devices. Meanwhile, single layer graphene film can transmit 97% of the visible light [102] together with its optimal electronic, optical and photonic properties is a suitable alternative for ITO save for its very high sheet resistance of up to 1000 Ω/square. Interestingly, chemical doping and hybridization with metal nanoparticles can reduce the sheet resistance of graphene for successful integration in OPV devices. Another set-back to graphene implementation as electrode in OPV devices is the non-uniform wetting of charge transporting materials on the surface due to hydrophobicity caused by the inert sp^2 hybridized carbon structure.

In the case of hydrophobicity, Jung et al. [103] synthesized a relatively smooth thin film of norepinephrine via polymerization to modify the surface of graphene prior to doping with poly(3,4-ethylenedioxythiophene)-poly(styrenesulfonate) (PEDOT:PSS) coating. They reported a PCE of 7.93% with the graphene-doped PEDOT:PSS comparable to the ITO doped reference device (8.73%). The same group developed a novel method for fabricating low-temperature ZnO nanoparticles combined with graphene as TCE in OPV device [104]. The ZnO nanoparticles uniformly covered the graphene in an n-doped blend to promote charge transfer from the photoactive layer to the electrode. Although, the PCE of ITO reference device was still higher at 9.13% compared to 8.16% and 7.41% for graphene-ZnO nanoparticles blend on glass and PET substrates respectively, the low-temperature and economic production as well as flexible application of the OPV device makes the graphene-based set-up more valuable.

Chen et al. [105] attempted to reduce the sheet resistance of graphene and improve the low work function by depositing PEDOT:PSS on monolayer graphene. They noted that PEDOT:PSS exhibit low electrical conductivity due to the insulating nature of PSS, so prior to deposition dimethyl sulfoxide (DMSO) was doped with PEDOT:PSS. The DMSO-doped PEDOT:PSS/graphene composite electrode exhibited enhanced light absorption, charge collection and flexibility evident in higher PCE of 4.39% and mechanical resistance to bending compared to the rigid ITO with 4.38% PCE. Meanwhile, Shin et al. [106] successfully reduced the high sheet resistance of graphene and reflectance on one hand by doping with Ag nanowires and increased the ultra violet-visible absorption and flexibility by co-doping with GQDs on the other. The GQDs-mixed Ag nanowires/graphene composite material was employed as TCE in OPV device with excellent bending stabilities of over 90% after 1000 bending cycles and PCE of 3.66%.

4.2   Hole transport layer (HTL)

Unlike PSCs, the most commonly used material for hole transport in OPV devices is poly(3,4-ethylenedioxythiophene):poly(styrenesulfonate) (PEDOT:PSS) [107] which unfortunately suffers from acidity and hygroscopicity [108]. Therefore, a suitable alternative is semiconducting metal oxides or graphene materials with advantageous excellent optical and electrical properties. However, metal oxides are not suitable for low-temperature and solution processable fabrication. This limitation could be dominated by blending with graphene which also maintains the intrinsic charge carrier mobility during production and improve the work function of graphene for application as HTL in OPV devices.

Zheng et al. [109] blend MoS_2 with graphene via a simple liquid-phase exfoliation method to ensure the oxygen-incorporated MoS_2 is contained in both 2H (trigonal prismatic D_{3h}) and 1T-(octahedral O_h) phases with tunable work function. The MoS_2-graphene composite was shown to have the lowest unoccupied molecular orbital (LUMO) level of *ca.* −3.9 eV and the highest occupied molecular orbital (HOMO) level of *ca.* −5.7 eV making it suitable as a p-type hole extraction layer material. The material was employed as an interlayer between PEDOT:PSS/ITO to improve hole transport via distribution of an electric field to extract charges from the absorbed excitons. Subsequently, the layer combination ITO/GMoS$_2$/PEDOT:PSS/active-layer/Ca/Ag gave a PCE of 9.4% which is higher than 9.0% for ITO/PEDOT:PSS/active-layer/Ca/Ag as well as better lifetime stability since the GMoS$_2$ layer act as protective tissue for the PEDOT:PSS against hygroscopisity. Meanwhile Dang et al. [110] removed the PEDOT:PSS layer and used graphene-MoO$_3$ as the HTL to achieve better hole mobility and higher work function in OPV device with PCE of 7.07% .

4.3   Electron acceptor and donor

The cathode and anode influences to a large extent the efficiency of OPV devices. The electron donor or cathode is associated with the HTL which makes electron available for the acceptor or anode. Eventually, both electrodes are crucial for efficient dissociation of excitons. Graphene quantum dots (GQDs) are small pieces of graphene sheets less than 100 nm size. Together with the intrinsic properties of graphene they possess low sheet resistance [111], tunable work function [112], and varying bandgaps [113]. Therefore, GQDs can be incorporated into the active layer of OPV devices as electron acceptor part of a binary layer or second electron acceptor in ternary solar cells.

Wu et al. [114] fabricated a binary (ITO/ZnO/P3HT:GQDs/MoO$_3$/Ag) and ternary (ITO/ZnO/P3HT:PCBM:GQDs/MoO$_3$/Ag) device to investigate the electronic energy states among the GQDs, the donor (P3HT) and acceptor (PCBM). They discovered that

Materials Research Foundations **88** (2021) 29-61

https://doi.org/10.21741/9781644901090-2

the HOMO energy levels of GQDs (–5.67 eV) matched with PCBM (–6.1 eV) and P3HT (–5.2 eV) to promote efficient exciton separation at their interfaces, that is, hole transport from PCBM to P3HT and electron transport from P3HT to PCBM . Also the oxygen-containing group on GQDs encourages high reactivity with functional groups. Wang et al. [115] prepared amino-functionalized GQDs as cathode interlayer, incorporating it with ZnO. The as-fabricated OPV device (ITO)/ZnO/AF-GQDs/poly[[4,8-bis[(2-ethylhexyl)oxy]benzo[1,2-b:4,5-b′]dithiophene -2,6-diyl][3-fluoro-2-[(2-ethylhexyl)carbonyl]thieno[3,4-b]thiophenediyl]](PTB7):[6,6]-phenyl-C71-butylic acid methyl ester (PC71BM)/MoO₃/Al gave 10.14% PCE with amino-functionalized GQDs (AF-GQDs) and 8.76% without it.

5. Graphene/silicon Schottky junction

Work function makes the whole difference in Schottky junction. Therefore, a metal can form a Schottky junction with any semiconductor provided they exhibit suitable work function difference. The zero band gap energy level of graphene pass it easily as a metal, coupled with its work function range 4.89 – 5.16 eV [116] can form Schottky junction with varieties of semiconductors; Si [117], GaAs [118], InP [119], and CdTe [120]. With the work function of Si ranging between 4.05 – 4.5 eV and other fascinating conductive properties, graphene/ silicon Schottky junction is very attractive. Unlike conventional p-n junction solar cells, the graphene in graphene/silicon Schottky junction acts as the transparent conductive electrode, the active layer for electron-hole separation, and hole transporting layer. This multi-functional graphene limits the number of components in graphene/silicon Schottky junction solar cells to; a top electrode layer for creating built-in potential to aid separation of exciton, a photoactive layer for absorbing photon to generate exciton, and a bottom electrode layer for collecting electrons/holes.

Li et al. [121] in 2010 first reported a PCE of 1.65% which has risen to over 15% within the decade. This rapid development is a result of adopting interfacial layer between the graphene and silicon to suppress electron recombination, doping the graphene to increase carrier concentration and conductivity, and application of antireflective coating on the silicon to increase light trapping and photon absorption. Shin et al. [122] attempted to improve the PCE of graphene/silicon Schottky junction solar cells by simply incorporating multilayer graphene nanomeshes and porous silicon to increase the bandgap and light absorption respectively. They reported an overall efficiency of 6.78% with a long-term stability of 1% in 1000 h usage. In the same vein, Rehman et al. [123] directly grew graphene on bare silicon. They showed that increase in graphene thickness increases the work function and absorption in the photoactive layer, and attempted to further elucidate the effect of this parameter in carrier recombination. Although, 5.51%

PCE was achieved, the efficiency was further increased to 9.18% when graphene was doped with HNO_3 to intensify the sp^2 and sp^3 hybrid and coated with antireflective polymethyl methacrylate (PMMA).

The poor bandgap energy of graphene can as well be improved by doping with a transition metal dichalcogenides with bandgap between 1-2 eV. Therefore, high crystalline MoS_2 monolayer was employed as passivation layer between graphene and silicon. The layer also function as hole transport and electron blocking at the interface to give a very high PCE of 15.8% after both the graphene and MoS_2 layer were coated with PMMA and TiO_2 was additionally coated on graphene as TCO [124].

Also, in a bid to reduce charge recombination, interfacial layers acting as insulators were added between the graphene and silicon [125–127]. Gnisci et al. [125] employed graphene-based derivatives as interfacial layers to achieve moderate PCE increase from 2.1% to 3.5% and 6.7% when doped with HNO_3. Rehman et al. [126] used Al_2O_3 and reported a high PCE of 8.4% from 3.6% without doping, while Alnuaimi et al. [127] used HfO_2 to increase the efficiency from 3.9% to 9.1% with PMMA coating.

Conclusion

From the foregoing, it is evident that the three popular third generation solar cells; that is, dye-sensitized, perovskite and organic photovoltaic are similar in architecture save for a few variations like dye-sensitized require a photosensitizer and electrolyte, perovskite material is the active layer for PSC and organic semiconductor for OPV. The similarity in their design is responsible for the rapid increase in power conversion efficiency of PSC as the dye was easily replaced with a perovskite material to prevent leakages and contamination.

Graphene is the championing material of the proposed success of the third generation solar cell due to its application in all aspects of the various architectures. This chapter has successfully reviewed application of graphene as semiconductor layer, electrolyte and counter-electrode in DSSCs; electrode, perovskite, electron and hole transporting layers in PSCs; and electrode, hole transporting layer and electron acceptor and donor in OPV; in addition to graphene/silicon Schottky junction in recent years increasing the PCE from 1.9% to over 22% with PSCs design.

Specifically, this 2-dimensional carbon allotrope decreases dark current in the semiconductor layer of DSSCs, minimize leakages in electrolyte when employed as gel and reduces the overall cost of the solar cell on substituting Pt as counter electrode. They improved the bending property of PSCs, mitigate grain boundaries effect in perovskite materials and increase charge transport in the ETL. Graphene has shown prominence as

Materials Research Forum LLC
https://doi.org/10.21741/9781644901090-2

electrode material in OPV due to its higher visible light absorption, and as HTL due to its improved work function. This tunable work function property of graphene certified it to easily substitute metals in metal/semiconductor Schottky junction.

Therefore, current researches and future perspectives is expected to develop a more stable design with lifespan comparable to commercially available silicon-based p-n junction, since recent researches have taken care of flexibility and cost.

References

[1] D. Gielen, F. Boshell, D. Saygin, M. D. Bazilian, N. Wagner, R. Gorini, The role of renewable energy in the global energy transformation, Energy Strateg Rev. 24 (2019) 38-50. https://doi.org/10.1016/j.esr.2019.01.006

[2] O.P. Avbenake, R.S. Al-Hajri, B.Y. Jibril, Catalytic upgrading of heavy oil using $NiCo/\gamma-Al_2O_3$ catalyst: Effect of initial atmosphere and water-gas shift reaction, Fuel 235 (2019) 736–743. https://doi.org/10.1016/j.fuel.2018.08.074

[3] BP Statistical Review of World Energy June 2016, http://www.bp.com/content/dam/ bp/pdf/energy-economics/statistical-review-2016/bp-statistical-review-of-world-energy-2016-full-report.pdf (accessed 12 January 2020)

[4] C.J. Cleveland, C. Morris, Photovoltaics, in: Handbook of energy, Volume II: chronologies, top ten lists, and word clouds draws, Elsevier Science (2014). https://doi.org/10.1016/B978-0-12-417013-1.00015-7

[5] D.M. Chapin, C.S. Fuller, G.L. Pearson, A new silicon p-n junction photocell for converting solar radiation into electrical power, J. Appl. Phys 25 (1954) 676–677. https://doi.org/10.1063/1.1721711

[6] D. A. Cusano, CdTe solar cells and photovoltaic heterojunctions in II–VI compounds, Solid State Electron. 6 (1963) 217–232. https://doi.org/10.1016/0038-1101(63)90078-9.

[7] G. Yu, J. Gao, J.C. Hummelen, F. Wudl, A.J. Heeger, Polymer photovoltaic cells: Enhanced efficiencies via a network of internal donor-acceptor heterojunctions, Science 270 (1995) 1789-1791. https://doi.org/10.1126/science.270.5243.1789

[8] J.M. Yun, J.S. Yeo, J. Kim, H.G. Jeong, D.Y. Kim, Y.J. Noh, S.S. Kim, B.C. Ku, S.I. Na, Solution-processable reduced graphene oxide as a novel alternative to PEDOT:PSS hole transport layers for highly efficient and stable polymer solar cells, Adv. Mater. 23 (2011) 4923–4928. https://doi.org/10.1002/adma.201102207

[9] Y. Areerob, K.Y. Cho, C.H. Jung, W.C. Oh, Synergetic effect of $La_2CdSnTiO_4$-WSe_2 perovskite structured nanoparticles on graphene oxide for high efficiency of dye sensitized solar cells, J. Alloys Compd. 775 (2019) 690-697. https://doi.org/10.1016/j.jallcom.2018.10.189

[10] Z. Pour-mohammadi, M. Amirmazlaghani, Asymmetric finger-shape metallization in Graphene-on-Si solar cells for enhanced carrier trapping, Mat. Sci. Semicon. Proc. 91 (2019) 13–21. https://doi.org/10.1016/j.mssp.2018.11.002

[11] B. Mazumder, Production of high purity silicon for solar cell and electronic applications by trichlorosilane process, T. Indian Ceram. Soc. 40 (1981) 155-159. https://doi.org/10.1080/0371750X.1981.10822539

[12] A. Ramos-Ruiz, J.V. Wilkening, J.A. Field, R. Sierra-Alvarez, Leaching of cadmium and tellurium from cadmium telluride (CdTe) thin-film solar panels under simulated landfill conditions, J. Hazard Mater. 336 (2017) 57–64. https://doi.org/10.1016/j.jhazmat.2017.04.052

[13] X.M. Li, T.S. Zhao, H.W. Zhu, Quantum dot and heterojunction solar cells containing carbon nanomaterials, in: W. Lu, J.B. Baek, L. Dai (Eds.), Carbon nanomaterials for advanced energy systems: Recent advancements in materials syntheses and device applications, John Wiley & Sons, Inc., 2014, pp. 237-266. https://doi.org/10.1002/9781118980989.ch7

[14] A.K. Geim, K.S. Novoselov, The rise of graphene, Nat. Mater. 6 (2007) 183–191. https://doi.org/10.1038/nmat1849

[15] J. Zhu, D. Yang, Z. Yin, Q. Yan, H. Zhang, Graphene and graphene-based materials for energy storage applications, Small 10 (2014) 3480–3498. https://doi.org/10.1002/smll.201303202

[16] K.P. Loh, S.W. Tong, J. Wu, Graphene and graphene-like molecules: Prospects in solar cells, J. Am. Chem. Soc. 138 (2016) 1095-1102. https://doi.org/10.1021/jacs.5b10917

[17] R. Garg, S. Elmas, T. Nann, M.R. Andersson, Deposition methods of graphene as electrode material for organic solar cells, Adv. Energy Mater. (2016) 1601393. https://doi.org/10.1002/aenm.201601393

[18] N.P.D. Ngidi, M.A. Ollengo, V.O. Nyamori, Heteroatom-doped graphene and its application as a counter electrode in dye-sensitized solar cells, Int J Energ Res. (2018) 1–33. https://doi.org/10.1002/er.4326

[19] D.H. Kweon, J.-B. Baek, Edge-functionalized graphene nanoplatelets as metal- free electrocatalysts for dye-sensitized solar cells, Adv. Mater. (2018) 1804440. https://doi.org/10.1002/adma.201804440

[20] C. Ge, Md.M Rahman, N.C.D. Nath, M.J. Ju, K.-M. Noh, J.-J Lee, Graphene-incorporated photoelectrodes for dye-sensitized solar cells, B. Korean Chem. Soc. 36 (2015) 762–771. https://doi.org/10.1002/bkcs.10140.

[21] Z. Yin, J. Zhu, Q. He, X. Cao, C. Tan, H. Chen, Q. Yan, H Zhang, Graphene-based materials for solar cell applications, Adv. Energy Mater. 4 (2014) 1300574. https://doi.org/10.1002/aenm.201300574

[22] T.H. Chowdhury, A. Islam, A.K.M. Hasan, M.A.M. Terdi, M. Arunakumari, S.P. Singh, M.K. Alam, I.M. Bedja, M.H. Ruslan, K. Sopian, N. Amin, M. Akhtaruzzaman, Prospects of graphene as a potential carrier-transport material in third-generation solar cells, Chem. Rec. 16 (2016) 614–632. https://doi.org/10.1002/tcr.201500206

[23] Y. Sun, W. Zhang, H. Chi, Y. Liu, C. Hou, D. Fang, Recent development of graphene materials applied in polymer solar cell, Renew. Sustain. Energ. Rev. 43 (2015) 973–980. https://doi.org/10.1016/j.rser.2014.11.040

[24] J. Ouyang, Applications of carbon nanotubes and graphene for third-generation solar cells and fuel cells, Nano. Materials Science 1 (2019) 77–90. https://doi.org/10.1016/j.nanoms.2019.03.004

[25] C.A. Ubani, M.A. Ibrahim, M.A.M. Teridi, K. Sopian, J. Ali, K.T. Chaudhary, Application of graphene in dye and quantum dots sensitized solar cell, Sol. Energy 137 (2016) 531–550. https://doi.org/10.1016/j.solener.2016.08.055

[26] F.W. Low, C.W. Lai, Recent developments of graphene-TiO_2 composite nanomaterials as efficient photoelectrodes in dye-sensitized solar cells: A review, Renew. Sust. Energ. Rev. 82 (2018) 103–125. https://doi.org/10.1016/j.rser.2017.09.024

[27] Y. Zhang, H. Li, L. Kuo, P. Dong, F. Yan, Recent applications of graphene in dye-sensitized solar cells, Curr. Opin. Colloid Interface Sci. 20 (2015) 406–415. https://doi.org/10.1016/j.cocis.2015.11.002

[28] T. Mahmoudi, Wang, Y., Hahn, Y.B., Graphene and its derivatives for solar cells application, Nano Energy 47 (2018) 51–65. https://doi.org/10.1016/j.nanoen.2018.02.047

[29] M.Z. Iqbal, A.-U. Rehman, Recent progress in graphene incorporated solar cell devices, Sol. Energy 169 (2018) 634–647. https://doi.org/10.1016/j.solener.2018.04.041

[30] K. Parvez, R. Li, K. Müllen, Graphene as transparent electrodes for solar cells, in: Feng, X., (Ed.), Nanocarbons for advanced energy conversion, Wiley-VCH Verlag GmbH & Co. KGaA., 2015, pp. 249-280. https://doi.org/10.1002/9783527680016.ch10

[31] M.A. Mat-Teridi, M.A. Ibrahim, N. Ahmad-Ludin, S.N.F.M. Nasir, M.Y. Sulaiman, K. Sopian, Graphene as sensitizer, in: Yusoff, A.R.M., (Ed), Graphene-based energy devices, Wiley-VCH Verlag GmbH & Co. KGaA., 2015, pp. 407-430. https://doi.org/10.1002/9783527690312.ch16

[32] J.Z. Wu, Graphene, in: Levy, D., Castellón, E., (Eds), Transparent conductive materials: materials, synthesis, characterization, applications, Wiley-VCH Verlag GmbH & Co. KGaA., 2018, pp. 165-192. https://doi.org/10.1002/9783527804603.ch3_2

[33] A. Kalluri, D. Debnath, B. Dharmadhikari, P Patra, Graphene quantum dots: synthesis and applications, in: C. V. Kumar (Ed.), Enzyme nanoarchitectures: Enzymes armored with graphene, Elsevier Inc., 2018, pp. 335-354. https://doi.org/10.1016/bs.mie.2018.07.002

[34] S.F. Adil, M. Khan, D. Kalpana, Graphene-based nanomaterials for solar cells, in: Z. L. Meidan, Y. M. Wang, (Eds.), Multifunctional photocatalytic materials for energy, Woodhead Publishing, 2018, pp. 127-152. https://doi.org/10.1016/B978-0-08-101977-1.00008-9

[35] L. Givalou, D. Tsichlis, F. Zhang, C.-S. Karagianni, M. Terrones, K. Kordatos, P. Falaras, Transition metal – graphene oxide nanohybrid materials as counter electrodes for high efficiency quantum dot solar Cells, Catal. Today (2019) In press. https://doi.org/10.1016/j.cattod.2019.03.035

[36] N. Balis, E. Stratakis, E. Kymakis, Graphene and transition metal dichalcogenide nanosheets as charge transport layers for solution processed solar cells, Mater. Today 19 (2016) 580-594. https://doi.org/10.1016/j.mattod.2016.03.018

[37] C. Ciceroni, A. Agresti, A. Di Carlo, F. Brunetti, Graphene oxide for DSSC, OPV and perovskite stability, in: M. Lira-Cantu (Ed.), The future of semiconductor oxides in next-generation solar cells. Elsevier Inc., 2018, pp. 503-531. https://doi.org/10.1016/B978-0-12-811165-9.00013-2

[38] E. Kymakis, D. Konios, Graphene oxide-like materials in organic and perovskite solar cells, in: M. Lira-Cantu (Ed.), The future of semiconductor oxides in next-generation solar cells. Elsevier Inc., 2018, pp. 357-394. https://doi.org/10.1016/B978-0-12-811165-9.00009-0

[39] R. Szostak, A. Morais, S.A. Carminati, S.V. Costa, P.E. Marchezi, A.F. Nogueira, Application of graphene and graphene derivatives/oxide nanomaterials for solar cells, in: M. Lira-Cantu (Ed.), The Future of semiconductor oxides in next- generation solar cells. Elsevier Inc., 2018, pp. 395-437. https://doi.org/10.1016/B978-0-12-811165-9.00010-7

[40] P.V. Kamat, Graphene-based nanoassemblies for energy conversion, J. Phys. Chem. Lett. 2 (2011) 242–251. https://doi.org/10.1021/jz101639v

[41] J.V. Milic´, N. Arora, M.I. Dar, S.M. Zakeeruddin, M. Grätzel, Reduced graphene oxide as a stabilizing agent in perovskite solar cells, Adv. Mater. Interfaces (2018) 1800416. https://doi.org/10.1002/admi.201800416

[42] S.K. Balasingam, Y. Jun, Recent progress on reduced graphene oxide-based counter electrodes for cost-effective dye-sensitized solar cells, Isr. J. Chem. 55 (2015) 955–965. https://doi.org/10.1002/ijch.201400213

[43] W.-R. Liu, Graphene-based energy devices, in: Yusoff, A.R.M., (Ed.), Graphene-based energy devices, Wiley-VCH Verlag GmbH & Co. KGaA., 2015, pp. 85-121. https://doi.org/10.1002/9783527690312.ch3

[44] L.C. Cotet, C.I. Fort, L.C. Pop, M. Baia, L. Baia, Insights into graphene-based materials as counter electrodes for dye-sensitized solar cells, in: M. Soroush, K. K.S. Lau (Eds.), Dye-sensitized solar cells: mathematical modelling and materials design and optimization, Academic Press, 2019, pp. 341-396. https://doi.org/10.1016/B978-0-12-814541-8.00010-0

[45] B. O'Regan, M. Grätzel, A low-cost, high-efficiency solar cell based on dye-sensitized colloidal TiO_2 films, Nature 353 (1991) 737–740. https://doi.org/10.1038/353737a0

[46] M. Grätzel, Solar energy conversion by dye-sensitized photovoltaic cells, Inorg. Chem. 44 (2005) 6841-6851. https://doi.org/10.1021/ic0508371

[47] T.N. Murakami, M. Grätzel, Counter electrodes for DSC: application of functional materials as catalysts, Inorg. Chim. Acta 361 (2008) 572–580. https://doi.org/10.1016/j.ica.2007.09.025

[48] W. Cho, Y. R. Kim, D. Song, H. W. Choi, Y. S. Kang, High-efficiency solid-state polymer electrolyte dye-sensitized solar cells with a bi-functional porous layer, J. Mater. Chem. A 2 (2014) 17746–17750. https://doi.org/10.1039/c4ta04064c

[49] S. Kment, F. Riboni, S. Pausova, L. Wang, L. Wang, H. Han, Z. Hubicka, J. Krysa, P. Schmuki, R. Zboril, Photoanodes based on TiO_2 and α-Fe_2O_3 for solar water splitting–superior role of 1D nanoarchitectures and of combined heterostructures, Chem. Soc. Rev. 46 (2017) 3716-3769. https://doi.org/10.1039/c6cs00015k

[50] U. Mehmood, S.H.A. Ahmad, A.U.H. Khan, A.A. Qaiser, Co-sensitization of graphene/TiO_2 nanocomposite thin films with ruthenizer and metal free organic photosensitizers for improving the power conversion efficiency of dye-sensitized solar cells (DSSCs), Sol. Energy 170 (2018) 47–55. https://doi.org/10.1016/j.solener.2018.05.051

[51] S. Muduli, W. Lee, V. Dhas, S. Mujawar, M. Dubey, K. Vijayamohanan, S.-H. Han, S. Ogale, Enhanced conversion efficiency in dye-sensitized solar cells based on hydrothermally synthesized TiO_2 −MWCNT nanocomposites, ACS Appl. Mater. Interfaces 1 (2009) 2030–2035. https://doi.org/10.1021/am900396m

[52] S. Sun, L. Gao, Y. Liu, Enhanced dye-sensitized solar cell using graphene-TiO_2 photoanode prepared by heterogeneous coagulation, Appl. Phys. Lett. 96 (2010) 083113. https://doi.org/10.1063/1.3318466

[53] C.S.N.O.A. Sreekala, J. Indiramma, K.B.S.P. Kumar, K. Sreelatha, M. Roy, Functionalized multi-walled carbon nanotubes for enhanced photocurrent in dyesensitized solar cells, J. Nanostructure Chem. 3 (2013) 19. https://doi.org/10.1186/2193-8865-3-19

[54] S.A. Kazmi, S. Hameed, A.S. Ahmed, M. Arshad, A. Azam, Electrical and optical properties of graphene-TiO_2 nanocomposite and its applications in dye sensitized solar cells (DSSC), J. Alloys Compd. 691 (2017) 659–665. https://doi.org/10.1016/j.jallcom.2016.08.319.

[55] K. Nemade, P. Dudhe, P. Tekade, Enhancement of photovoltaic performance of polyaniline/graphene composite-based dye-sensitized solar cells by adding TiO_2 nanoparticles, Solid State Sci. 83 (2018) 99–106. https://doi.org/10.1016/j.solidstatesciences.2018.07.009

[56] S.-Q. Fan, C. Kim, B. Fang, K.-X. Liao, G.-J. Yang, C.-J. Li, J.-J. Kim, J. Ko, Improved efficiency of over 10% in dye-sensitized solar cells with a ruthenium

Materials Research Forum LLC

https://doi.org/10.21741/9781644901090-2

complex and an organic dye heterogeneously positioning on a single TiO_2 electrode, J. Phys. Chem. C 115 (2011) 7747–7754. https://doi.org/10.1021/jp200700e

[57] S. Mathew, A. Yella, P. Gao, R. Humphry-Baker, B.F.E. Curchod, N. Ashari-Astani, I. Tavernelli, U. Rothlisberger, M.K. Nazeeruddin, M. Grätzel, Dye-sensitized solar cells with 13% efficiency achieved through the molecular engineering of porphyrin sensitizers, Nat. Chem. 6 (2014) 242–247. https://doi.org/10.1038/nchem.1861

[58] K.S.K. Lo, W.W.F. Leung, Dye-sensitized solar cells with shear-exfoliated graphene, Sol. Energy 180 (2019) 16–24. https://doi.org/10.1016/j.solener.2018.12.077

[59] L. Chen, Y. Zhou, W. Tu, Z. Li, C. Bao, H. Dai, T. Yu, J. Liu, Z. Zou, Enhanced photovoltaic performance of a dye-sensitized solar cell using graphene–TiO_2 photoanode prepared by a novel in situ simultaneous reduction-hydrolysis technique, Nanoscale 5 (2013) 3481–3485. https://doi.org/10.1039/c3nr34059g

[60] Z. He, G. Guai, J. Liu, C. Guo, J.S.C. Loo, C.M. Li, T.T. Yang Tan, Nanostructure control of graphene-composited TiO_2 by a one-step solvothermal approach for high performance dye-sensitized solar cells, Nanoscale 3 (2011) 4613–4616. https://doi.org/10.1039/c1nr11300c

[61] A.A. Madhavan, S. Kalluri, D.K. Chacko, T.A. Arun, S. Nagarajan, K.R.V. Subramanian, A.S. Nair, S.V. Nair, A. Balakrishnan, Electrical and optical properties of electrospun TiO_2-graphene composite nanofibers and its application as DSSC photo-anodes, RSC Adv. 2 (2012) 13032–13037. https://doi.org/10.1039/c2ra22091a

[62] S.N. Sadikin, M.Y.A. Rahman, A.A. Umar, T.H.T. Aziz, Improvement of dye sensitized solar cell performance by utilizing graphene-coated TiO_2 films photoanode, Superlattice Microst. 128 (2019) 92–98. https://doi.org/10.1016/j.spmi.2019.01.014

[63] D.K. Kumar, D. Suazo-Davila, D. García-Torres, N.P. Cook, A. Ivaturi, M.-H. Hsu, A.A. Martí, C.R. Cabrera, B. Chen, N. Bennett, H.M. Upadhyaya, Low-temperature titania-graphene quantum dots paste for flexible dye-sensitised solar cell applications, Electrochim. Acta 305 (2019) 278-284. https://doi.org/10.1016/j.electacta.2019.03.040.

[64] C.-H. Tsai, P.Y Chuang, H.L. Hsu, Adding graphene nanosheets in liquid electrolytes to improve the efficiency of dye-sensitized solar cells, Mater. Chem. Phys. 207 (2018) 154-160. https://doi.org/10.1016/j.matchemphys.2017.12.059

[65] S. Rehman, M. Noman, A.D. Khan, A. Saboor, M.S. Ahmad, H.U. Khan, Synthesis of polyvinyl acetate/graphene nanocomposite and its application as an electrolyte in

dye sensitized solar cells, Optik 202 (2020) 163591.
https://doi.org/10.1016/j.ijleo.2019.163591

[66] L. Liu, Y. Wu, F. Chi, Z. Yi, H. Wang, W. Li, Y. Zhang, X. Zhang, An efficient quasi-solid-state dye-sensitized solar cell with gradient polyaniline-graphene/PtNi tailored gel electrolyte, Electrochim. Acta 316 (2019) 125-132.
https://doi.org/10.1016/j.electacta.2019.05.115

[67] K.C. Sun, A.A. Arbab, I.A. Sahito, M.B. Qadir, B.J. Choi, S.C. Kwon, S.Y. Yeo, S.C. Yi, S.H. Jeong, A PVdF-based electrolyte membrane for a carbon counter electrode in dyesensitized solar cells, RSC Adv. 7 (2017) 20908-20918.
https://doi.org/10.1039/C7RA00005G

[68] B. Siwach, D. Mohan, K.K. Singh, A. Kumar, M. Barala, Effect of carbonaceous counter electrodes on the performance of ZnO-graphene nanocomposites based dye sensitized solar cells, Ceram. Int. 44 (2018) 21120-21126.
https://doi.org/10.1016/j.ceramint.2018.08.151

[69] K. Robinson, G.R.A. Kumara, R.J.G.L.R. Kumara, E.N. Jayaweera, R.M.G. Rajapakse, SnO2/ZnO composite dye-sensitized solar cells with graphene-based counter electrodes, Org. Electron. 56 (2018) 159-162.
https://doi.org/10.1016/j.orgel.2018.01.040

[70] D.H. Seo, M. Batmunkh, J. Fang, A.T. Murdock, S. Yick, Z. Han, C.J. Shearer, T.J. Macdonald, M. Lawn, A. Bendavid, J.G. Shapter, K.K. Ostrikov, Ambient air synthesis of multi-layer CVD graphene films for low-cost, efficient counter electrode material in dye-sensitized solar cells, Flat. Chem. 8 (2018) 1–8.
https://doi.org/10.1016/j.flatc.2018.02.002

[71] C.A. Tseng, C.P. Lee, Y.J. Huang, H.W. Pang, K.C. Ho, Y.T Chen, One-step synthesis of graphene hollow nanoballs with various nitrogen-doped states for electrocatalysis in dye-sensitized solar cells, Mater. Today Energy 8 (2018) 15-21.
https://doi.org/10.1016/j.mtener.2018.02.006

[72] X. Meng, C. Yu, X. Song, J. Iocozzia, J. Hong, M. Rager, H. Jing, S. Wang, L. Huang, J. Qiu, Z. Lin, Scrutinizing defects and defect density of selenium- doped graphene for high-efficiency triiodide reduction in dye-sensitized solar cells, Angew. Chem. Int. Ed. 57 (2018) 4682-4686. https://doi.org/10.1002/anie.201801337

[73] R.S. Ganesh, K. Silambarasan, E. Durgadevi, M. Navaneethan, S. Ponnusamy, C.Y. Kong, C. Muthamizhchelvan, Y. Shimura, Y. Hayakawa, Metal sulfide nanosheet–nitrogen-doped graphene hybrids as low-cost counter electrodes for dye- sensitized

solar cells, Appl. Surf. Sci. 480 (2019) 177–185.
https://doi.org/10.1016/j.apsusc.2019.02.251

[74] C.K. Kim, H.M. Kim, M. Aftabuzzaman, I.-Y. Jeon, S.H. Kang, Y.K. Eom, J.B. Baek, H.K. Kim, Comparative study of edge-functionalized graphene nanoplatelets as metal free counter electrodes for highly efficient dye-sensitized solar cells, Mater. Today Energy 9 (2018) 67-73. https://doi.org/10.1016/j.mtener.2018.05.003

[75] V. Murugadoss, P. Panneerselvam, C. Yan, Z. Guo, S. Angaiah, A simple one-step hydrothermal synthesis of cobalt-nickel selenide/graphene nanohybrid as an advanced platinum free counter electrode for dye sensitized solar cell, Electrochim. Acta 312 (2019) 157-167. https://doi.org/10.1016/j.electacta.2019.04.142

[76] A.H. Alami, K. Aokal, D. Zhang, A. Taieb, M. Faraj, A. Alhammadi, J.M. Ashraf, B. Soudan, J. El Hajjar, M. Irimia-Vladu, Low-cost dye-sensitized solar cells with ball-milled tellurium-doped graphene as counter electrodes and a natural sensitizer dye, Int. J. Energy Res. (2019) 1–10. https://doi.org/10.1002/er.4684

[77] B. Zhou, X. Zhang, P. Jin, X. Li, X. Yuan, J. Wang, L. Liu, Synthesis of In$_{2.77}$S$_4$ nanoflakes/graphene composites and their application as counter electrode in dye-sensitized solar cells, Electrochim. Acta 281 (2018) 746-752. https://doi.org/10.1016/j.electacta.2018.06.031

[78] Y. Areerob, J.Y. Cho, W.K. Jang, K.Y. Cho, W.-C. Oh, An alternative of NiCoSe doped graphene hybrid La$_6$W$_2$O$_{15}$ for renewable energy conversion used in dye-sensitized solar cells, Solid State Ionics 327 (2018) 99–109. https://doi.org/10.1016/j.ssi.2018.10.026

[79] M.U. Rahman, F. Xie, Y. Li, X. Sun, M. Wei, Grafting cobalt sulfide on graphene nanosheets as a counterelectrode for dye-sensitized solar cells, J. Alloys Compd. 808 (2019) 151701. https://doi.org/10.1016/j.jallcom.2019.151701

[80] B. Pang, S. Lin, Y. Shi, Y. Wang, Y. Chen, S. Ma, J. Feng, C. Zhang, L. Yu, L. Dong, Synthesis of CoFe$_2$O$_4$/graphene composite as a novel counter electrode for high performance dye-sensitized solar cells, Electrochim. Acta 297 (2019) 70-76. https://doi.org/10.1016/j.electacta.2018.11.170

[81] A. Kojima, K. Teshima, Y. Shirai, T. Miyasaka, Organometal halide perovskites as visible-light sensitizers for photovoltaic cells, J. Am. Chem. Soc. 131 (2009) 6050–6051. https://doi.org/10.1021/ja809598r

[82] J. M. Kim, C. W. Jang, J. H. Kim, S. Kim, S.-H. Choi, Use of AuCl$_3$-doped graphene as a protecting layer for enhancing the stabilities of inverted perovskite solar

cells, Appl. Surf. Sci. 455 (2018) 1131–1136.
https://doi.org/10.1016/j.apsusc.2018.06.068

[83] H. Zhou, Q. Chen, G. Li, S. Luo, T. Song, H. Duan, Z. Hong, J. You, Y. Liu, Y. Yang, Interface engineering of highly efficient perovskite solar cells, Science 345 (2014) 542–546. https://doi.org/10.1126/science.1254050

[84] Best Research-Cell Efficiency https://www.nrel.gov/pv/assets/pdfs/best-research-cell-efficiencies-190416.pdf. (Accessed 12 January 2020)

[85] C.W. Jang, J.M. Kim, S.-H. Choi, Lamination-produced semi-transparent/flexible perovskite solar cells with doped-graphene anode and cathode, J. Alloys Compd. 775 (2019) 905-911. https://doi.org/10.1016/j.jallcom.2018.10.190

[86] S. Kim, H.S. Lee, J.M. Kim, S.W. Seo, J.H. Kim, C.W. Jang, S.-H. Choi, Effect of layer number on flexible perovskite solar cells employing multiple layers of graphene as transparent conductive electrodes, J. Alloys Compd. 744 (2018) 404-411. https://doi.org/10.1016/j.jallcom.2018.02.136

[87] S. Kim, S.H. Shin, S.-H. Choi, N-i-p-type perovskite solar cells employing n-type graphene transparent conductive electrodes, J. Alloys Compd. 786 (2019) 614-620. https://doi.org/10.1016/j.jallcom.2019.01.372

[88] D.H. Shin, J.M. Kim, S.H. Shin, S.-H. Choi, Highly-flexible graphene transparent conductive electrode/perovskite solar cells with graphene quantum dots-doped PCBM electron transport layer, Dyes Pigments 170 (2019) 107630. https://doi.org/10.1016/j.dyepig.2019.107630

[89] X. Gan, S. Yang, J. Zhang, G. Wang, P. He, H. Sun, H. Yuan, L. Yu, G. Ding, Y. Zhu, Graphite-N doped graphene quantum dots as semiconductor additive in perovskite solar cells, ACS Appl. Mater. Interfaces 11 (2019) 37796−37803. https://doi.org/10.1021/acsami.9b13375

[90] Y. Li, W.W.-F. Leung, Introduction of graphene nanofibers into the perovskite layer of perovskite solar cells, Chem.Sus.Chem. 11 (2018) 2921-2929. https://doi.org/10.1002/cssc.201800758

[91] F. Giordano, A. Abate, J.P.C. Baena, M. Saliba, T. Matsui, S.H. Im, S.M. Zakeeruddin, M.K. Nazeeruddin, A. Hagfeldt, M. Graetzel, Enhanced electronic properties in mesoporous TiO_2 via lithium doping for high-efficiency perovskite solar cells, Nat. Commun. 7 (2016) 10379. https://doi.org/10.1038/ncomms10379

Materials Research Forum LLC
https://doi.org/10.21741/9781644901090-2

[92] S. Sidhik, A. Cerdan-Pasaran, D. Esparza, T. Lopez-Luke, R. Carriles, E. De La Rosa, Improving the optoelectronic properties of mesoporous TiO_2 by cobalt doping for high-performance hysteresis free perovskite solar cells, ACS Appl. Mater. Interfaces 10 (2018) 3571-3580. https://doi.org/10.1021/acsami.7b16312

[93] E. Edri, S. Kirmayer, A. Henning, S. Mukhopadhyay, K. Gartsman, Y. Rosenwaks, G. Hodes, D. Cahen, Why lead methylammonium tri-iodide perovskite-based solar cells require a mesoporous electron transporting scaffold (but not necessarily a hole conductor), Nano Lett. 14 (2014) 1000-1004. https://doi.org/10.1021/nl404454h

[94] H. Xia, Z. Ma, Z. Xiao, W. Zhou, H. Zhang, C. Du, J. Zhuang, X. Cheng, X. Liu, Y. Huang, Interfacial modification using ultrasonic atomized graphene quantum dots for efficient perovskite solar cells, Org. Electron. 75 (2019) 105415. https://doi.org/10.1016/j.orgel.2019.105415

[95] M. Zhang, L. Bai, W. Shang, W. Xie, H. Ma, Y. Fu, D. Fang, H. Sun, L. Fan, M. Han, Facile synthesis of water-soluble, highly fluorescent graphene quantum dots as a robust biological label for stem cells, J. Mater. Chem. 22 (2012) 7461–7467. https://doi.org/10.1039/C2JM16835A

[96] X. Meng, X. Cui, M. Rager, S. Zhang, Z. Wang, J. Yu, Y.W. Harn, Z. Kang, B.K. Wagner, Y. Liu, C. Yu, J. Qiu, Z. Lin, Cascade charge transfer enabled by incorporating edge-enriched graphene nanoribbons for mesostructured perovskite solar cells with enhanced performance, Nano Energy 52 (2018) 123–133. https://doi.org/10.1016/j.nanoen.2018.07.028

[97] S. Sidhik, S.S. Panikar, C.R. Perez, T.L. Luke, R. Carriles, S.C. Carrera, E.D.L. Rosa, Interfacial engineering of TiO_2 by graphene nanoplatelets for high efficiency hysteresis-free perovskite solar cells, ACS Sustain. Chem. Eng. 6 (2018) 15391-15401. https://doi.org/10.1021/acssuschemeng.8b03826

[98] D. Shen, W. Zhang, F. Xie, Y. Lia, A. Abate, M. Wei, Graphene quantum dots decorated TiO_2 mesoporous film as an efficient electron transport layer for high-performance perovskite solar cells, J. Power Sources 402 (2018) 320–326. https://doi.org/10.1016/j.jpowsour.2018.09.056

[99] Y.H. Cao, Z.Y. Deng, M.Z. Wang, J. Bai, S.H. Wei, H.J. Feng, Interface engineering of graphene/$CH_3NH_3PbI_3$ heterostructure for novel P-I-N structural perovskites solar cells, J. Phys. Chem. C 122 (2018) 17228-17237. https://doi.org/10.1021/acs.jpcc.8b04042

[100] N.S. Sariciftci, L. Smilowitz, A.J. Heeger, F. Wudl, Photoinduced electron transfer from a conducting polymer to Buckminsterfullerene, Science 258 (1992) 1471-1476. https://doi.org/10.1126/science.258.5087.1474.

[101] L.J. Meng, F. Placido, Annealing effect on ITO thin films prepared by microwave enhanced dc reactive magnetron sputtering for telecommunication applications, Surf. Coat. Tech. 166 (2003) 44-50. https://doi.org/10.1016/S0257-8972(02)00767-3

[102] J.W. Suk, A. Kitt, C.W. Magnuson, Y. Hao, S. Ahmed, J. An, A.K. Swan, B.B. Goldberg, R.S. Ruoff, Transfer of CVD-grown monolayer graphene onto arbitrary substrates, ACS nano 5 (2011) 6916-6924. https://doi.org/10.1021/nn201207c

[103] S. Jung, H. Kim, J. Lee, G. Jeong, H. Kim, J. Park, H. Park, Bio-inspired catecholamine-derived surface modifier for graphene-based organic solar cells, ACS Appl. Energy Mater. 1 (2018) 6463-6468. https://doi.org/10.1021/acsaem.8b01396

[104] S. Jung, J. Lee, J. Seo, U. Kim, Y. Choi, H. Park, Development of annealing free, solution-processable inverted organic solar cells with N-doped graphene electrodes using zinc oxide nanoparticles, Nano Lett. 18 (2018) 1337-1343. https://doi.org/10.1021/acs.nanolett.7b05026

[105] Y. Chen, Y.Y. Yue, S.R. Wang, N. Zhang, J. Feng, H.-B. Sun, Thermally-induced wrinkles on PH1000/graphene composite electrode for enhanced efficiency of organic solar cells, Sol. Energy Mater. Sol. Cells 201 (2019) 110075. https://doi.org/10.1016/j.solmat.2019.110075

[106] D.H. Shin, S.W. Seo, J.M. Kim, H.S. Lee, S.H. Choi, Graphene transparent conductive electrodes doped with graphene quantum dots-mixed silver nanowires for highly flexible organic solar cells, J. Alloys Compd. 744 (2018) 1-6. https://doi.org/10.1016/j.jallcom.2018.02.069

[107] Y.H. Kim, C. Sachse, M.L. Machala, C. May, L. Mueller-Meskamp, K. Leo, Highly conductive PEDOT:PSS electrode with optimized solvent and thermal post-treatment for ITO-free organic solar cells, Adv. Funct. Mater. 21 (2011) 1076–1081. https://doi.org/10.1002/adfm.201002290

[108] E. Voroshazi, B. Verreet, T. Aernouts, P. Heremans, Long-term operational lifetime and degradation analysis of P3HT: PCBM photovoltaic cells, Sol. Energy Mater. Sol. Cells 95 (2011) 1303–1307. https://doi.org/10.1016/j.solmat.2010.09.007

[109] X. Zheng, H. Zhang, Q. Yang, C. Xiong, W. Li, Y. Yan, R.S. Gurney, T. Wang, Solution-processed Graphene-MoS$_2$ heterostructure for efficient hole extraction in

organic solar cells, Carbon 142 (2019) 156-163.
https://doi.org/10.1016/j.carbon.2018.10.038

[110] Y. Dang, Y. Wang, S. Shen, S. Huang, X. Qu, Y. Pang, S.R.P. Silva, B. Kanga, G. Lu, Solution processed hybrid Graphene-MoO₃ hole transport layers for improved performance of organic solar cells, Org. Electron. 67 (2019) 95–100. https://doi.org/10.1016/j.orgel.2019.01.013

[111] M. M. Li, W. Ni, B. Kan, X. J. Wan, L. Zhang, Q. Zhang, G. K. Long, Y. Zuo, Y. S. Chen, Graphene quantum dots as the hole transport layer material for high-performance organic solar cells, Phys. Chem. Chem. Phys. 15 (2013) 18973-18978. https://doi.org/10.1039/C3CP53283F

[112] Z. M. Luo, G. Q. Qi, K. Y. Chen, M. Zou, L. H. Yuwen, X. W. Zhang, W. Huang, L. H. Wang, Microwave-assisted preparation of white fluorescent graphene quantum dots as a novel phosphor for enhanced white-light-emitting diodes, Adv. Funct. Mater. 26 (2016) 2739-2744. https://doi.org/10.1002/adfm.201505044

[113] J. Liu, G. H. Kim, Y. H. Xue, J. Y. Kim, J. B. Baek, M. Durstock, L. M. Dai, Graphene oxide nanoribbon as hole extraction layer to enhance efficiency and stability of polymer solar cells, Adv. Mater. 26 (2014) 786-790. https://doi.org/10.1002/adma.201302987

[114] W. Wu, J. Zhang, W. Shen, M. Zhong, S. Guo, Graphene quantum dots band structure tuned by size for efficient organic solar cells, Phys. Status Solidi 216 (2019) 1900657. https://doi.org/10.1002/pssa.201900657

[115] S. Wang, Z. Li, X. Xu, G. Zhang, Y. Li, Q. Peng, Amino-Functionalized graphene quantum dots as cathode interlayer for efficient organic solar cells: quantum dot size on interfacial modification ability and photovoltaic performance, Adv. Mater. Interfaces 6 (2019), 1801480. https://doi.org/10.1002/admi.201801480

[116] J.S. Park, J.K. Kim, J. Cho, T.T. Seong, Review-group III-nitride-based ultraviolet light-emitting diodes: ways of increasing external quantum efficiency, ECS J. Solid State SC 6 (2017) 42-52. https://doi.org/10.1149/2.0111704jss

[117] A. Suhail, G. Pan, D. Jenkins, K. Islam, Improved efficiency of graphene/Si Schottky junction solar cell based on back contact structure and DUV treatment, Carbon 129 (2018) 520-526. https://doi.org/10.1016/j.carbon.2017.12.053

[118] X. Li, S. Lin, X. Lin, Z. Xu, P. Wang, S. Zhang, H. Zhong, W. Xu, Z. Wu, W. Fang, Graphene/h-BN/GaAs sandwich diode as solar cell and photodetector, Opt. Express 24 (2016) 134-145. https://doi.org/10.1364/OE.24.000134

[119] P. Wang, X. Li, Z. Xu, Z. Wu, S. Zhang, W. Xu, H. Zhong, H. Chen, E. Li, J. Luo, Q. Yu, S. Lin, Tunable graphene/indium phosphide heterostructure solar cells, Nano Energy 13 (2015) 509-517. https://doi.org/10.1016/j.nanoen.2015.03.023

[120] S. Lin, X. Li, S. Zhang, P. Wang, Z. Xu, H. Zhong, Z. Wu, H. Chen, Graphene/CdTe heterostructure solar cell and its enhancement with photo-induced doping, Appl. Phys. Lett. 107 (2015) 191106. https://doi.org/10.1063/1.4935426

[121] X. Li, H. Zhu, K. Wang, A. Cao, J. Wei, C. Li, Y., Jia, Z., Li, X., Li, D., Wu, Graphene-on-silicon Schottky junction solar cells, Adv. Mater. 22 (2010) 2743–2748. https://doi.org/10.1002/adma.200904383

[122] D.H. Shin, J.H. Kim, D.H. Jung, S.-H. Choi, Graphene-nanomesh transparent conductive electrode/porous-Si Schottky-junction solar cells, J. Alloys Compd. 803 (2019) 958-963. https://doi.org/10.1016/j.jallcom.2019.06.264

[123] M.A. Rehman, S.B. Roy, I. Akhtar, M.F. Bhopal, W. Choi, G. Nazir, M.F. Khan, S. Kumar, J. Eom, S.-H. Chun, Y. Seo, Thickness-dependent efficiency of directly grown graphene based solar cells, Carbon 148 (2019) 187-195. https://doi.org/10.1016/j.carbon.2019.03.079

[124] J. Ma, H. Bai, W. Zhao, Y. Yuan, K. Zhang, High efficiency graphene/MoS$_2$/Si Schottky barrier solar cells using layer controlled MoS$_2$ films, Sol. Energy 160 (2018) 76–84. https://doi.org/10.1016/j.solener.2017.11.066

[125] A. Gnisci, G. Faggio, L. Lancellotti, G. Messina, R. Carotenuto, E. Bobeico, P.D. Veneri, A. Capasso, T. Dikonimos, N. Lisi, The role of graphene-based derivative as interfacial layer in graphene/n-Si Schottky barrier solar cells, Phys. Status Solidi A 216 (2018) 1800555. https://doi.org/10.1002/pssa.201800555

[126] M.A. Rehman, I. Akhtar, W. Choi, K. Akbar, A. Farooq, S. Hussain, M.A. Shehzad, S.H. Chun, J. Jung, Y. Seo, Influence of an Al$_2$O$_3$ interlayer in a directly grown graphene-silicon Schottky junction solar cell, Carbon 132 (2018) 157-164. https://doi.org/10.1016/j.carbon.2018.02.042

[127] A. Alnuaimi, I. Almansouri, I. Saadat, A. Nayfeh, High performance graphene silicon Schottky junction solar cells with HfO$_2$ interfacial layer grown by atomic layer deposition, Sol. Energy 164 (2018) 174–179. https://doi.org/10.1016/j.solener.2018.02.020

Materials for Solar Cell Technologies I
Materials Research Foundations **88** (2021) 62-85

Materials Research Forum LLC
https://doi.org/10.21741/9781644901090-3

Chapter 3

Carbon Nanomaterials Beyond Graphene for Solar Cell and Electrochemical Sensing

Fethi Achi[1,*], Abdellah Henni[2], Sabah Menaa[1], Amira Bensana[3]

[1]Laboratory of valorization and promotion of Saharan resources, University of Kasdi Merbah, Ouargla, Algeria

[2] Laboratory of Dynamics, Interactions and Reactivity Systems DIRS, Departement of Process Engineering University of Kasdi Merbah, Ouargla, Algeria

[3] Laboratory of Chemical Process Engineering, Departement of process engineering, Faculty of Technology, Ferhat Abbas University Sétif-1-, Setif 19000, Algeria

* achifethi@hotmail.fr

Abstract

Carbon-based nanomaterials have different structures with excellent physical and electronic properties. Graphene and carbon nanomaterials are widely used in sensing areas due to its high positive effect on the response of modified electrodes. Their presence increases sensitivities and gives the lower detection limits and enhances the analytical performance of biosensors for food safety and environmental monitoring. In addition, carbon nanomaterials play an important role for the good exploitation of solar energy by developing new structures of silicon-based photovoltaic cells. In this work we report the effect of the most recent graphene and carbon nonmaterial used for electrochemical detection of substances. This chapter also presents an overview of solar cell synthesis using graphene and carbon nanomaterials.

Keywords

Carbon, Nanomaterials, Graphene, Solar Cells, Reduced Graphene Oxide, Food Safety, Environmental Monitoring

Contents

Carbon Nanomaterials Beyond Graphene for Solar Cell and Electrochemical Sensing...62

1.   Introduction..63

2.   **Solar cells based on carbon nanomaterials** **64**

    2.1   Dye-sensitized solar cells based on carbon nanomaterials 65

    2.3   Organic solar cells based on carbon nanomaterials 67

    2.4   Perovskite solar cells based on carbon nanomaterials 69

    2.5   All-carbon solar cells 70

3.   **Carbon based-materials for the construction of biosensors** **71**

    3.1   Functionalizing graphene 71

    3.2   Functionalizing carbon nanotubes 72

    3.2.1   Ionic liquids for functionalizing carbon nanotubes 72

    3.3   3-dimensional porous structure 73

    3.3.1   Reduced graphene oxide (rGO)based biosensors 73

Conclusions **74**

References **75**

1.   Introduction

In recent years, energy has been the subject of real social debate (liberalization of markets, price increases, emission of greenhouse gases, renewable energy, global warming, etc.). The industrialized world is totally dependent on energy in all its forms. This is the engine of our dynamism, the support of our indolence, the basis of industrial and agricultural productivity and access to consumer goods.

Access to renewable and sustainable energy sources is humanity's greatest challenge for the 21st century. All of the activities of modern society are very energy dependent. In general, all countries (and in particular the developing countries) aspire to absolute industrial and economic growth which cannot be achieved without having access to important sources of energy. Some countries have natural energy resources, but most countries have no resources and must rely on the import of energy carriers such as oil and carbon. Furthermore, petroleum-based energy sources have negative consequences for the environment. It is therefore essential to develop alternative energy sources to polluting resources.

The earth is subjected to solar irradiation with an average power of 174,000 Tera Watt (TW) and thus, each year, the earth receives approximately 3,850 Zetta joules (ZJ) of energy from the sun [1].

Materials Research Forum LLC
https://doi.org/10.21741/9781644901090-3

The most widespread and efficient current exploitation of this energy is ensured by the use of photovoltaic (PV) devices. Silicon-based photovoltaic cells are the most common these days. Despite the fact that these cells are the most energy efficient (24% efficiency) these are not very accessible by populations because of their high production cost. Therefore, the development of new structures and new materials is necessary for more attractive PV cells.

Figure 1 [2] published by the National Renewable Energy Laboratory (NREL) shows trends by family of technologies, from the oldest to emerging solutions and the evolution of the efficiency of different technologies. These results illustrate the efficiencies achieved for the best cells in the laboratory.

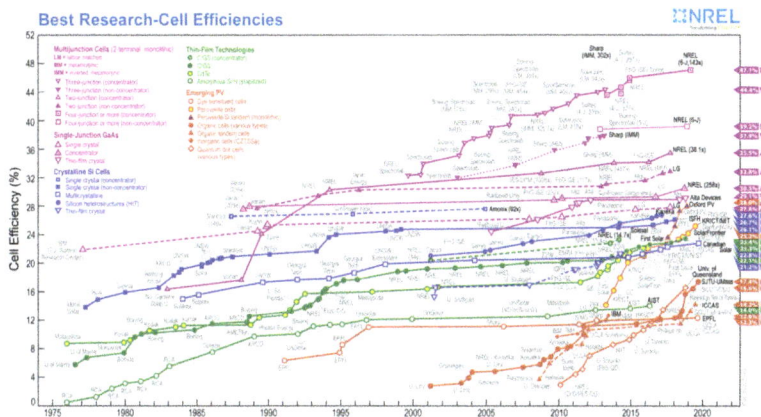

Figure 1 Efficiency of different types of photovoltaic technologies according to years [2].

2.   Solar cells based on carbon nanomaterials

Semiconductor nanocrystals have been widely studied because of their unique optical and physical properties, which are a function of their small size, from 1 to 10 nm. These inorganic nanoparticles are often used in solar cells [3,4].

In order to spread these applications and to reinforce the performances of these different optoelectronic systems a new approach has been adopted by the scientific world which consists in using hybrid structures of inorganic nanoparticles / organic nanomaterials types.

Several studies have been carried out on nanocomposites such as metallic nanoparticles or semiconductor/carbon nanotubes [5,6], considering the fascinating properties of single and multi-walled carbon nanotubes. Subsequently, there has been research to replace nanotubes (coiled graphene sheet) by graphene which has similar properties, but much larger surfaces [7,8]. In addition, the cost of producing graphene sheets in large quantities is much lower than that of carbon nanotubes. These characteristics make graphene a good alternative to replace carbon nanotubes in nanocomposite systems.

Furthermore, graphene has certain advantages as a support for the dispersion of catalyst particles (TiO_2), semiconductor particles (CdSe) or metallic particles (Pt), such as: a large specific surface, an electrical conductivity high, good transparency and low manufacturing cost (considering chemical methods). On the other hand, it is known in the literature that due to the van der Waals interactions between the graphene layers, the latter tend to aggregate. To control this phenomenon, it is possible to proceed by electrostatic stabilization or by chemical functionalization [9].

These two methods have proven to be effective in keeping graphene in individual sheets, hence the advantage of using them. Recently, the number of appearing researches on PV cells containing in its composition carbon nanomaterials is huge.

2.1   Dye-sensitized solar cells based on carbon nanomaterials

Around 1991, a new class of dye-sensitized solar cells (DSSCs) were tested by Professor Michael Grätzel [10]. These cells are very accessible because of their low cost and the simplicity of the manufacturing process. In addition, their energy efficiencies [11] remain quite low compared to those of silicon-based cells. Numerous researches have been reported to improve the oxide properties such as doping, hetero structuring and alloying [12–18] for DSSCs.

Due to their exceptional properties (optical, mechanical and physical) and their good electrical contact with the TiO_2 used in dye cells, carbon nanomaterials are promising materials in the improvement of dye cells. In recent years, the use of carbonaceous materials (carbon nanotubes, graphene, carbon nanocornets, etc.) in the field of DSSC has aroused more and more interest. For the choice of materials, cost is an important parameter. Carbon nanomaterials have very good electrical conductivity. These physical properties can improve the transport of electrons in the active layer. However, the optimal amount to use must be determined. Excessive use of highly conductive nanomaterials may create a short circuit in the device. Otherwise, the effect of this addition of carbons would be insignificant and difficult to distinguish.

The "nanocarbon" can indeed be used in all the components of the cell (against electrode, electrolyte, photo electrode, etc.). Generally, the use of carbon nanostructures and nano-objects in DSSCs has aimed at the following objectives [19]:

- Improving the collection and the transport of charges;
- Reduce the recombination of charges in the cell;
- Increase the diffusion coefficient and the catalytic efficiency of charge generation (against electrode);
- Lower the cost of the counter electrode by using Pt-free electrodes;
- Broaden the absorption spectrum by using new dyes based on "nanocarbons".

The first attempt to replace Pt with a Carbon nanomaterials (mixture of graphite and black carbon) was carried out in 1996 by Kay et al. [11]. The cell showed a yield of 6.7%.

Since this study, different types of carbonaceous materials for example carbon black [20], mesoporous carbon [21], graphene [22], composites based on graphene [23,24], and nanotubes of carbon [25], have been widely used (Figure 2 [26-28]). The yields of cells based on these materials are comparable to those of cells based on Pt electrodes.

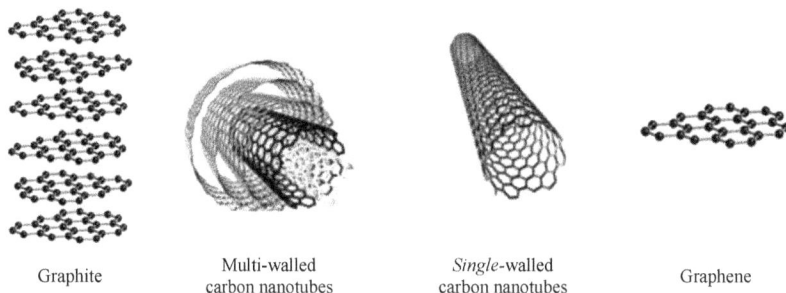

| Graphite | Multi-walled carbon nanotubes | *Single*-walled carbon nanotubes | Graphene |

Figure 2 Structure of graphite, carbon nanotubes and graphene [26–28].

Different types of carbon nanotubes CNTs (Multi-sheets [29] or single-sheets [30], pure [31] or treated [32]) have been added to the active layer in order to improve the efficiency of charge collection, since transport electrons can then be made directly by the nanotubes. In addition, the presence of CNTs also makes it possible to reduce the recombination of charges in the cell. This reduction comes from the reduction in the number of TiO_2 interfaces crossed during the transport of electrons. On the other hand, electronic traps located on the surface of TiO_2 nanoparticles, the presence of CNTs reduces charge recombination. An analysis of the literature has shown that the methods of synthesis of nanocomposites and the properties of materials lead to an optimal concentration range of CNTs in the active layer of DSSC (from 0.01wt% to 0.3 wt% [29]) .

This increase comes mainly from a better collection of charges and the reduction of recombination. However, the open circuit voltage decreases slightly. This decrease in V_{oc} has also been observed by other authors [33,34]. It can be associated with a change in the Fermi level of TiO_2 towards more negative potentials, because the edge of the E_{CB} conduction band of CNTs is generally weaker than that of TiO_2 [31].

In all cases, once the optimal concentration of CNTs is reached, the cell yield increases compared to a reference cell. For example, Dembele et al. [35] have shown that the yield of the liquid cell increases from 7% to 9% when 0.01% of CNT is added within the porous layer.

A polythiophene/graphene composite was developed by the University of Tezpur in the department of chemical sciences [36]. This composite has shown highly efficient as a counter electrode in DSSCs (Figure 3 [37]).

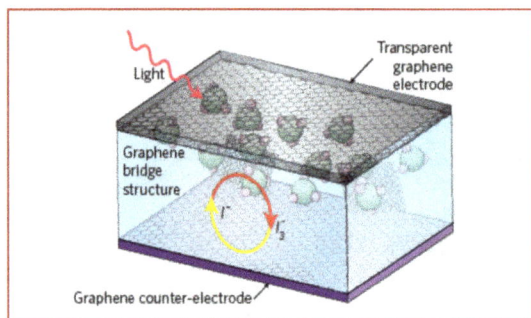

Figure 3 Schematic of DSSC with graphene [37].

2.3   Organic solar cells based on carbon nanomaterials

As observed for 30 years in organic semiconductor materials, the photovoltaic effect has experienced a great boom in the last decade. The basic photovoltaic cell consists of one or more active layers surrounded by electrodes. Each layer of the stack must meet certain criteria. The performance of organic solar cells (OSC) depends on many parameters, such as absorption, charge transport, length of excitonic diffusion, states of interfaces, etc.

There are different structures of organic photovoltaic cells, we find the Schottky type structure, the bilayer structure (p-n heterojunction) and the interpenetrating network structure.

Initially giving very low conversion efficiency values ($<10^{-5}$%), this particular application of organic semiconductors began to draw attention in 1986 when Tang et al. [38] showed that yields close to one percent were achievable. The cell, then made up of a bilayer of molecules evaporated under vacuum, reaches 0.95% conversion efficiency. Other materials and structures have since been developed specifically for this application and tested, leading to a record value of 7.9%, owned by the SOLARMER ENERGY INC [39]. The materials used are a fullerene derivative called PCBM and an alternating copolymer.

Unlike cells based on inorganic materials, these solar cells offer the advantage of being able to be deposited in large areas, at high speed, by conventional printing techniques. They also pave the way for light, nomadic and flexible applications at low cost. On the other hand, they currently have lifetimes considered to be lower than those of inorganic cells, and lower conversion yields [40].

To improve their stability and efficiency of organic solar cells, the carbon nanomaterials have been incorporated into the layers of solar cells. For over 10 years research has been carried out using carbon and polymer composites to develop high performance functional materials [41]. Different approaches show that the use of single-walled carbon nanotubes improves the PV properties by structuring the bulk-heterojunction active layer [42,43]. However, the use of CNTs in organic PVs is not limited to single-walled carbon nanotubes as demonstrated by incorporation of multi-walled carbon nanotubes MWCNTs functionalized PCBM in polymer solar cells [44]. Single-walled carbon nanotubes were first used as the acceptor in polymer solar cells by Kymakis and Amaratunga [45] who the performance of the solar cells was increased. A heterojunction based on one single-walled carbon nanotubes (SWCNTs) and P$_3$HT was estimated to yield a high V$_{OC}$ of 0.5 V with a 3% of efficiency. The authors attributed that the performance of this device is 50 – 100 times higher than those obtained in CNT-P$_3$HT blends [46,47].

Graphene derivatives have been used as electron transport both and hole layers in organic solar cell (Figure 4) [37]. On the other hand, graphene also has been used as hole transport layers in combination with other nanomaterials as bilayers to improve the extraction of holes [48]. To improve the charge carrier transport in active layers for ternary or binary blends, the rGO have been used as electron-acceptors [49,50].

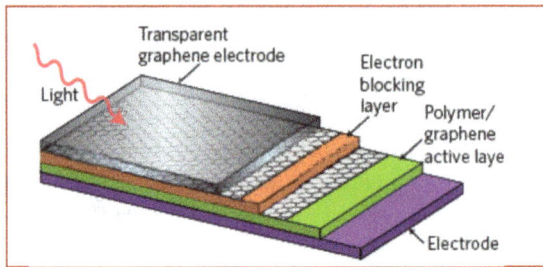

Figure 4 Schematic of organic solar cells with graphene [37].

2.4   Perovskite solar cells based on carbon nanomaterials

Snaith and Nicholas team [51] is the first to integrate graphene into the mesoporous TiO_2 layer of a perovskite solar cell (PSC). The formation of the mesoporous TiO_2 layer usually requires sintering of around 500°C. The authors lowered the annealing temperature to 150°C. A cell composed of an electron transport layer made only of TiO_2 sintered at 500°C led to a yield of 14.1%, and to a yield of only 10% when sintered at 150 °C.

On the other hand, when the cell is composed of a TiO_2/graphene composite layer, it has a yield of 15.6% after sintering at 150°C [51]. The researchers attribute these results to the significant reduction in serial resistance of the PV cell in the presence of graphene (Figure 5)[52], but also to an increase in the recombination resistance of the electron-hole pairs of the active layer.

Among the studies carried out, some have varied the graphene concentration. As in the case of DSSCs, it has been observed that the yield of perovskite cells increases with the graphene content up to a certain concentration beyond which the performance decreases [53–55].

According to Sidhik et al. [56] the presence of graphene nanoplates modifies the optical and passive gap the surface defects of TiO_2, thus making it possible to reduce the phenomena of charge recombination and to improve the transport of photo-generated electrons. Beneficial effects have also been reported in the literature for compact layers of TiO_2 [57].

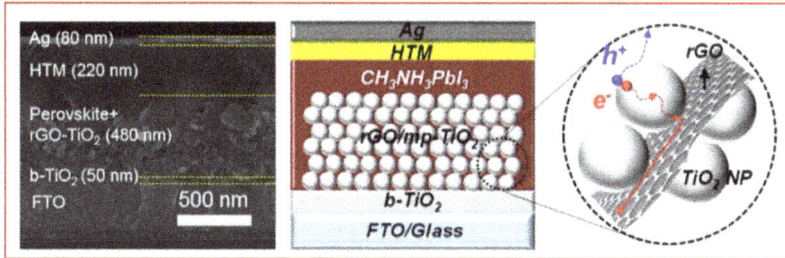

Figure 5 SEM image in section of the perovskite photovoltaic cell and the diagram representing the stacking of the various layers of the photovoltaic cell and illustration of the charge transfer from a TiO$_2$ nanoparticle to a reduced graphene oxide sheet developed by Han et al. [52].

2.5   All-carbon solar cells

Recently, researchers develop a flexible and robust PV cell based entirely on carbon materials [58]. These solar cells are based on semiconductor layers with different carbon allotropes like anode and cathode. Using carbon instead of more expensive materials (such as silicon, ITO or silver), this PV cell takes advantage of carbon's exceptional mechanical, electrical and optical properties-transparency, flexibility, tunability and stability at high temperatures. The results obtained can be used as an organic PV cell with characteristics particularly suited to difficult environmental conditions (Figure 6) [59].

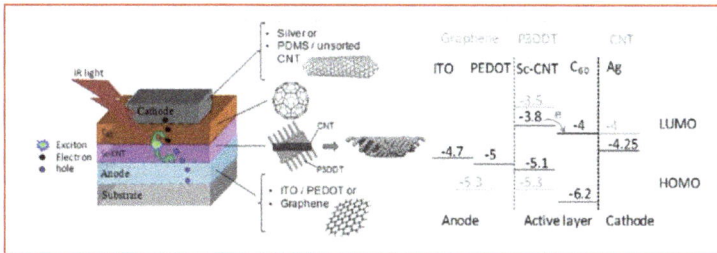

Figure 6 The structure of all carbon solar cell with rGO or ITO/PEDOT as the anode and doped SWNTs or Ag as the cathode [59].

The power conversion efficiency (PCE) of PV devices that use graphene-based composites have been summarized in Table 1 [51, 60-67].

Table 1 The power conversion efficiency (PCE) of works used rGO in solar cells [51, 60-67].

Type	Configuration of solar cells	PCE (%)	Ref.
PSC	FTO/graphene/TiO$_2$/CH$_3$NH3PbI3-xClx/Spiro-OMeTAD/Au	15.6	[51]
DSSC	Glass/rGO/TiO$_2$/dye/spiro-OMeTAD/Au	0.26	[60]
DSSC	FTO/TiO$_2$/dye/Co(III)/(II) mediated electrolyte/graphene	9.3	[61]
OSC	Quartz/rGO/PEDOT:PSS/ CuPc:C60/BCP/Ag	0.85	[62]
OSC	PET/rGO/PEDOT:PSS/CuPc:C60/BCP/Al	1.18	[63]
OSC	Au-graphene/PEDOT:PSS/ P3HT:PCBM/ZnO/ITO	3.04	[64]
ISC	Glass/graphene/ZnO/CdS/CdTe/graphite paste	4.17	[65]
PSC	FTO/TiO$_2$/CH$_3$NH$_3$PbX3/Spiro-OMeTAD/PEDOT:PSS/PDMS/PMMA/graphene	12.37	[66]
All-Carbon Solar Cells	FTO/rGO-TiO$_2$/rGO-CH$_3$NH$_3$PbI$_3$/rGO-Spiro-OMeTAD/Ag	16.5	[67]

3. Carbon based-materials for the construction of biosensors

3.1 Functionalizing graphene

The use of graphene as dispersing agent with cellulose microfibers (CMF) can improve the surface area of electrode surface with high porosity [68, 69]. Functionalizing graphene can prevent its aggregation and provides a more nano interface to immobilize biomaterials. Hence, variety of functional groups have been evaluated for use as immobilizing ligands to prepare graphene biosensing platforms such as non-covalent functionalization of graphene oxide sheets (GO) with 1-pyrenebutanoic acid succinimidyl ester (PASE) (LOD=24 nM) or with using graphene-silk peptide (Gr–SP) nanosheets that displays a good stability (93.6 % 1 month) [70]. Moreover, highly sensitive tyrosinase biosensor for the determination of phenol (113.1 µA/µM) was prepared using 1-formylpyrene (1-FP) to functionalize reduced graphene oxide (rGO) [71]. The synthesize of graphene nanoplatelets (GNPs) by hydrothermal method and titanium dioxide nanotubes (TNTs) leads to construct biosensing platform with good stability (87.7 % 4 weeks) and low detection limit (0.055 µM) [72].

Materials for Solar Cell Technologies I
Materials Research Foundations **88** (2021) 62-85

Materials Research Forum LLC
https://doi.org/10.21741/9781644901090-3

3.2 Functionalizing carbon nanotubes

The use of carbon nanotubes in biosensing platforms for the detection of phenolic compounds has been intensely investigated (Table 2) [73-84]. Their excellent electrochemical properties and their unique structure enable large area biosystems applications that offer a great opportunity to enhance analytical performance of biosensing phenolic compounds [75-80].

Table 2 Electrochemical sensors for the detection of phenols [73-84].

E-Matrix/Electrode	Sensitivity	L.R.	L.O.D.	pH	Km	Stability (%)	Ref.
Tyr/AuNPs/MWCNTs/SPCE	131 nA/µM	10 – 200 nM	2.94 nM	6.5	4.9 µM	95 % 7 days	[73]
GA/Tyr/MWCNTs/SPE	52.5 nA/µM	0.05 –1 µM	0.14µM	6.5	68 mM	75 % 30 days	[74]
Tyr-Diazonium-MWCNTs/BDD	1.81 µA/µM.cm²	0.01 –100 nM	10 pM	7.2	N.R.	80 % 7 weeks	[75]
Tyr/MWCNTs-ZnO-Nafion/GCE	0.77 A/M	0.06 – 32 µM	47 nM	7.0	18 µM	88 % 2 weeks	[76]
MWCNTs-Tyr-MNPs/SPE	4.8 µA/mM.cm²	10 – 80 µM	7.61 µM	6.5	178.52 µM	N.R.	[77]
Tyr/SWNTs/LDHs/GCE	115.46 mA/M.cm²	0.155 – 10.6 µM	155 nM	6.0	12 µM	53 % 30 days	[78]
Tyr-SF-MWCNTs-CoPc/GCE	0.71 µA/µM	0.05 – 3 µM	30 nM	7.4	12µM	75 % 30 days	[79]
SWCNTs-AuNPs/Tyr/SPE	0.374 µA/µM	0.083 – 24 µM	45 nM	7.0	20.1 µM	62 % 2 months	[80]
Tyr/1.FP/rGo/SPE	113.1 µA/mM	0.5 –150 µM	0.17 µM	7.0	N.R.	85 % 1 month	[81]
GA/Tyr/MWCNTs/GCE	0.2261 µA/µM	0.4 – 10µM	0.2 µM	7.5	N.R.	72 % 1 month	[82]
Tyr/Poly(GMA)-g-MWCNTs/ITO	2.925 µA/µM	0.01 – 0.08 mM	N.R.	7.0	41 µM	N.R.	[83]
Tyr-IL-MWCNTs-DHP/GCE	32.8 mA/M	4.9 – 1100 µM	0.58 µM	7.0	190 µM	95 % 30 days	[84]

Therefore, immobilizing tyrosinase with multi-walled carbon nanotubes can be carried out with cross-linking method using different electrodes such as glassy carbon or screen-printed electrodes [74]. Moreover, functionalization of carbon nanotubes before enzyme immobilization is of great importance [81-84]. Hence, variety of materials were used such as hybrid hydrotalcite-like materials that provide good analytical performance for catechol biosensor (LOD=43 nM) compared to magnetic nanoparticles (MNPs) [77] (LOD= 7.61 µM).

3.2.1 Ionic liquids for functionalizing carbon nanotubes

Ionic liquids are good solvents for many inorganic and organic materials due to their high thermal stability [85]. Some of these solvents including, imidazole bromide or (1-butyl-3-methylimidazolium chloride) have attracted a great attention to functionalize multi-walled carbon nanotubes (MWCNTs) used in biosensing platform (Figure 7) [83].

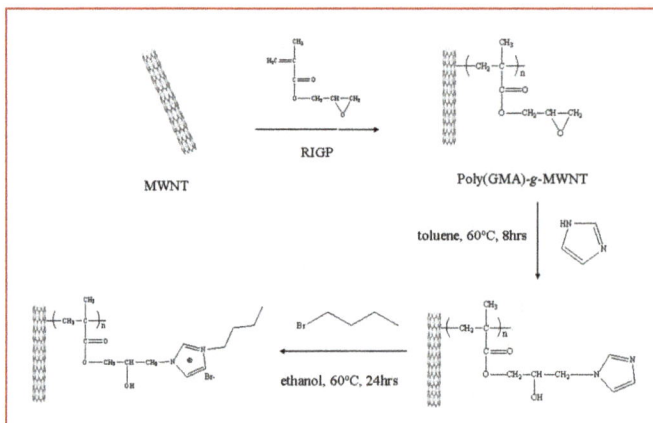

Figure 7 Synthesis procedure of the MWNTs ionic liquid for biosensors [83].

3.3   3-dimensional porous structure

New sensitive electrodes based on 3-dimensional porous structure were obtained using drop-casting single-walled carbon nanotubes (SWCNTs)(Figure8) [86], or by mixing graphene oxide with polypyrrole nanofibers and zinc oxide–copper oxide p–n junction heterostructures [87]. Electrodeposition method of graphene quantum dots (GQDs) [88], laser scribed graphene (LSG) [89], hydrothermal method [90] and chemical synthesis process [91] are the most important methods applied to obtain two and three-dimensional porous structure of electrochemical sensing platforms of phenolic compounds.

3.3.1 Reduced graphene oxide (rGO)based biosensors

Reduced graphene oxide (rGO) was largely used to modify the surface of electrochemical transducers. This material has an excellent ability to combine with various of polymeric films such as poly(phenol red), poly(3.4 ethylenedioxy-thiophene) [92] or with using 5-amino 8-hydroxy quinoline (AHQ) (Table 3) [87-91, 93-98]. Accordingly, reduced graphene oxide was largely used to prepare an electrochemical sensors for the detection of phenolic compounds and neurotransmitters using glassy carbon electrode modified with poly(3.4 ethylenedioxythiophene). The surface rugosity and electroactive area of reduced graphene oxide are factors improving electroanalytical sensing characteristics [94].

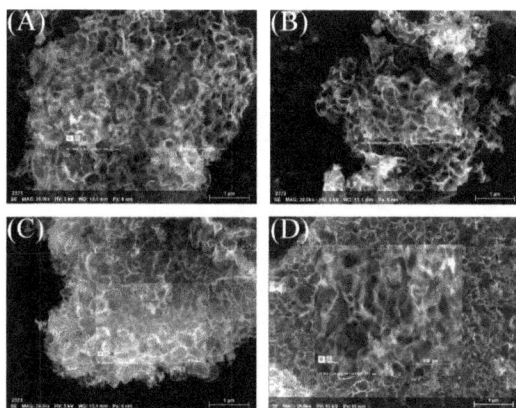

Figure 8 Element mapping images of as-synthesized three-dimensional interconnected porous carbon materials obtained at 500 (A), 600 (B), 700 (C) and 800°C (D) [86].

Table 3 Electrochemical sensors for the detection of phenolic compounds and neurotransmitters [87-91,93-98].

Modified electrode	Sensitivity	Linear range	LOD	pH	Stability	Ref.
3DCuO-ZnO/PPy/rGo/GCE	0.041 µA/µM	0.04 – 420 µM	0.012 µM	7.0	97.05 % 2 weeks	[87]
GQDs/GCE	0.6145 µA/µM	0.4 – 100 µM	50 nM	7.0	91.6 % 2 weeks	[88]
PEDOT/LSG	0.220 µA/µM	1 – 150 µM	0.33 µM	7.4	93.8 % 3 days	[89]
2Dg-C3N4/CuO/CuO/GCE	0.316 µA/µM.cm²	2 nM – 71.1 µM	0.1 nM	7.2	97.2 % 1 month	[90]
PEDOT/CeO₂/MWCNTs/GCE	3.669 µA/µM	0.1 – 10 µM	0.03 µM	7.0	89.5 % 50 days	[91]
Poly(AHQ)/rGo/GCE	13.3 A/M.cm²	0.1 – 1.4 µM	32.7 nM	7.0	98 % 30 days	[93]
CrGo/GCE	86.62 nA/µM	1 – 1000 µM	55 nM	7.0	N.R.	[94]
MWCNTs-CPE/PDDA-rGo	4.144 nA/µM	0.05 – 120 µM	0.016 µM	7.5	93.18 % 30 days	[95]
(Ppy)/ZIF67/Nafion/GCE	1.656 µA/µM.cm²	0.08 – 500 µM	0.0308 µM	6.5	89.7 % 2 months	[96]
N-rGo-180-8/NH₃/GCE	1.82 µA/µM	0.5 – 150 µM	410 nM	7.4	74.4 % 14 days	[97]
rGo/Bi₂S₃/GCE	2.0461 µA/µM	0.01 – 40 µM	12.3 nM	6.0	91.6 % 30 days	[98]

Conclusions

In summary, carbon–based photovoltaic (PV) active layers constitute a promising novel direction for photostable, efficient, solution–processable, thin–film, solar cells that are amenable to large–scale manufacturing. This material is not limited to nanotubes and graphene, but rather span a vast array of suitable carbon compounds.

On the other hand, the composition of carbon molecules is suitable for a healthy diagnosis in vivo sensing of various living molecules in biology. Therefore, carbon based-materials have an excellent electrochemical property with unique structure that enable large area biosystems applications and offering a great opportunity to enhance analytical performance of biosensing platforms.

References

[1] P.L. Wagner, V. Smil, General energetics: energy in the biosphere and civilization, Geogr. Rev. 83 (1993) 110-112. http://doi.org/10.2307/215395

[2] Best research-cell efficiency chart, Photovoltaic Research, www.nrel.gov/pv/cell-efficiency.html (accessed May 18, 2020)

[3] M.H. Sayed, E.V.C. Robert, P.J. Dale, L. Gütay, Cu_2SnS_3 based thin film solar cells from chemical spray pyrolysis, Thin Solid Films. 669 (2019) 436-439. http://doi.org/10.1016/j.tsf.2018.11.002

[4] M.S. Hossain, K.S. Rahman, M.R. Karim, M.O. Aijaz, M.A. Dar, M.A. Shar, H. Misran, N. Amin, Impact of CdTe thin film thickness in $Zn_xCd_{1-x}S$/CdTe solar cell by RF sputtering, J. Sol. Energy. 180 (2019) 559–566. http://doi.org/10.1016/j.solener.2019.01.019

[5] M.A. Correa-Duarte, J. Pérez-Juste, A. Sánchez-Iglesias, M. Giersig, L.M. Liz-Marzán, Aligning Au nanorods by using carbon nanotubes as templates, Angew. Chem. Int. Ed. 44 (2005) 4375–4378. http://doi.org/10.1002/anie.200500581

[6] J. Guerra, M.A. Herrero, Hybrid materials based on Pd nanoparticles on carbon nanostructures for environmentally benign C-C coupling chemistry, Nanoscale. 2 (2010)1390–1400. http://doi.org/10.1039/c0nr00085j

[7] R. Muszynski, B. Seger, P. V. Kamat, Decorating graphene sheets with gold nanoparticles, J. Phys. Chem. C. 112 (2008) 5263–5266. http://doi.org/10.1021/jp800977b

[8] J. Wu, S. Bai, X. Shen, L. Jiang, Preparation and characterization of graphene/CdS nanocomposites, Appl. Surf. Sci. 257 (2010) 747–751. http://doi.org/10.1016/j.apsusc.2010.07.058

[9] G. Williams, B. Seger, P. V. Kamt, TiO_2-graphene nanocomposites. UV-assisted photocatalytic reduction of graphene oxide, ACS Nano. 2 (2008) 1487–1491. http://doi.org/10.1021/nn800251f

[10] B. O'Regan, M. Grätzel, A low-cost, high-efficiency solar cell based on dye-sensitized colloidal TiO₂ films, Nature. 353 (1991) 737–740. http://doi.org/10.1038/353737a0

[11] A. Kay, M. Grätzel, Low cost photovoltaic modules based on dye sensitized nanocrystalline titanium dioxide and carbon powder, Sol. Energy Mater. Sol. Cells. 44 (1996) 99–117. http://doi.org/10.1016/0927-0248(96)00063-3

[12] D. Selloum, A. Henni, A. Karar, A. Tabchouche, N. Harfouche, O. Bacha, S. Tingry, F. Rosei, Effects of Fe concentration on properties of ZnO nanostructures and their application to photocurrent generation, Solid State Sci. 92 (2019) 76–80. http://doi.org/10.1016/j.solidstatesciences.2019.03.006

[13] A. Henni, A. Merrouche, L. Telli, A. Karar, F.I. Ezema, H. Haffar, Optical, structural, and photoelectrochemical properties of nanostructured In-doped ZnO via electrodepositing method, J. Solid State Electr. 20 (2016) 2135–2142. http://doi.org/10.1007/s10008-016-3190-y

[14] A. Henni, A. Merrouche, L. Telli, A. Karar, Studies on the structural, morphological, optical and electrical properties of Al-doped ZnOnanorods prepared by electrochemical deposition, J. Electroanal. Chem. 763 (2016) 149–154. http://doi.org/10.1016/ j.jelechem.2015.12.037

[15] Y. Bouznit, A. Henni, Characterization of Sb doped SnO₂ films prepared by spray technique and their application to photocurrent generation, Mater. Chem. Phys. 233 (2019) 242–248. http://doi.org/10.1016/j.matchemphys.2019.05.072

[16] A. Mahroug, B. Mari, M. Mollar, I. Boudjadar, L. Guerbous, A. Henni, N. Selmi, Studies on structural, surface morphological, optical, luminescence and Uvphotodetection properties of sol–gel Mg-doped ZnO thin films, Surf. Rev. Lett. 26 (2019) 1850167. http://doi.org/10.1142/S0218625X18501676

[17] A. Henni, N. Harfouche, A. Karar, D. Zerrouki, F.X. Perrin, F. Rosei, Synthesis of graphene–ZnO nanocomposites by a one-step electrochemical deposition for efficient photocatalytic degradation of organic pollutant, Solid State Sci. 98 (2019) 106039. http://doi.org/10.1016/j.solidstatesciences.2019.106039

[18] A. Henni, A. Merrouche, L. Telli, S. Walter, A. Azizi, N. Fenineche, Effect of H₂O₂ concentration on electrochemical growth and properties of vertically oriented ZnOnanorods electrodeposited from chloride solutions, Mat. Sci. Semicon. Proc. 40 (2015) 585–590. http://doi.org/10.1016/j.mssp.2015.07.046

Materials Research Forum LLC
https://doi.org/10.21741/9781644901090-3

[19] R.D. Costa, F. Lodermeyer, R. Casillas, D.M. Guldi, Recent advances in multifunctional nanocarbons used in dye-sensitized solar cells, Energy Environ. Sci.7 (2014)1281–1296. http://doi.org/10.1039/c3ee43458c

[20] T.N. Murakami, S. Ito, Q. Wang, M.K. Nazeeruddin, T. Bessho, I. Cesar, P. Liska, R. Humphry-Baker, P. Comte, P. Péchy, M. Grätzel, Highly efficient dye-sensitized solar cells based on carbon black counter electrodes, J. Electrochem. Soc.153 (2006) A2255–A2261. http://doi.org/10.1149/1.2358087

[21] M.E. Plonska-Brzezinska, A. Lapinski, A.Z. Wilczewska, A.T. Dubis, A. Villalta-Cerdas, K. Winkler, L. Echegoyen, The synthesis and characterization of carbon nano-onions produced by solution ozonolysis, Carbon. 49 (2011) 5079–5089. http://doi.org/10.1016/j.carbon.2011.07.027

[22] L. Kavan, J.H. Yum, M. Grätzel, Optically transparent cathode for dye-sensitized solar cells based on graphene nanoplatelets, ACS Nano. 5 (2011) 165–172. http://doi.org/10.1021/nn102353h

[23] G. Wang, S. Zhuo, W. Xing, Graphene/polyaniline nanocomposite as counter electrode of dye-sensitized solar cells, Mater. Lett. 69 (2012) 27–29. http://doi.org/10.1016/j.matlet.2011.11.086

[24] R. Bajpai, S. Roy, N. Kulshrestha, J. Rafiee, N. Koratkar, D.S. Misra, Graphene supported nickel nanoparticle as a viable replacement for platinum in dye sensitized solar cells, Nanoscale. 4 (2012) 926–930. http://doi.org/10.1039/c2nr11127f

[25] D. Noureldine, T. Shoker, M. Musameh, T.H. Ghaddar, Investigation of carbon nanotube webs as counter electrodes in a new organic electrolyte based dye sensitized solar cell, J. Mater. Chem. 22 (2012) 862–869. http://doi.org/10.1039/C1JM15055C

[26] S. Iijima, Helical microtubules of graphitic carbon, Nature. 354 (1991) 56–58. http://doi.org/10.1038/354056a0

[27] S. Iijima, T. Ichihashi, Single-shell carbon nanotubes of 1-nm diameter, Nature. 363 (1993) 603–605. http://doi.org/10.1038/363603a0

[28] A.K. Geim, K.S. Novoselov, The rise of graphene, Nat. Mater. 6 (2007) 183–191. http://doi.org/10.1038/nmat1849

[29] Y.F. Chan, C.C. Wang, B.H. Chen, C.Y. Chen, Dye-sensitized TiO_2 solar cells based on nanocomposite photoanode containing plasma-modified multi-walled carbon nanotubes, Prog. Photovoltaics. 21(2013)47–57. http://doi.org/10.1002/pip.2174

[30] X. Dang, H. Yi, M.H. Ham, J. Qi, D.S. Yun, R. Ladewski, M.S. Strano, P.T. Hammond, A.M. Belcher, Virus-templated self-assembled single-walled carbon

Materials Research Forum LLC
https://doi.org/10.21741/9781644901090-3

nanotubes for highly efficient electron collection in photovoltaic devices, Nat. Nanotechnol. 6 (2011) 377–384. http://doi.org/10.1038/nnano.2011.50

[31] P. Du, L. Song, J. Xiong, N. Li, L. Wang, Z. Xi, N. Wang, L. Gao, H. Zhu, Dye-sensitized solar cells based on anatase TiO_2/multi-walled carbon nanotubes composite nanofibers photoanode, Electrochim. Acta. 87(2013) 651–656. http://doi.org/10.1016/j.electacta.2012.09.096

[32] W. Guo, Y. Shen, L. Wu, Y. Gao, T. Ma, Performance of dye-sensitized solar cells based on $MWCNT/TiO_{2-x}N_x$ nanocomposite electrodes, Eur. J. Inorg. Chem. 2011 (2011) 1776–1783. http://doi.org/10.1002/ejic.201001241

[33] Z. Lin, A. Orlov, R.M. Lambert, M.C. Payne, New insights into the origin of visible light photocatalytic activity of nitrogen-doped and oxygen-deficient anatase TiO_2, J. Phys. Chem. B. 109 (2005) 20948–20952. http://doi.org/10.1021/jp053547e

[34] W. Guo, Y. Shen, G. Boschloo, A. Hagfeldt, T. Ma, Influence of nitrogen dopants on N-doped TiO_2 electrodes and their applications in dye-sensitized solar cells, Electrochim. Act. 56 (2011) 4611–4617. http://doi.org/10.1016/j.electacta.2011.02.091

[35] K.T. Dembele, G.S. Selopal, C. Soldano, R. Nechache, J.C. Rimada, I. Concina, G. Sberveglieri, F. Rosei, A. Vomiero, Hybrid carbon nanotubes-TiO_2 photoanodes for high efficiency dye-sensitized solar cells, J. Phys. Chem. C. 117 (2013) 14510–14517. http://doi.org/10.1021/jp403553t

[36] H. Yan, J. Wang, B. Feng, K. Duan, J. Weng, Graphene and Ag nanowires co-modified photoanodes for high-efficiency dye-sensitized solar cells, Sol. Energy. 122 (2015) 966–975. http://doi.org/10.1016/j.solener.2015.10.026

[37] F. Bonaccorso, Z. Sun, T. Hasan, A.C. Ferrari, Graphene photonics and optoelectronics, Nat. Photonics. 4 (2010) 611–622. http://doi.org/10.1038/nphoton.2010.186

[38] C.W. Tang, Two-layer organic photovoltaic cell, Appl. Phys. Lett. 48 (1986)183–185 http://doi.org/10.1063/1.96937

[39] G. Zhao, Y. He, Z. Xu, J. Hou, M. Zhang, J. Min, H.Y. Chen, M. Ye, Z. Hong, Y. Yang, Y. Li, Effect of carbon chain length in the substituent of PCBM-like molecules on their photovoltaic properties, Adv. Funct. Mater. 20 (2010) 1480–1487. http://doi.org/10.1002/adfm.200902447

[40] A. Mishra, M.K.R. Fischer, P. Bäuerle, Metal-free organic dyes for dye-sensitized solar cells: from structure: property relationships to design rules., Angew. Chem. Int. Ed. 48 (2009) 2474–2499. http://doi.org/10.1002/anie.200804709

[41] B. Ratier, J.M. Nunzi, M. Aldissi, T.M. Kraft, E. Buncel, Organic solar cell materials and active layer designs-improvements with carbon nanotubes: A review, Polym. Int. 61 (2012) 342–354. http://doi.org/10.1002/pi.3233

[42] H. Derbal-Habak, C. Bergeret, J. Cousseau, J.M. Nunzi, Improving the current density J_{sc} of organic solar cells P3HT:PCBM by structuring the photoactive layer with functionalized SWCNTs, Sol. Energy Mater. Sol. Cells. 95 (2011) S53–S56. http://doi.org/10.1016/ j.solmat.2010.12.047

[43] R. Radbeh, E. Parbaile, M. Chakaroun, B. Ratier, M. Aldissi, A. Moliton, Enhanced efficiency of polymeric solar cells via alignment of carbon nanotubes, Polym. Int. 59 (2010) 1514–1519. http://doi.org/10.1002/pi.2916

[44] Y.S. Jung, Y.H. Hwang, A. Javey, M. Pyo, PCBM-grafted MWNT for enhanced electron transport in polymer solar cells, J. Electrochem. Soc. 158 (2011) A237–A240 http://doi.org/10.1149/1.3530197

[45] E. Kymakis, G.A.J. Amaratunga, Single-wall carbon nanotube/conjugated polymer photovoltaic devices, Appl. Phys. Lett. 80 (2002) 112–114. http://doi.org/10.1063/1.1428416

[46] E. Kymakis, G.A.J. Amaratunga, Carbon nanotubes as electron acceptors in polymeric photovoltaics, Rev. Adv. Mater. Sci. 10 (2005) 300–305

[47] J. Liu, Y. Xue, Y. Gao, D. Yu, M. Durstock, L. Dai, Hole and electron extraction layers based on graphene oxide derivatives for high-performance bulk heterojunction solar cells, Adv. Mater. Technol. 24 (2012) 2228–2233. http://doi.org/10.1002/adma.201104945

[48] S. Chen, X. Yu, M. Zhang, J. Cao, Y. Li, L. Ding, G. Shi, A graphene oxide/oxygen deficient molybdenum oxide nanosheet bilayer as a hole transport layer for efficient polymer solar cells, J. Mater. Chem. A. 3 (2015)18380–18383. http://doi.org/10.1039/c5ta04823k

[49] D. Romero-Borja, J.L. Maldonado, O. Barbosa-García, M. Rodríguez, E. Pérez-Gutiérrez, R. Fuentes-Ramírez, G. De La Rosa, Polymer solar cells based on P3HT:PC_{71}BM doped at different concentrations of isocyanate-treated graphene, Synth. Met. 200 (2015) 91–98. http://doi.org/10.1016/j.synthmet.2014.12.029

[50] M.M. Stylianakis, D. Konios, G. Kakavelakis, G. Charalambidis, E. Stratakis, A.G. Coutsolelos, E. Kymakis, S.H. Anastasiadis, Efficient ternary organic photovoltaics incorporating a graphene-based porphyrin molecule as a universal electron cascade material, Nanoscale. 7 (2015)17827–17835. http://doi.org/10.1039/c5nr05113d

[51] J.T.W. Wang, J.M. Ball, E.M. Barea, A. Abate, J.A. Alexander-Webber, J. Huang, M. Saliba, I. Mora-Sero, J. Bisquert, H.J. Snaith, R.J. Nicholas, Low-temperature processed electron collection layers of graphene/TiO$_2$ nanocomposites in thin film perovskite solar cells, Nano Lett. 14 (2014)724–730. http://doi.org/10.1021/nl403997a

[52] G.S. Han, Y.H. Song, Y.U. Jin, J.W. Lee, N.G. Park, B.K. Kang, J.K. Lee, I.S. Cho, D.H. Yoon, H.S. Jung, Reduced graphene oxide/mesoporous TiO$_2$ nanocomposite based perovskite solar cells, ACS Appl. Mater. Interfaces. 7 (2015) 23521–23526. http://doi.org/10.1021/acsami.5b06171

[53] T. Umeyama, D. Matano, J. Baek, S. Gupta, S. Ito, V. Subramanian, H. Imahori, Boosting of the performance of perovskite solar cells through systematic introduction of reduced graphene oxide in TiO$_2$ Layers, Chem. Lett. 44 (2015) 1410–1412.http://doi.org/10.1246/cl.150651

[54] A. Agresti, S. Pescetelli, B. Taheri, A.E. Del Rio Castillo, L. Cinà, F. Bonaccorso, A. Di Carlo, Back cover: Graphene–perovskite solar cells exceed 18 % efficiency: A stability study (ChemSusChem 18/2016), ChemSusChem. 9 (2016) 2716–2716. http://doi.org/10.1002/cssc.201601092

[55] A. Agresti, S. Pescetelli, A.L. Palma, A.E. Del Rio Castillo, D. Konios, G. Kakavelakis, L. Cinà, E. Kymakis, F. Bonaccorso, A. Di Carlo, Graphene interface engineering for perovskite solar modules: 12.6% power conversion efficiency over 50 cm^2 active area, ACS Energy Lett. 2 (2017)279–287. http://doi.org/10.1021/acsenergylett.6b00672

[56] S. Sidhik, S.S. Panikar, C.R. Pérez, T.L. Luke, R. Carriles, S.C. Carrera, E. De La Rosa, Interfacial engineering of TiO$_2$ by graphene nanoplatelets for high-efficiency hysteresis-free perovskite solar cells, ACS. Sustain. Chem. Eng. 6 (2018) 15391–15401. http://doi.org/10.1021/acssuschemeng.8b03826

[57] P. Yang, Z. Hu, X. Zhao, D. Chen, H. Lin, X. Lai, L. Yang, cesium-containing perovskite solar cell based on graphene/TiO$_2$ electron transport layer, Chemistry Select 2 (2017) 9433–9437. http://doi.org/10.1002/slct.201701479

[58] M.P. Ramuz, M. Vosgueritchian, P. Wei, C. Wang, Y. Gao, Y. Wu, Y. Chen, Z. Bao, Evaluation of solution-processable carbon-based electrodes for all-carbon solar cells, ACS Nano. 6 (2012) 10384–10395. http://doi.org/10.1021/nn304410w

[59] M.S. Arnold, J.D. Zimmerman, C.K. Renshaw, X. Xu, R.R. Lunt, C.M. Austin, S.R. Forrest, Broad spectral response using carbon nanotube/organic semiconductor/C$_{60}$ photodetectors, Nano Lett. 9 (2009) 3354–3358. http://doi.org/10.1021/nl901637u

Materials for Solar Cell Technologies I
Materials Research Foundations **88** (2021) 62-85

Materials Research Forum LLC
https://doi.org/10.21741/9781644901090-3

[60] Z. Yin, S. Wu, X. Zhou, X. Huang, Q. Zhang, F. Boey, H. Zhang, Electrochemical deposition of ZnOnanorods on transparent reduced graphene oxide electrodes for hybrid solar cells, Small. 6 (2010) 307–312. http://doi.org/10.1002/smll.200901968

[61] L. Kavan, J.H. Yum, M.K. Nazeeruddin, M. Grätzel, Graphene nanoplatelet cathode for Co(III)/(II) mediated dye-sensitized solar cells, ACS Nano. 5 (2011) 9171–9178. http://doi.org/10.1021/nn203416d

[62] H. Park, P.R. Brown, V. Bulović, J. Kong, Graphene as transparent conducting electrodes in organic photovoltaics: Studies in graphene morphology, hole transporting layers, and counter electrodes, Nano Lett. 12 (2012) 133–140. http://doi.org/10.1021/nl2029859

[63] L. Gomez De Arco, Y. Zhang, C.W. Schlenker, K. Ryu, M.E. Thompson, C. Zhou, Continuous, highly flexible, and transparent graphene films by chemical vapor deposition for organic photovoltaics, ACS Nano. 4 (2010) 2865–2873. http://doi.org/10.1021/nn901587x

[64] S. Li, Y. Luo, W. Lv, W. Yu, S. Wu, P. Hou, Q. Yang, Q. Meng, C. Liu, H.M. Cheng, Vertically aligned carbon nanotubes grown on graphene paper as electrodes in lithium-ion batteries and dye-sensitized solar cells, Adv. Energy Mater. 1 (2011) 486–490. http://doi.org/10.1002/aenm.201100001

[65] H. Bi, F. Huang, J. Liang, Y. Tang, X. Lü, X. Xie, M. Jiang, Large-scale preparation of highly conductive three dimensional graphene and its applications in CdTe solar cells, J. Mater. Chem.21 (2011) 17366–17370. http://doi.org/10.1039/c1jm13418c

[66] P. You, Z. Liu, Q. Tai, S. Liu, F. Yan, Efficient semitransparent perovskite solar cells with graphene electrodes, Adv. Mater. 27 (2015) 3632–3638. http://doi.org/10.1002/adma.201501145

[67] N. Balis, A.A. Zaky, C. Athanasekou, A.M. Silva, E. Sakellis, M. Vasilopoulou, T. Stergiopoulos, A.G. Kontos, P. Falaras, Investigating the role of reduced graphene oxide as a universal additive in planar perovskite solar cells, J. Photochem. Photobiol. 386 (2020) 112141. http://doi.org/10.1016/j.jphotochem.2019.112141

[68] X. Liu, R. Yan, J. Zhu, J. Zhang, X. Liu, Growing TiO_2 nanotubes on graphene nanoplatelets and applying the nanonanocomposite as scaffold of electrochemical tyrosinase biosensor, Sensor Actuat. B-Chem. 209 (2015) 328–335. http://doi.org/10.1016/j.snb.2014.11.124

[69] L. Fritea, A. Le Goff, J.L. Putaux, M. Tertis, C. Cristea, R. Săndulescu, S. Cosnier, Design of a reduced-graphene-oxide composite electrode from an

electropolymerizable graphene aqueous dispersion using a cyclodextrin-pyrrole monomer. Application to dopamine biosensing, Electrochim. Acta. 178 (2015) 108–112. http://doi.org/10.1016/j.electacta.2015.07.124

[70] Y. Qu, M. Ma, Z. Wang, G. Zhan, B. Li, X. Wang, H. Fang, H. Zhang, C. Li, Sensitive amperometric biosensor for phenolic compounds based on graphene–silk peptide/tyrosinase composite nanointerface, Biosens. Bioelectron. 44 (2013) 85–88. http://doi.org/10.1016/j.bios.2013.01.011

[71] Z. Hua, Q. Qin, X. Bai, X. Huang, Q. Zhang, An electrochemical biosensing platform based on 1-formylpyrene functionalized reduced graphene oxide for sensitive determination of phenol, RSC. Advances. 6 (2016) 25427–25434. http://doi.org/10.1039/C5RA27563F

[72] X. Liu, R. Yan, J. Zhu, J. Zhang, X. Liu, Growing TiO_2 nanotubes on graphene nanoplatelets and applying the nanonanocomposite as scaffold of electrochemical tyrosinase biosensor, Sensor Actuat. B-Chem. 209 (2015) 328–335. http://doi.org/10.1016/j.snb.2014.11.124

[73] F.R. Caetano, E.A. Carneiro, D. Agustini, L.C.S. Figueiredo-Filho, C.E. Banks, M.F. Bergamini, L.H. Marcolino-Junior, Combination of electrochemical biosensor and textile threads: A microfluidic device for phenol determination in tap water, Biosens. Bioelectron. 99 (2018) 382–388. http://doi.org/10.1016/j.bios.2017.07.070

[74] G. Alarcón, M. Guix, A. Ambrosi, M.T. Ramirez Silva, M.E. Palomar Pardave, A. Merkoçi, Stable and sensitive flow-through monitoring of phenol using a carbon nanotube based screen printed biosensor, Nanotechnology. 21 (2010) 245502. http://doi.org/10.1088/0957-4484/21/24/245502

[75] N. Zehani, P. Fortgang, M. S. Lachgar, A. Baraket, M. Arab, S.V. Dzyadevych, R. Kherrat, N. Jaffrezic-Renault, Highly sensitive electrochemical biosensor for bisphenol A detection based on a diazonium-functionalized boron-doped diamond electrode modified with a multi-walled carbon nanotube-tyrosinase hybrid film, Biosens. Bioelectron. 74 (2015) 830–835. http://doi.org/10.1016/j.bios.2015.07.051

[76] J.M. Lee, G.-R. Xu, B.K. Kim, H.N. Choi, W.-Y. Lee, Amperometric tyrosinase biosensor based on carbon nanotube-doped sol-gel-derived zinc oxide-nafion composite films, Electroanalysis. 23 (2011) 962–970. http://doi.org/10.1002/elan.201000556

[77] B. Pérez-López, A. Merkoçi, Magnetic nanoparticles modified with carbon nanotubes for electrocatalytic magnetoswitchable biosensing applications, Adv. Funct. Mater. 21 (2011) 255–260. http://doi.org/10.1002/adfm.201001306

[78] H. Wang, Y. Qin, K. Chen, H. Xue, the phenol biosensor based on LDHS/SWNTS hybrid materials, Int. J. Electrochem. Sci. 11 (2016) 777–79

[79] H. Yin, Y. Zhou, J. Xu, S. Ai, L. Cui, L. Zhu, Amperometric biosensor based on tyrosinase immobilized onto multiwalled carbon nanotubes-cobalt phthalocyanine-silk fibroin film and its application to determine bisphenol A, Anal. Chim. Acta. 659 (2010) 144–150. http://doi.org/10.1016/j.aca.2009.11.051

[80] Y. Li, D. Li, W. Song, M. Li, J. Zou, Y. Long, Rapid method for on-site determination of phenolic contaminants in water using a disposable biosensor, Front. Environ. Sci. Eng. 6 (2012) 831–838. http://doi.org/10.1007/s11783-012-0393-z

[81] Z. Hua, Q. Qin, X. Bai, X. Huang, Q. Zhang, An electrochemical biosensing platform based on 1-formylpyrene functionalized reduced graphene oxide for sensitive determination of phenol, RSC. Adv. 6 (2016) 25427–25434. http://doi.org/10.1039/C5RA27563F

[82] J. Ren, T.F. Kang, R. Xue, C.N. Ge, S.Y. Cheng, Biosensor based on a glassy carbon electrode modified with tyrosinase immmobilized on multiwalled carbon nanotubes, Microchim. Acta. 174 (2011) 303–309. http://doi.org/10.1007/s00604-011-0616-1

[83] K.-I. Kim, H.-Y. Kang, J.-C. Lee, S.-H. Choi, Fabrication of a multi-walled nanotube (MWNT) ionic liquid electrode and its application for sensing phenolics in red wines, Sensors. 9 (2009) 6701–6714. http://doi.org/10.3390/s90906701

[84] F.C. Vicentini, B.C. Janegitz, C.M.A. Brett, O. Fatibello-Filho, Tyrosinase biosensor based on a glassy carbon electrode modified with multi-walled carbon nanotubes and 1-butyl-3-methylimidazolium chloride within a dihexadecylphosphate film, Sensor Actuat. B Chem. 188 (2013) 1101–1108. http://doi.org/10.1016/j.snb.2013.07.109

[85] M. Erbeldinger, A.J. Mesiano, A.J. Russell, Enzymatic catalysis of formation of Z-aspartame in ionic liquid-an alternative to enzymatic catalysis in organic solvents, Biotechnol. Prog. 16 (2000) 1129–1131. http://doi.org/10.1021/bp000094g

[86] Y. Xiang, L. li, H. liu, Z. Shi, Y. Tan, C. Wu, Y. Liu, J. Wang, S. Zhang, One-step synthesis of three-dimensional interconnected porous carbon and their modified electrode for simultaneous determination of hydroquinone and catechol, Sensor Actuat. B Chem. 267 (2018) 302–311. http://doi.org/10.1016/j.snb.2018.04.051

[87] Kh. Ghanbari, S. Bonyadi, An electrochemical sensor based on reduced graphene oxide decorated with polypyrrole nanofibers and zinc oxide–copper oxide p–n junction heterostructures for the simultaneous voltammetric determination of ascorbic acid,

dopamine, paracetamol, and tryptophan, New J. Chem. 42 (2018) 8512–8523. http://doi.org/10.1039/C8NJ00857D

[88] S. Zheng, R. Huang, X. Ma, J. Tang, Z. Li, X. Wang, J. Wei, J. Wang, A highly sensitive dopamine sensor based on graphene quantum dots modified glassy carbon electrode, Int. J. Electrochem. Sci. 13 (2018) 5723–5735. http://doi.org/10.20964/2018.06.19

[89] G. Xu, Z.A. Jarjes, V. Desprez, P.A. Kilmartin, J. Travas-Sejdic, Sensitive, selective, disposable electrochemical dopamine sensor based on PEDOT-modified laser scribed graphene, Biosens. Bioelectron. 107 (2018) 184–191. http://doi.org/10.1016/j.bios.2018.02.031

[90] J. Zou, S. Wu, Y. Liu, Y. Sun, Y. Cao, J.P. Hsu, A.T. Shen Wee, J. Jiang, An ultra-sensitive electrochemical sensor based on 2D g-C_3N_4/CuO nanocomposites for dopamine detection, Carbon. 130 (2018) 652–663. http://doi.org/10.1016/j.carbon.2018.01.008

[91] A. Üğe, D. KoyuncuZeybek, B. Zeybek, An electrochemical sensor for sensitive detection of dopamine based on MWCNTs/CeO_2 -PEDOT composite, J. Electroanal. Chem. 813 (2018) 134–142. http://doi.org/10.1016/j.jelechem.2018.02.028

[92] V. Serafín, L. Agüí, P. Yáñez-Sedeño, J.M. Pingarrón, A novel hybrid platform for the preparation of disposable enzyme biosensors based on poly(3,4-ethylenedioxythiophene) electrodeposition in an ionic liquid medium onto gold nanoparticles-modified screen-printed electrodes, J. Electroanal. Chem. 656 (2011) 152–158. http://doi.org/10.1016/j. jelechem.2010.11.038

[93] V.M.A. Mohanan, A.K. Kunnummal, V.M.N. Biju, Selective electrochemical detection of dopamine based on molecularly imprinted poly(5-amino 8-hydroxy quinoline) immobilized reduced graphene oxide, J. Mater. Sci. 53 (2018) 10627–10639. http://doi.org/10.1007/s10853-018-2355-8

[94] D.P. Rocha, R.M. Dornellas, R.M. Cardoso, L.C.D. Narciso, M.N.T. Silva, E. Nossol, E.M. Richter, R.A.A. Munoz, Chemically versus electrochemically reduced graphene oxide: Improved amperometric and voltammetric sensors of phenolic compounds on higher roughness surfaces, Sensor Actuat. B Chem. 254 (2018) 701–708.http://doi.org/ 10.1016/j.snb.2017.07.070

[95] Z. Liu, M. Jin, J. Cao, R. Niu, P. Li, G. Zhou, Y. Yu, A. van den Berg, L. Shui, Electrochemical sensor integrated microfluidic device for sensitive and simultaneous quantification of dopamine and 5-hydroxytryptamine, Sensor Actuat. B Chem. 273 (2018) 873–883. http://doi.org/10.1016/j.snb.2018.06.123

[96] W. Zhang, D. Duan, S. Liu, Y. Zhang, L. Leng, X. Li, N. Chen, Y. Zhang, Metal-organic framework-based molecularly imprinted polymer as a high sensitive and selective hybrid for the determination of dopamine in injections and human serum samples, Biosens. Bioelectron. 118 (2018) 129–136. http://doi.org/10.1016/j.bios.2018.07.047

[97] P. Wiench, Z. González, R. Menéndez, B. Grzyb, G. Gryglewicz, Beneficial impact of oxygen on the electrochemical performance of dopamine sensors based on N-doped reduced graphene oxides, Sensor Actuat. B Chem. 257 (2018) 143–153. http://doi.org/10.1016/j.snb.2017.10.106

[98] X. Yan, Y. Gu, C. Li, B. Zheng, Y. Li, T. Zhang, Z. Zhang, M. Yang, Morphology-controlled synthesis of Bi_2S_3 nanorods-reduced graphene oxide composites with high-performance for electrochemical detection of dopamine, Sensor Actuat. B Chem. 257 (2018) 936–943. http://doi.org/10.1016/j.snb.2017.11.037

Materials for Solar Cell Technologies I
Materials Research Foundations **88** (2021) 86-128

Materials Research Forum LLC
https://doi.org/10.21741/9781644901090-4

Chapter 4

New Generation Transparent Conducting Electrode Materials for Solar Cell Technologies

Sandeep Pandey[1], Manoj Karakoti[1], Amit Kumar[2,3], Sunil Dhali[1], Aniket Rana[2], Kuldeep K. Garg [2,3], Rajiv K Singh[2,3], Nanda Gopal Sahoo[*1]

[1]Photovoltaic Metrology Section, Advanced Materials and Devices Metrology Division, CSIR-National Physical Laboratory, Dr. K.S. Krishnan Marg, New Delhi 110012, India

[2]Photovoltaic Metrology Section, Advanced Materials and Devices Metrology Division, CSIR-National Physical Laboratory, Dr. K.S. Krishnan Marg, New Delhi 110012, India

[3]Academy of Scientific and Innovative Research (AcSIR), Ghaziabad, Uttar Pradesh 201002, India

ngsahoo@yahoo.co.in

Abstract

Transparent conducting electrodes (TCEs) play a vital role for the fabrication of solar cells and pivoted almost 50% of the total cost. Recently several materials have been identified as TCEs in solar cell applications. Still, indium tin oxide (ITO) based TCEs have dominated the market due to their outstanding optical transparency and electrical conductivity. However, inadequate availability of indium has increased the price of ITO based TCEs, which attracts the researchers to find alternative materials to make solar technology economical. In this regard, various kinds of conducting materials are available and synthesized worldwide with high electrical conductivity and optical transparency in order to find alternative to ITO based electrodes. Especially, new generation nanomaterials have opened a new window for the fabrication of cost effective TCEs. Carbon nanomaterials such as graphene, carbon nanotubes (CNTs), metal nanowires (MNWs) and metal mesh (MMs) based electrodes especially attracted the scientific community for fabrication of low cost photovoltaic devices. In addition to it, various conducting polymers such as poly (3, 4-ethylene dioxythiophene): poly (styrenesulfonate) (PEDOT:PSS) based TCEs have also showed their candidacy as an alternative to ITO based TCEs. Thus, the present chapter gives an overview on materials available for the TCEs and their possible use in the field of solar cell technology

Materials for Solar Cell Technologies I   Materials Research Forum LLC
Materials Research Foundations **88** (2021) 86-128   https://doi.org/10.21741/9781644901090-4

Keywords

Solar Cells, Transparent Conducting Electrodes, Indium Tin Oxide, Carbon Nanomaterials, Metal Nanowires, Metal Mesh, Poly (3, 4-ethylene dioxythiophene): poly (styrenesulfonate) PEDOT: PSS

Contents

New Generation Transparent Conducting Electrode Materials for Solar Cell Technologies **86**

1.   Introduction...**87**
        1.1   Fundamentals for the evaluation of TCEs 90
2.   Materials for transparent electrodes **91**
        2.1   Indium tin oxide-based TCEs...92
        2.2   Fluorine doped indium tin oxide 93
        2.3   Carbon-based nanomaterials...93
        2.3.1 Graphene-based TCEs 94
        2.3.2 Carbon nanotube based TCEs...98
        2.3.3 Carbon nanosheets based TCEs.......................................100
        2.4   Metal nanowires (MNWs) based TCEs............................100
        2.5   Metal meshes based TCEs...102
        2.6   PEDOT: PSS based TCEs 107
        2.7   Other TCE materials...110
Conclusion and remarks..**111**
Acknowledgements...**112**
References **112**

1.   Introduction

The rapid growth of population and demand for clean energy by promising green energy sources have played an essential position in the sustainable growth of environment. Today, the need for green energy rises exponentially, especially in developing countries, where still the fulfillment of energy demands somewhat depends upon the limited source of fossil fuels. In this regard, solar energy is considered as the most promising resource for the production of a large amount of green energy through photovoltaic (PV) technology to meet the increasing demand of electricity [1]. In recent years, research was focused on the development of PV technology as reports on the forecast of the world electricity power supply stated that solar cell technologies will deliver about 345GW and 1081GW by 2020 and 2030, respectively [2]. Presently, there are approximate 24 models of solar cell technologies available using different kinds of materials and methods to get the best device performance [3]. However, in all these solar cell technologies, transparent conducting electrodes (TCEs) play an important role governing the performance of the solar cells depending on the properties of the TCEs. Primarily, the optoelectronic properties of a TCEs depend upon the sheet resistance (R_{sheet}) and optical transmittance. The value of sheet resistance and permissible range of optical transparency determines the role of TCEs and suitability for solar cell applications. Generally, TCEs with high transmittance and lower sheet resistance value are considered as the most excellent which is believed to depend upon the nature of the materials and fabrication techniques. Till date several TCE materials have been identified and used in various kinds of solar cell technologies. Among TCE materials, metal oxide and carbon nanomaterials take significant attraction for the development of versatile TCEs. The ITO based transparent electrodes dominate the market due to a very balanced optical transmittance in UV-visible region (T~85-90%) and lower sheet resistance (R_{sheet} ~ 10-20 ohm/sq) which makes it suitable for various optoelectronic applications [4]. Even though ITO based substrates dominate the market, the present generation of photovoltaic avoids the use of ITO based substrates due to limited availability and high price of indium. Even though the utilization of flexible substrates of ITO showed limited interest for flexible photovoltaic applications due to extremely brittle nature and high risk of cracking during mechanical exposure, which minimizes the conductivity of the substrate and thereby leading device degradation [5-6]. Furthermore, the brittle nature of the ITO substrate showed poor bending strain tolerance and relatively low cyclic flexibility, which makes them inadequate for real-life applications. Although, the market demand of TCEs for the ITO substrates, development of large scale ITO based low-cost thin-film organic/hybrid solar cell technology is still a dream for practical applications, as such, the developments with these substrates are limited to the laboratory premises. Moreover, the fabrication process of ITO substrates via a process of sputtering require high vacuum and temperature process to produce uniform crystalline ITO substrates with high electrical conductivity (R_{sheet}~10 ohm/sq) and optical transmittance (T>85 %). The high cost of sputtering has made it less effective when compared with the TCEs made with vacuum free roll to roll printing technique. So, the wide scope of ITO based substrates has become limited when it comes to the aspect of commercialization. It is usually stated that

Materials for Solar Cell Technologies I　　　　　　　　　　　Materials Research Forum LLC
Materials Research Foundations **88** (2021) 86-128　　　　https://doi.org/10.21741/9781644901090-4

any substitute of TCEs should possesses a R_{sheet} < 100 ohm/sq with the permissible transparency of T>90 % in visible region [7-8].

In this regard, solution-processable conducting materials take significant attraction for the production of low-cost TCEs for optoelectronic applications. Especially, graphene and carbon nanotubes gained significant attraction for the advancement of affordable TCEs for photovoltaic applications. The extraordinary properties of carbon nanomaterials such as tremendous electrical and physical properties, superior conductivity with high mobility, fit them as the candidate of TCEs [9-13]. However, a CNT based network as TCE suffered from the huge contact resistance of tube-tube junctions, as a result practicability of only CNT based TCEs does not show much success rate for photovoltaic applications. Another candidate of TCE materials i.e. metal nanowires are also of significant interest because of high optical transmittance and low sheet resistance. The optoelectronic performance of silver nanowire has shown excellent results with R_{sheet} ~ 10-20 ohm/sq) with 90% optical transparencies [14-15]. However, cost-benefit analysis of silver nanowire-based TCEs for photovoltaic applications still becomes a subject of debate.

Recently, graphene and graphene based composites have been proposed as most capable candidates as TCEs for cost-effective photovoltaic due to their extraordinary assets such as elevated conductivity i.e. low sheet resistance, great optical transmittance, good thermal, electrical and mechanical properties [16-18]. Graphene is regarded as a single atomic sheet of graphite where all carbon atoms are bonded together as sp^2 hybridized carbon atoms in a honeycomb lattice. The remarkable properties of the graphene such as high intrinsic electron mobility (2×10^5 cm^2/V-s) facilitate superior electrical conductivity, high thermal conductivity (~5×10^3 W/m-K) enables applications in power electronics, extraordinary mechanical strength fulfils the demand of flexible electronics and unique structure gives it chemical stability for the harsh environments [19-20]. Though, the and optical transparency of the graphene nanosheets showed variation with number of layers and doping material. An ideal single layer of undoped graphene sheets showed a significant sheet resistance of ~ 31 ohm/sq, which makes graphene as a promising material for TCEs. The predicted upper limit of few-layer undoped graphene films exhibits ~ 337 ohm/sq with 90 % optical transparency, which is far beyond the industrial demands [21]. However, for the doped graphene sheets, the value of sheet resistance falls to = 62.4/N Ω-sq^{-1}, where N represents number of layers in the graphene thin film [22]. However, large scale fabrications of TCEs with the graphene as transparent conducting material have been suffering from conductivity loss, when applied to large area optoelectronic applications [23-24]. Thus, the performance control of a large-area device fabricated with these wonder materials often showed mediocre performance in

Materials Research Forum LLC
https://doi.org/10.21741/9781644901090-4

comparison to their small scale analogues. Thus, one of the important aspects which should be possessed by a TCEs material is to show transcending properties, when expanded to large scale optoelectronic applications. In this regard, metal meshes and metal nanowires (MNWs) based TCEs have shown excellent properties. These metal meshes and MNWs could be used as (i) subcomponents of the TCEs to improve the electrical conductivity or (ii) standalone TCEs. Accordingly, metal meshes and MNWs based TCEs compete with the ITO based TCEs to replace them in photovoltaic applications because of the excellent sheet resistance value and unbeatable optical transmittance [25-27]. Among MNWs, silver nanowires (AgNWs) showed wonderful performance in optoelectronic applications because of excellent T- trade-off in comparison to ITO based TCEs. In addition to MNW based TCEs, metal meshes also exhibit outstanding T- trade-off compared to ITO based TCEs. The flexible mechanical robustness of the metal meshes opens a new door for the flexible and low-cost scalable optoelectronics. Additionally, higher transparency of metal meshes in comparison to ITO, particularly in near infra-red (NIR) region enhances the importance of the metal meshes as TCEs in solar cell applications [28-29]. The special fabrication techniques of metal meshes make it completely invisible to the human eye. More importantly, these metal meshes can be fabricated to large scale by printing techniques, thereby, excluded the high costing sputtering process, which demands a high value of vacuum. This process generally requires temperature dependent process in the range of 100-120 °C or post sintering process with low energy values in the temperature range of ≤150–200 °C, thereby highly recommended by the researchers working in the field of the TCEs [30]. However, the figure of merit (FoM) of each kind of the TCEs decides the superiority of the TCEs suitable for photovoltaic applications.

1.1   Fundamentals for the evaluation of TCEs

The performance of the solar cell depends upon the transparency and electrical conductivity of its TCEs. Therefore, to appraise the performance of the solar cell, evaluation of the TCEs must be done by maintaining T- swap-balance between the electrical conductivity and optical transmittance. So, performance of TCEs decides quality of the TCEs, which could be evaluated from figure of merit (FoM). The figure of merit (FoM) acts as a performance indicator for TCEs in most of the optoelectronic applications and is evaluated from its optical and electrical constants. Haacke [31] proposed the first formulation to evaluate FoM for TCEs, based on the value of optical transmittance at 550nm and sheet resistance (). The FoM (Φ_{TC}) can be calculated from the following equations:

$$\Phi_{TC} = T_{550nm}^{10}/(\Omega^{-1}) \tag{1}$$

In addition to the above equation, Dressel and Grunner [32] also derived another formulation for the evaluation of FoM of TCEs based by developing a correlation between σ_{dc} and σ_{opt}, where σ_{dc} and σ_{opt} represents direct current conductivity and optical conductivity, respectively. According to Dressel and Grunner the FoM (σ_{dc}/σ_{opt}) for TCE can be evaluated from the following expression:

$$\frac{\sigma_{dc}}{\sigma_{opt}} = 188.5/\ (T_{550nm}^{-1/2}-1) \tag{2}$$

The high value of the σ_{dc}/σ_{opt} indicates better transmittance for a given or better for T_{550nm}. So, an elevated value of FoM is often preferred for excellent TCEs. From the industrial aspects, a TCE with FoM value ~35 (corresponding to T> 90% and <100 Ω-sq^{-1}) is considered to be suitable for optoelectronic applications [33-34].

2. Materials for transparent electrodes

The present era of the photovoltaic technology depends upon the choice of the TCEs, which decides the overall performance of the solar cells. In this regard, several materials have been synthesized which can be used for the wide range of the optoelectronic applications. Among this, the widely used materials are semiconducting oxides such as indium tin oxide and fluorine-doped tin oxide (FTO)-based TCEs. The popularity of the ITO and FTO based electrode gained attention of the researchers from the last decade. However, affordable cost-benefit analysis of these TCEs does not meet with the consumers demands. In this regard, researchers identified several other conducting materials which can be used as the alternative of ITO and FTO. Generally, the synthesis and fabrication of TCE materials other than ITO and FTO require low-cost technology which can easily meet with the conservation of energy, economy and ecology" (EEE). About ten years back, researchers moved towards the vacuum-free fabrication processes to save the high demand for energy. Fortunately, researchers discovered several new materials which can be synthesized easily in a laboratory environment and fabrication process involves solution-processable printing techniques such as roll to roll printing, doctor blade coating, spin coating etc. These materials include graphene and graphene composite type carbon nanomaterials, which have taken significant attraction due to their promising optoelectronic properties. On the other hand, metal nanowires (MNWs) and metal meshes are also potential candidate of TCEs. Therefore, synthesis and fabrication

Materials for Solar Cell Technologies I   Materials Research Forum LLC
Materials Research Foundations **88** (2021) 86-128   https://doi.org/10.21741/9781644901090-4

techniques of these materials must be understood to confine the applicability of these materials for the solar cell applications.

2.1 Indium tin oxide-based TCEs

Till date, Indium tin oxide a tin-doped In_2O_3 based wide bandgap n-type semiconductor is regarded as the best promising electrode available in the market due to its high electrical conductivity and high optical transparency in the visible region. The high FoM for ITO based TCEs makes it suitable for various kinds of optoelectronic applications. However, there are two face sides shown by the ITO electrodes. At one side while the ITO showed two-fold advantages regarding the electrical conductivity and excellent transparency, at the same time, the other face side of ITO showed several disadvantages which render the limitations to its uses, while considering the cost-benefit analysis by the opinion of circular economy with industrial symbiosis. The second face of the ITO based TCEs includes high fabrication cost, which demands the process of physical vapor deposition (PVD) by the high vacuum of spurting process. Therefore, alternatives to ITO electrodes were drastically investigated in the last decade. In this regard, nanomaterials have shown tremendous remarkable properties to use them as an alternative to the ITO based electrodes. Interestingly, the search of nanomaterials leads researchers towards the synthesis of ITO nanoparticles and their utilization for the low cost and affordable TCEs. Especially, ITO nanoparticle based porous coating has shown promising behaviour as a candidate of TCEs due to their high optical transparency. The main advantage associated with these ITO nanoparticles is their solution-processable ability for the fabrication of conductive electrodes, however, this technique couldn't receive much attraction for flexible substrates due to their post thermal heating step at high temperature, as the plastic substrate can bear upto a certain limit of temperature [35]. Recently a group of researchers has shown the fabrication of porous ITO nanoparticles based TCEs via solution process method followed by sintering at 300°C. The TCEs thus coated with porous ITO nanoparticles exhibits relatively low resistivity of 5×10^{-3} ohm-cm, with the sheet resistance and optical of 356 ohm/sq and 93% respectively. However, several other studies also conducted by various researchers to lower the sintering temperature up to 130°C. The sheet resistance of thus reported porous ITO nanoparticle-based nanocomposite was found to > 400 ohm/sq with optical transparency of 85%-90% [36-38]. However, the limitations of solution-processable method for the fabrication of porous ITO nanoparticle-based flexible TCEs leads the researchers for a simple vacuum coating process to produce almost transparent ITO based TCEs with novel nanoarray structure. The electrode thus fabricated possesses a sheet resistance of 230 ohm/sq and showed completely transparent behavior for the entire visible region [39]. However, still, the cost-effective issues and limited sources of indium poses a challenge to researchers, therefore, the research for alternatives to ITO for sustainable ecology and economy continues.

2.2 Fluorine doped indium tin oxide

Recently, the processes to find the alternatives of ITO based TCEs have gained much attention because of the high cost of indium in ITOs. Some studies found ITO as the most expensive component used during the device fabrication including the active layer, interlayer, top-electrode and device encapsulation [40]. Therefore, researchers generally look for the alternative doping materials which can replace indium from tin oxide. In this regard, fluorine is found to be the most suitable candidate to replace indium from tin oxide. Also, one of the major drawbacks related to the use of ITO based TCEs in the devices is the diffusion of indium in the active layer and even in the top electrode of the devices, which leads the instability of the devices [41-42]. Especially, FTO based TCEs plays a vital role in those cases where high-temperature requirement becomes a necessary condition to fabrication the interfacial layers in the devices. It has noted that FTO is more thermally stable than ITO and these have been reported by a study in which sheet resistance of the ITO changes from 18 ohm/sq to 53 ohm/sq with gradually rise in temperature up to 450°C during the procedure of thermal annealing. Therefore, FTO based TCEs generally suited in those cases, where high temperature is needed for annealing purpose. For instance, in the case of TiO_2 as one of the interlayer, FTO is used rather than ITO. Further, FTO with sheet resistance 7.5 ohm/sq is widely available in the market and found to more mature than ITO for optoelectronic applications [43]. Generally, it has been observed that TCEs with lower sheet resistance showed much better fill factor in the OSCs than those TCEs having higher sheets resistance. This has been depicted by a study where OSCs were fabricated by using FTO, ITO with the sheet resistance of 8 ohm/sq and 20 ohm/sq respectively. As expected OSCs with FTO as TCE showed high fill factor than the OSCs with ITO as TCEs. However, because of the high surface roughness within the FTO based TCEs, high leakage current often observed within the OSCs.

2.3 Carbon-based nanomaterials

Carbon nanomaterials are the class of the nanomaterials that are considered to be promising replacement of metal oxide-based TCEs (Figure 1). Among carbon nanomaterials, graphene and carbon nanotubes (CNTs) are extensively used during the last decades by the researchers due to extraordinary electrical and mechanical properties. Especially, high electrical conductivity and elevated mobility make them perfect candidates in the field of TCEs. Further, the high optical transparency in the visible region of the electromagnetic spectrum enhances their candidacy for a variety of the

Materials for Solar Cell Technologies I   Materials Research Forum LLC
Materials Research Foundations **88** (2021) 86-128   https://doi.org/10.21741/9781644901090-4

optoelectronic applications, where TCEs are used as the main component of the device structure [44-46]. It has been noted that a perfect TCEs should have enough sheet resistance and optical transparency that it compliances the merits of the conductive electrode. Further, the manufacturing cost should not exceed when going to large-area fabrication.

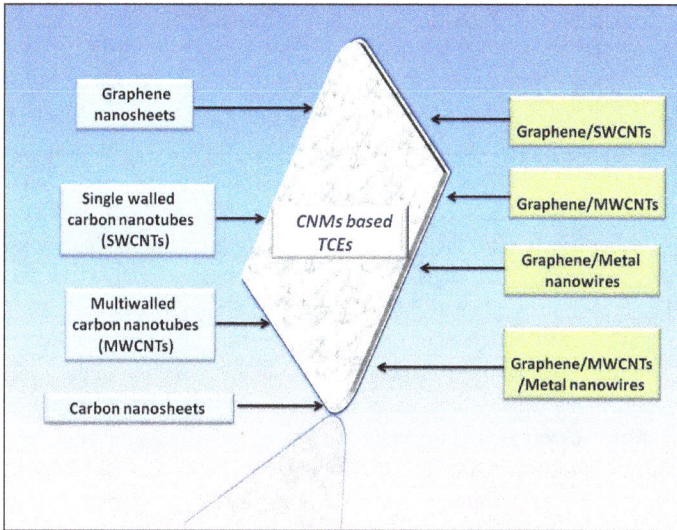

Figure 1   *Different kinds of CNMs for TCEs*

In this regard, still, the ITO based TCEs has dominated over the market because of promising optical transparency and superior electrical conductivity. However, the limited availability of the indium and critical process of fabrication with the high cost of manufacturing represent makes them poor choice for large scale applications. In this regard, graphene, CNTs and their composite materials showed superior properties for industrial symbiosis and cost-benefit analysis. About a decade ago, the fabrication of the TCEs with graphene and CNT understood as the costly process due to limited synthesis process. But nowadays, a variety of the synthesis routes for graphene and CNTs have opened new windows for the fabrication of cost effective and high area based TCEs.

2.3.1 Graphene-based TCEs

Graphene, a two-dimensional carbon nanomaterial with sp^2 hybridized carbon atoms crammed in structured hexagonal honeycomb type lattice showed outstanding properties such as high electrical, mechanical and optical properties, which makes it superior alternative to substitute metal oxide type TCEs commonly used in solar cells, flexible displays, and similar kind optoelectronic devices. However, still the large area production of TCEs with graphene facing problems due to mass-scale production of highly conducting graphene nanosheets. Therefore, presently, researchers are focusing on the mass-scale production techniques for graphene, which can be directly implemented for the wide scope TCEs. Another problem faced by researchers in the selection of an appropriate insulator substrate on which the graphene nanosheets has to be transferred. There are several techniques have been reported in previous years to fabricate TCEs with graphene as transparent conducting material. Among them, CVD processed and solution-processed graphene electrodes are widely used. Solution-processed fabrication of graphene-based electrodes is regarded as one of the best methods for the large area transparent electrode due to their ease of fabrication. Usually, solution-processed fabrication technique takes two steps, in which the first step generally takes the top-down approach i.e. breaking of graphite bundles into individual graphene flake, while the second step involves the fabrication of thin films of the graphene nanosheets into the substrates. Modified chemical exfoliation and ultrasonic exfoliation in the aqueous phase are the basic methods which come under the top-down approach. Once the exfoliation takes place, stable suspensions of the graphene flakes can be prepared after the process of the purification. However, in the case of the ultrasonic process, sometimes surfactant is needed to make the suspension stable. The other methods which follow bottom-up approach include Langmuir-Blodgett film technique [47-48], spin coating [49-51], filtration transfer [52], rod coating [53] and liquid-air interface self-assembly [54]. It is generally noted that large area graphene sheets must be needed in order to get highly efficient technological applications such as photovoltaic applications, which could achieved by some efficient transfer techniques applicable to both rigid and flexible substrates. [55]. Therefore, a precise clear view of the synthesis and fabrication process is regarded as one of the important parameters while going for the large-area fabrication of the graphene-based TCEs. For example, chemical oxidation of the graphite by following the modified Hummers method protocol could lead to an effective dispersion of the graphene oxide in the aqueous media of the water or some other solvent. The spin coating or roll coating of thus prepared graphene oxide dispersion in the insulators substrates like glass or flexible PET could reveal the large-area films of the graphene oxide sheets with single or multilayer coating of the graphene oxide sheets. However, coating of graphene oxide sheets over the desired substrate behaves as insulator due to presence of various oxygen-containing groups, therefore reduction of the thin films of the graphene oxide sheets is extremely needed to maintain conductivity of the graphene sheets, which is done by the either chemical reduction by introducing suitable reducing

Materials for Solar Cell Technologies I   Materials Research Forum LLC
Materials Research Foundations **88** (2021) 86-128   https://doi.org/10.21741/9781644901090-4

agents or by thermal reduction technique under the inert atmosphere of N_2 or Ar [56-62]. Further, another point that should be noted while processing the solution casting technique via chemical oxidation method is the unwanted presence of structural defects arises during the vigorous methods. The reviews on the oxidation and reduction suggest that none of the methods could perfectly maintain the originality of the graphene sheets [63]. The explanatory report could be understood by reading the work reported by a group of researchers where minimum sheet resistance and optical transmittance were reported to 840 ohm/sq and 78%, with FOM value of 1.8 respectively [64]. Further a group of researchers showed the thin film of solution processed GO on glass and plastic substrates [65-66]. However, none of these methods has passed the barrier of the grain boundaries. Especially chemically method was found to be unsuitable for to remove the barriers of the grain boundaries. To avoid some the major drawbacks of the chemical oxidation method to produce graphene oxide suspension, which includes vigorous oxidation and reduction, other liquid phase methods were developed on order to make less defected graphene nanosheets. These methods include liquid-phase exfoliation of graphite by using various organic solvents such as N-methyl pyrrolidone (NMP), N, N-dimethylacetamide (DMA), γ-butyrolactone, dimethylformamide (DMF) [67-68]. However, tremendous work has been done to make large scale graphene-based thin films for optoelectronic applications. In this regard, electrochemical exfoliation of the graphite widely used, where a group of researcher has shown the fabrication of the thin films of the graphene with the $= 1.35 \times 10^5$ ohm/sq with nearly 70% optical transmittance [69]. The large value of the sheet resistance and low optical transmittance could be used for photovoltaic applications. In this regard, chemical vapour deposition (CVD) technique has to emerge as one of the best promising methods for the fabrication of defect-free thin films of graphene nanosheets. The growth of the graphene films was conducted by using Cu or Ni as catalytic substrate, results in high-quality thin films of the graphene comparable with highly oriented pyrolytic graphite (HOPG). This technique can make graphene films quickly, but in comparison with the solution-processed graphene films, costs of production are remaining higher. But to confine industrial aspects and cost-benefit analysis, low-pressure CVD found sophisticated for optoelectronic applications. The primary research on the CVD based thin film of graphene sheets was done by a group of researchers who interestingly showed the deposition of graphene films over the polycrystalline Ni films, which is later on shifted to substrates of glass and plastic by using poly [methymethaacrylate] (PMMA), or polydimethylsiloxane (PDMS). The TCEs thus formed showed and optical transmittance of 770-1000 ohm/sq and 90%, 20, 280 ohm/sq and 76% with the FOM values of 3.5-4.5 and 4.1 respectively [70]. However, another group of researchers showed the superiority of the copper-based substrates over the Ni-based polycrystalline substrates. The film grown over the copper substrates was found to be a single layer within a repetitive manner, while in the case of the Ni-based substrates films grown on the substrates was found to be multilayer with the variation of a graphene layer on repetitive experiments. The sheet resistance and optical transmittance of the graphene films grown over the Cu substrates exhibit 350 ohm/sq and 90% respectively, with the FOM value nearly 10 [71]. The basic reason behind this is the homogeneity issues, which seriously affected the charge carrier mobility [72-73]. Because of the high homogeneity, CVD based graphene showed good electronic properties, which makes them suitable for the application of solar cells [74]. In order to get uniformity and homogeneity for the graphene films, a group of researchers synthesized large area monolayer and bilayer graphene films over the Cu-Ni alloy foils by using hydrogen and methane gas as precursor source for the synthesis. This group showed that quality and the thickness of the graphene films could be controlled by changing the temperature and cooling rate [75]. Further, real field application of CVD processed graphene TCEs was demonstrated by a group of researchers where a breakthrough was depicted by fabricating a 30-inch wide touch screen panel of graphene-based TCEs through the transfer of CVD processed graphene into the PET substrate via roll to roll technique. The sheet resistance and optical transparency of this TCE were found to be 30 ohm/sq and 90% respectively, with the FOM value of 118 [76]. The excellent FOM value of the graphene-based TCEs was found because of low defect density and homogeneity of the graphene sheets. The Raman spectrum of the CVD grown graphene was also found similar with the mechanically exfoliated graphene sheets from HOPG, whereas the graphene sheets obtained from the solution-processed technique showed the high intensity of D band, showed enhanced defect regions in the graphene sheets. Further, another factor reveals importance of the grain size for the graphene sheets. A published work focused on the grain size depicted that increased grain size in the graphene nanosheets enhanced the charge mobility, therefore reduces sheet resistance [77]. However, still, the debatable question has arisen about the superiority of the methods available for the fabrication of TCEs based on graphene. While one way CVD processed graphene thin films showed excellent results, but the cost-effective analysis is still marking a debatable question for the wide scope of the CVD processed thin film of graphene sheets. On the other way, solution-processable graphene thin films provide cost-benefit in terms of economic aspects, but the average performance of thus fabricated graphene thin films limited the scope of the solution-processed thin films of the graphene for the optoelectronic applications. In this regard, many researchers have proposed graphene-based hybrid TCEs, which includes the doping of the other nanomaterials such as CNT, metal nanowires and metal oxides. Many researchers incorporated CNT as a dopant to make graphene-based hybrid TCEs. One of researchers group has mixed GO and CNT in the reducing media of anhydrous hydrazine and spin-coated it on top of glass electrode followed by the doping of the thionyl chloride depicted considerable sheet resistance and optical transmittance of 240 ohm/sq and 86% with the FOM value of 10 [78], while another researcher group reported sheet resistance and optical transmittance of 735 ohm/sq and 90 % transmittance by incorporating CNT with CVD processed graphene sheets [79]. Another approach has also demonstrated by using metal nanowire as conducting filler materials to fabricate graphene-based hybrid TCE. Recently a group of researchers showed the addition of AgNWs as conducting filler in the matrix of the graphene exhibit variably high FOM value of 182, corresponding to the sheet resistance and optical transmittance of 33 ohm/sq and 94% respectively [80]. However, the presence of CNTs and metal nanowires also imposed tube-tube junction resistance, which somewhat affects charge carrier dynamics of devices.

2.3.2   Carbon nanotube based TCEs

Carbon nanotubes (CNTs) are another class of carbon nanomaterials, which can be considered as the cylindrical form of the graphene nanosheets. Depending upon the walls on the tubes, CNTs are generally classified according to the number of walled present on the carbon tubes. The most versatile form of carbon nanotubes present as single-walled carbon nanotubes (SWCNTs) or as multiwalled carbon nanotubes (MWCNTs) [81]. Because of their optical transparency, electrical conductivity and flexibility, CNTs are also regarded as an alternative of ITO based conductive electrodes. The open window of flexibility for optoelectronic applications makes it the most viable alternative to ITO based electrodes. However, several issues related with the CNTs as a candidate of TCEs still needed several advancements to use them as an alternative of ITO such poor adhesion to the substrate, which reduces the wide-scale applications of CNT based TCEs. Till date, several methods have been designed and optimized to fabricate CNT based electrodes, this method includes fabrication by direct CVD, spin coating, deposition by vacuum filtration, inkjet printing, spray and dip coating, and [82-86]. These methods successfully employed for both SWCNTs and MWCNTs thin films. SWCNTs depicted significant attention in the present decades due to their better electrochromic properties, more resistivity towards the corrosion leading chemicals and wide potential window for various optoelectronic applications, especially for solar cell applications. To reveal the viability of the SWCNTs as TCEs for solar application, several fabrication techniques have been demonstrated by various researchers [87]. A group of researchers reported the fabrication of optical transparent SWCNT based conducting electrode, by using a simple CVD based reactor with incorporating organic solvents. The thin film of this SWCNT electrode showed comparable parameters with ITO based electrode [88]. While in one of the methods, the researcher uses surfactants to fabricate thin films of the SWCNTs via spin coating. This method includes the use of oligothiophene-terminated poly(ethylene

Materials for Solar Cell Technologies I   Materials Research Forum LLC
Materials Research Foundations **88** (2021) 86-128   https://doi.org/10.21741/9781644901090-4

glycol), a non-ionic amphiphilic surfactant, which help the system to achieve a highly dense network of SWCNT due to the charge repulsion factors. Further to improve the sheet resistance of the thin film of the SWCNT, treatment with HNO_3 and $SOCl_2$ was done followed by the deep cleaning with dichloromethane and distilled water. The final sheet resistant and optical transmittance of the thus fabricated thin film of SWCNT was found to be 56 ohm/sq and 71 % respectively, at the wavelength of 550 nm [89]. Proceeding, the work towards the flexible electrode preparation, another group of researchers has fabricated the SWCNTs thin film on both glass and PET substrate. Nafion and water media at the ratio of 50:50 was used to optimize the results to get significant sheet resistant value of 500-600 ohm/sq [90]. Further, the main work regarding the fabrication of SWCNTs was focused towards the flexible TCEs in recent years. In this regard, one of the groups has demonstrated the utility of the SWCNT as a flexible electrode material. This group demonstrated fabrication of transparent SWCNT electrodes on the flexible silicon substrate by using spray coating method and obtained optical transmittance in the range of 65% to 85% and 1 to 8.5 k-ohm/sq sheet resistance for thin-film coatings, while low sheet resistance of the 200 ohm/sq was obtained for thick films by compromising optical transmittance [91]. Though, industrial applications of SWCNTs must be reliable to the utilization of fruitful properties of the SWCNTs. In this regard, a group of researchers fabricated SWCNTs based conducting electrodes by using the printing technique. SWCNT obtained from arc discharge method were fabricated on the plastic substrate. TCEs thus fabricated showed 200 ohm/sq sheet resistance and 85% optical transmittance of at 550 nm. The same SWCNT based TCEs also used as an anode for the fabrication polymer fullerene bulk heterojunction solar cells. OPV cell fabricated on SWCNT/plastic electrode showed almost similar PCE of 2.5%, as obtained for ITO based OPVs, which showed 3% PCE for the same type architecture [92]. In this way, while one way SWCNTs showed extensive properties to use them as TCEs, at the same time cost-benefit constraints limited the use of SWCNTs as replacement of ITO based TCEs, until a cost-effective and eco-friendly mass scale productive technique has developed. However, some of the researchers also reported the viability of MWCNTs as a candidate of TCEs. Recently, a group of researchers reported MWCNTs based flexible TCE by using PET substrate. Sodium dodecyl sulphate (SDS) along with MWCNTs was allowed for ultrasonication bath to prepare MWCNT based TCEs. TCEs thus fabricated with MWCNTs at the optimum concentration of 1.2mg/ml showed 180 ohm/sq sheet resistance with an optical transmittance of 85% at 550nm [93], while another group of researchers showed a semitransparent cells with free-standing MWCNTs as top transparent electrodes for n-i-p type configuration. The cell thus fabricated has shown very low leakage current, up to 60% fill factor and promising 1.5 % efficiency with long term durability and stability [94].

2.3.3 Carbon nanosheets based TCEs

Recently carbon nanosheets based TCEs also attracts the researchers because of their high electrical and optical properties. It has been reported that carbon nanosheets could be used as TCEs in OPVs. In this regard, a group of researchers has showed pitch converted carbon nanosheets as TCEs in OPVs. Thin films of carbon nanosheets were fabricated via solution processing technique, in which pitch solution was prepared in dimethlyformamide and spin coated on quartz substrates followed by carbonization process to covert the pitch solution into carbon nanosheets. Thus, fabricated carbon nanosheets based TCEs were used as anode for in OSC and showed nearly 1.7 % efficiency under AM 1.5 G and 100 nW cm^{-2} illumination [95].

2.4 Metal nanowires (MNWs) based TCEs

The search for alternatives to the ITO based TCEs lead the researchers towards the growth of metal-based TCEs. Metals exhibit excellent electrical conductivity. However, opaque nature in its bulk form becomes a crucial issue to use them as material for the transparent electrode. Therefore, research moves towards the synthesis of metal-based nanomaterials to utilize their fruitful properties for the TCEs applications. In this regard, metal nanowires (MNWs) showed excellent properties to use them as TCEs because of their unbeatable electrical conductivity and optical transparency in the visible region. Two kinds of approaches for the synthesis of metal nanowires are generally employed, which includes the bottom-up and top-down approach, respectively. The bottom-up approach generally employ for the solution processes to assemble the metal nanowires in the form of metal films followed by post-treatment to weld the contacting metal nanowires for better electrical performance, while top-down approach required the deposition of metal films in the form of nanowires onto a template without wire-wire junction resistance for fully connected metal electrodes. The first solution-processed metal nanowire system was fabricated on the top of polyethene terephthalate (PET) substrate by using Ag nanowires synthesized by the chemical reduction of silver nitrate [96]. Figure (2) [97] depicted the dispersion of the Ag nanowires via solution process approach for making flexible Ag nanowire-based TCE.

Further, locally joined metal nanowires need to be well jointed for good electrical conductivity. To join these metal nanowires well, several techniques have been adopted such as post-annealing treatment at high temperature, electrochemically deposition of another thin film of other conducting materials, plasmonic welding and high-pressure cold welding method. Yet, the annealing method at high temperature (200°C) weld the metal nanowires and joint them well to give high electrical conductivity, but for the flexible TEs, the method is not suited well, because at this temperature most of the

Materials for Solar Cell Technologies I   Materials Research Forum LLC
Materials Research Foundations **88** (2021) 86-128   https://doi.org/10.21741/9781644901090-4

flexible substrates get crumbling and lose their transparency. Among these methods, the welding process by cold pressing at high pressure has shown significant development for the good electrical and optical performance of the Ag nanowires. A group of researchers has shown cold pressing at 25 MPa for 5 s for the welding of Ag nanowires and reported low sheet resistance of 8.6 ohm/sq with 80% optical transmittance [98]. However, improvement in the sheet resistance and optical transmittance again showed by another group of researchers which showed the spray coating process for the fabrication of Ag nanowires as TCE. The Ag nanowires based TCEs thus fabricated with this process exhibited sheet resistance 18.9 ohm/sq and 94% optical transmittance, which is better than the previously reported results [99]. In addition to the fabrication process, the value of and optical transmittance also decided by length of Ag nanowires. It has been reported that long nanowires exhibited low sheet resistance and high optical transmittance in comparison to the short nanowires. In this regard, a group of researchers has developed multi cycling growing technique to fabricate long Ag nanowires network with an approximate length of 100μm exhibited of 9 ohm/sq with 89 % optical transparency, which is found much higher than Ag nanowire with average length) 20μm [100]. Apart from the Ag nanowire, other highly conducting metals such as Gold (Au) and Copper (Cu) were also employed for nanowire-based TCEs. Copper has great electrical conductivity with lesser prices in comparison to the Ag and Au metals, therefore regarded as another alternative to ITO as candidate of TCEs. Several studies have been done for the growth of Cu nanowires. Among them, the simplest growth of the Cu nanowire depicted by the studies done by the group of researchers, which showed the growth of Cu nanowire from spherical seed of the Cu in the aqueous solution of NaOH, Cu (NO)$_3$, ethylenediamine and hydrazine with the average length of 10μm and 90nm in diameter. The fabricated TCEs based on these Cu nanowires showed good sheet resistance of 15 ohm/sq, but relatively low optical transmittance (65%), which was found relatively low in comparison to Ag-based TCEs [101-103]. The low value of optical transmittance may depicted here because of the short length of the Cu nanowires and proved by the another study by a group of researcher which showed that Cu nanowires based TCEs with average length of 28.4 μm long and 75nm thick NWs exhibited sheet resistance of 60 ohm/sq and 94.4 % optical transmittance, which is close to the Cu nanowires with average length of 40 μm and 16.2nm in diameter exhibit a sheets resistance of 51.5 ohm/sq at the transmittance of 93.1% [104]. However, the main problems that concern with the Cu nanowires are the issues related to the stability of Cu nanowire in air. Reports depicted that Cu nanowire is unstable in air and the resistance of the Cu nanowire increases by one magnitude at the temperature of 85°C. To resolve this issue, the protective layer of the Ni was fabricated. However, still, research is going on to develop another way for the fabrication of Cu nanowire-based TCEs. Further, Au nanowires were also developed during the last decade. However, the synthesis of Au nanowire is not easier as in the case of Ag and Cu. In this regard, a group of researchers has synthesized Au nanowires by reducing the solution of $HAuCl_4$ with oleyamine to produce Au nanowires with a length of few microns and an average diameter of 1.6 nm [105]. The TCEs fabricated by using these Au nanowires exhibited a sheet resistance of 400 ohm/sq and approximately 96.5 % transmittance [106]. So, briefly, still, the Ag-based nanowires dominated over Cu and Au based nanowires for the fabrication of effective TCEs for optoelectronic applications. So preferably, Ag nanowire-based conductive electrodes widely used as an anode for OSCs. A study done by a group of researchers used Ag nanowire as TCE to substitute metal grids in P3HT: PCBM based OSCs showed almost 19% elevated short circuit current, however approx 10% lower PCE in comparison to the ITO based OSCs with same active layer materials [107]. Therefore, the overall study for the optoelectronic properties of the metal nanowire-based TCEs showed that Ag, Cu and Au could become a promising alternative to the ITO based TCEs.

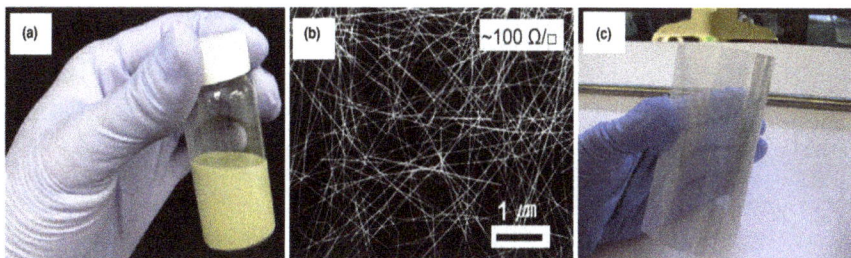

Figure 2   (a) Dispersion of Ag nanowires (b) SEM image of the on the top of the flexible substrate, (c) Ag nanowire-based flexible transparent conductive electrode (Reprinted (adapted) with permission from (Reference [97]). Copyright © 2010, American Chemical Society)

2.5   Metal meshes based TCEs

The innovative ideas and intensive research for the finding of the alternative of ITO based TCEs have shifted the direction of the researchers towards an entirely new concept of TCEs, which includes the incorporation of metal meshes as conducting network of the transparent electrode for optoelectronic applications. Metal meshes based TCEs showing interesting feature for TCEs because of their good electrical conductivity, mechanical robustness, optical transparency and significantly cost-effective fabrication process, which makes them affordable and compatible with large scale optoelectronic applications. These metal meshes based TCEs show their utility for wide-scale optoelectronic applications including third generation solar cells, therefore depicted new windows in the field of energy conversion and storage devices. The TCEs based on the MNW and Metal meshes showed similarity in many aspects, but their fabrication process is different. The synthesis of MNW can be easily done with solution process methods and other previously reported methods [108-109]. However, high-quality TCEs via solution processing method is still a challenge. On the other hand, fabrication of metal meshes based TCE includes different fabrication techniques such as thermal deposition, sputtering, laser direct writing, electrodeposition, lithography, printing and amalgamation of two methods to fabricate hybrid electrodes [110-121]. Till date, different classes of metal meshes based transparent conducting electrodes have been fabricated, which includes the TCEs based on Ag meshes, Au meshes, Ni meshes, Cu meshes, Mo meshes, Pt meshes and combination of two more metal meshes [122-133]. The properties of the metal mesh-based electrodes vary for metal to metal and type of fabrication technique used for the fabrication process. Generally, the printing technique is one of the versatile method used for the large scale metal mesh-based TCEs. Printing process involves simple and sophisticated vacuum free roll to roll process. Hence, the printing method is regarded as the best cost-effective method in comparison to the physical and chemical methods. Further, the electrical and optical properties of the metal mesh-based TCEs depends upon mesh patterns, which in turn of it depends upon the effect of aspect ratio (AR), pitch size thickness and line width. The AR is described as the ratio of thickness/width, which must be increased to increase electrical conductivity and optical transmittance. The general strategy to increase the AR of the metal mesh-based electrodes includes enhancement in the number of metal lines without affecting optical transparency and the height of the metal lines to increase the conductivity in the mesh network. Principally, AR of the metal mesh-based TCEs can be controlled by the different kinds of patterns and shape geometries of the metal network. Different kinds of geometries and shapes had been adopted for the fabrication of metal mesh-based TCEs. The general shapes adopted for metal mesh-based TCEs includes lines, grids, pyramids, brick walls, honeycomb and circular shape [134-139]. However, researchers are also looking for alternate shapes and geometries for metal mesh-based electrodes which could optimize results than previously adopted shapes and designs. One of the key points that should be addressed here is about the possibility of the electrical short-circuiting between the metal mesh network and the top electrode, therefore to neglecting to possibility of electric short circuit, researchers have attempted some successful efforts by depositing an additional conduction passive film on the apex of the metal mesh electrode [140-141] or by introducing the metal mess into the plastic elastomeric substrate by the deposition of an extra conducting film [142-143]. Performance of the metal-based TCEs depends upon the metal layer, which in turn of it, depends upon the geometry and dimensions of the line width (w), pitch size (p), height (h), web width (a), and kind of face along with its topography i.e. embossed or embedded (Figure 3 (a-d)) [144].

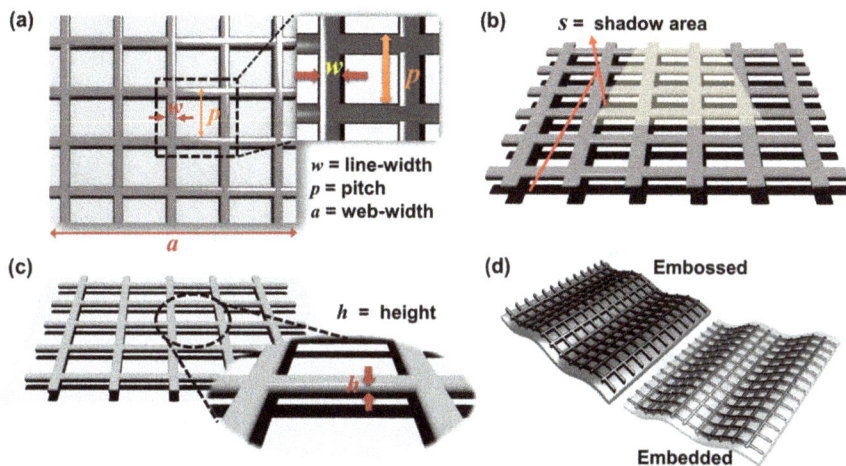

Figure 3   Square grid network of metal mesh-based TCE showing (a) the line width, pitch size and web-width, (b) shadow area during illumination beneath the grid network, (c) the height of the metal grids, (d) flexible substrates in embossed and embedded form (Reprinted (adapted) with permission from (Reference [144]). Copyright © 2013, Royal Society of Chemistry)

For instance, increasing line width (w) enhances the conductivity of the metal mesh electrode, but it also enhances the shadow area (Figure 4(a)) [144], which depicted unfavourable condition for the fabrication of the optoelectronic devices. Especially, this condition will greatly affect the solar cell devices, where a broad spectrum of light intensity is needed to get the best device performance. However, if the line width reduces, it enhances the sheet resistance, which again creates a problematic scenario for a perfect metal mesh TCE for optoelectronic application. Therefore, the researcher suggests the deposition of a thicker line of metal grids to compensate for this issue. Working on the suggestion, a group of researchers worked on Au based metal mesh electrodes and deposited a thicker Au mesh and effectively minimized sheet resistance by a factor of three [145]. Similarly, pitch (p) of the metal grid also plays an important role in the applicability of the metal grids for optoelectronic applications. An increased p within the metal grid network enhances optical transmittance at the cost of sheet resistance and vice versa (Figure 4(b)) [144].

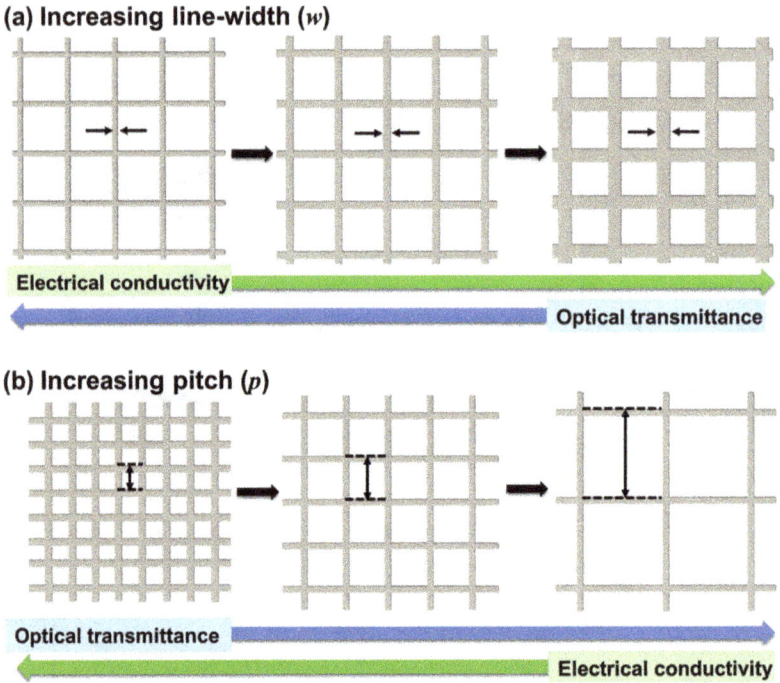

Figure 4   *Effect of line width (w) and pitch (p) in metal-based TCEs (a) increasing line –width (w), (b) increasing pitch (p) (Reprinted (adapted) with permission from (Reference [144]). Copyright © 2013, Royal Society of Chemistry)*

The larger value of p imposed the performance of TCEs for optoelectronic applications due to higher diffusion pathway for the better charge collection for the charge carriers and therefore increased the chances of charge recombination. Further, thickness of the metal mesh gridlines is regarded as another important factor that greatly affects the performance of metal mesh TCEs. Higher thickness of the metal grid lines will reduce sheet resistance, although it enhances the surface roughness of the TCEs, which affects the performance of it, while used for optoelectronic applications. This can be understood by the study performed by a group of researchers who fabricated Ni-based metal mesh with different height of the metal gridlines. The study showed that metal grid with 5 nm line thickness showed low sheet resistance of 5 ohm/sq, but the optical transmittance was found to be much lower, nearly opaque, while metal grid line with the thickness of 2 nm showed high optical transmittance > 80%, but the sheet resistance highly increased and showed a value of nearly 950 ohm/sq, which is much higher for optoelectronic applications [146]. So, generally a thin ultra-smooth layer of conducting polymer such as PEDOT: PSS is generally employed to remove the barrier of surface roughness. Also, sometimes, the metal mesh grids are embedded within a plastic substrate to enhance the surface smoothness (Fig 3(d)) [144]. In addition to these factors, the choice of materials used for the fabrication of metal mesh conducting electrode also showed variable results for different material. After discussing the fabrication techniques, factors affecting the execution of the metal mesh-based TCEs for solar cell application were discussed and depicted by the various group of researchers. A TCE is to understand as one of the important parts of solar cells, as it directly dictates the performance of the solar cells. To analyze the consistency of the metal mesh-based electrode in organic solar cell, a group of research investigated Ag, Au and Cu based metal mesh electrodes as alternative of ITO based electrodes and reported almost similar results with ITO based electrodes because of lower sheet resistance < 30 ohm/sq and high optical transmittance> 80%. Further, another group of researchers reported the diethylene glycol (DEG) treated PEDOT: PSS as a passivation layer for the smoothness of the Ag metal mesh electrode and used this hybrid TCE as an alternative of the ITO based TCE in OSCs [147]. The sheet resistance of DEG-PEDOT: PSS-Ag based metal mesh TCEs was found to be nearly 70 ohm/sq, which somewhat compete with ITO based electrodes. Further, continuing the work of Ag metal mesh electrodes, another group of researchers depicted a flexible Ag electrode with honeycomb geometry fashion as current collecting mesh via screen printing technique by adjusting line width restricted to 160 μm and showed very low sheet resistance of 1 ohm/sq [148]. Further, this group also demonstrated the fabrication of OSC with an active area of 4 cm^2 by using PEDOT: PSS/Ag honeycomb structure hybrid TCEs as an alternative of ITO based TCEs and demonstrated PCE nearly 1.82 %, which is found as almost two fold more than ITO based OSC with a PCE of 0.95 %. Again, continuing to this work, other group researcher reported an enhanced PCE of 5.85% with an active area 1.21 cm^2 for a flexible OSC with PEDOT: PSS/Ag hybrid transparent conducting electrode [149]. Proceeding to same work, another group also implemented PEDOT: PSS/Ag-based hybrid TCE for small molecule OSCs [150]. The power conversion efficiency obtained by using this TCEs was found to nearly 2.8%, for an active area of 0.08 cm2, while 2.9 % PCE was achieved for ITO based TCEs for the same device architecture. In addition to the conductive polymers, graphene nanosheets were also employed along with Ag metal mesh TCEs [151]. However, cost-benefit analysis of the graphene assisted Ag metal mesh TCE lags behind than the PEDOT: PSS based Ag metal mesh TCEs. So, whatever be the route of fabrication will be implemented for the search of alternative TCEs, a cost-benefit analysis will only suggest their applicability in marketing applications (Figure 5) [144].

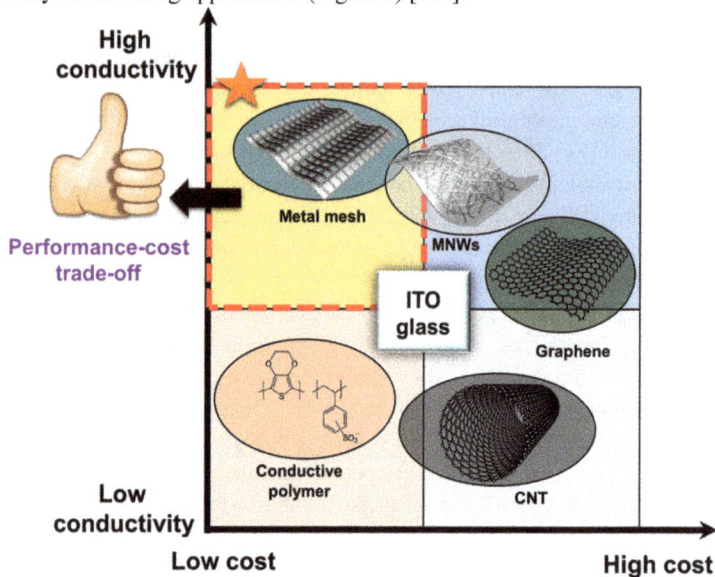

Figure 5   Comparison of the performance-fabrication cost between commercial available ITO and other available alternative TCEs (Reprinted (adapted) with permission from (Reference [144]). Copyright © 2013, Royal Society of Chemistry)

2.6   PEDOT: PSS based TCEs

The demand of flexible and cost-effective TCEs have diverted the attraction of researchers towards the eco-friendly conductive polymers due to their excellent electrical conducting properties, optical transparency and outstandingly their ease of fabrication process for the large scale TCE applications. In this regard, poly (3, 4-ethylene dioxythiophene): poly (styrenesulfonate) called PEDOT: PSS (Figure 6) has attracted the researchers due to its promising properties for various optoelectronic applications such as excellent electrical conductivity and optical transparency, tunable work function (WF), and high flexibility with stretchable tendency [152-154]. These properties make it viable

Materials for Solar Cell Technologies I   Materials Research Forum LLC
Materials Research Foundations **88** (2021) 86-128   https://doi.org/10.21741/9781644901090-4

for many applications such as third generation solar cells, OLEDs, OTFTs, health monitors, thermoelectric, flexible artificial intelligence robotics and touch sensors.

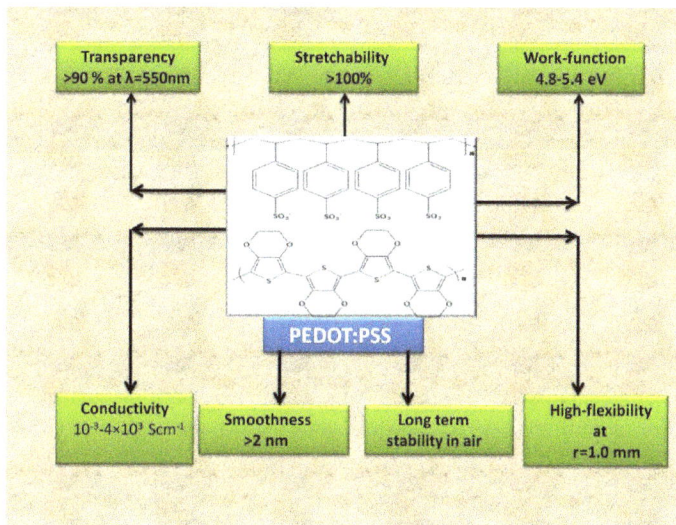

Figure 6   Chemical structure and properties of the PEDOT: PSS

In the last fifteen years, extensive research has been done in the field of the PEDOT: PSS, which awarded it as the most promising conducting polymer for various rigid and flexible electronics. Basically, PEDOT: PSS is regarded as a polymer electrolyte with highly conjugating PEDOT as positive charges and insulating PSS moiety with negative charges. Generally, the chemical structure and practical demonstration of PEDOT moiety showed that it is insoluble in water, which imposes the barrier for its applicability. Therefore, it is attached chemically with another moiety i.e. PSS. The PSS moieties not only make PEDOT dispersible to water, but it also maintains the stability of the PEDOT via coulombic interactions. The beautiful deep blue colour of PEDOT: PSS and cost-effective fabrication process such as spin coating, blade coating, spray coating, dip coating, inkjet printing, screen printing open a new door for various optoelectronic applications. Remarkably, in comparison to the scalable fabrication of MNW and CNT based TCEs, PEDOT: PSS based TCEs offers very low prices with almost the same properties. The TCEs fabricated with PEDOT: PSS offers very high surface smoothness and uniformity regardless of glass and plastic substrates [155-156]. Recently, PEDOT: PSS are widely used as TCEs, as an alternative for ITO electrode for organic solar cell applications. A group of researchers used modified PEDOT: PSS based TCEs as an alternative of ITO in OSCs. The doping treatments were done before use PEDOT: PSS as TCEs by using glycerol and sorbitol respectively to decrease the sheet resistance from 1.5×10^5 ohm/sq to 1.0×10^3 ohm/sq. The resulting device showed a PCE of 3.0%, while the device based on ITO with same architecture showed a PCE of 5.4% [157]. However, the study reveals the possibility of PEDOT: PSS as TCEs for futuristic OSCs. Another group of researchers doped PEDOT: PSS with ethylene glycol to convert mixed PEDOT: PSS into the linear matrix and reported a PCE of 1.5% with MEH-PPV based OSCs [158]. It is generally seen that PEDOT: PSS based conducting electrodes showed better performance when doped with some suitable reagent. In this regard a group of researchers showed the performance of PEDOT: PSS based TCEs doped with 5% dimethylsulfoxide (DMSO). The DMSO doped PEDOT: PSS based TCEs showed high conductivity than the TCEs fabricated with PEDOT: PSS alone. TCEs doped with DMSO showed conductivity of 470 S/cm^{-1}, which showed benchmark conductivity for optoelectronic applications [159]. However, this value of conductivity lags from the ITO based electrode, which showed a charming conductivity of 1050 S/cm^{-1} for plastic substrates and 6740 S/cm^{-1} for glass-based electrodes. Modification of PEDOT: PSS based TCEs leads the wide use of them as anode for various kinds of OSCs. A group of researchers showed the fabrication of flexible and stretchable PEDOT: PSS based TCEs modified with the additive treatment of DMSO and fluorosurfactant (Zonyl FS-300) and used them in P3HT: PCBM based organic solar cells. The PCE thus achieved for this modified PEDOT: PSS based TCEs was found to be comparable to that of ITO based device [160]. Lightweight and ultrathin OSCs were also fabricated by using modified PEDOT: PSS based TCEs with 0.2 μm thickness depicted the high PCE of 4.2% for P$_3$HT: PCBM based flexible organic solar cells. The modified PEDOT: PSS based electrode used for this study demonstrated a remarkable of 100 ohm/sq, while DMSO and Zonyl-FS300 were used as the additive for the modification of PEDOT: PSS based electrode [161]. However, the solvent modification in such extent showed only moderate modification in the conductivity of PEDOT: PSS based TCEs. The breakthrough in this regard was demonstrated a study done by the group of researchers, which showed how effectively soaking in acidic media of the PEDOT: PSS showed relatively high conductivity just comparable to the ITO based TCEs. The study revealed how the conductivity of the PEDOT: PSS based thin films enhanced by deep soaking of films into the 1M H$_2$SO$_4$ followed by heating at 160 °C. The thin films of PEDOT: PSS thus treated with H$_2$SO$_4$ depicted conductivity of 3000 S/cm, which is high enough to the previously reported the thin films of the PEDOT: PSS, which showed conductivity in the range of

500-1000 S/cm. Further, OSCs fabricated by using these TCEs showed an enhanced PCE of 3.5%, which is analogous with OSCs based on the ITO as TCE [162]. This work is further carry forward by another group of researchers, which reported the effect of soaking of thin films of PEDOT: PSS with 100% concentrated H_2SO_4. The thin films of PEDOT: PSS soaked with concentrated H_2SO_4 showed an optimum sheet resistance of 46.1 ohm/sq and 90 % optical transmittance at 550 nm, due to the structural alignment of the polymeric chains of PEDOT into the highly crystalline nanofibrils. Further, OSCs fabricated by using these TCEs showed a high PCE of 6.6% [163]. So, highly acidic treatment of PEDOT: PSS films leading towards the high conductivity area, but its strongly affects the substrate morphology, thereby optical transmittance greatly effects by acid treatment. For instance, it is reported that acid treatment greatly affects the flexibility and transparency of the plastic substrates, therefore reduces the optoelectronic performance of the flexible PEDOT: PSS based TCEs [164-165]. Therefore, while one way the acid treatment enhanced the conductivity of the PEDOT: PSS flexible TCEs, on the hand acid treatment, greatly affect the flexibility and transmittance of these TCEs. So, to find out the appropriate solution to this issue, a group researcher showed an effective way to remove the issues related to the corrosive nature of acid. This group established a facile transfer-printing for acid treated PEDOT: PSS based flexible conducting electrodes on polyethylene naphthalate (PEN), which exhibited a low of 45 ohm/sq and >90% optical transparency. The PTB7: PCBM based OSCs fabricated with transfer printed PEDOT: PSS based anodes showed an optimum PCE of 7.7%, which resemblance with ITO based OSC of same active layer materials [166].

2.7   Other TCE materials

One of the transparent conducting materials that used as an alternative to the ITO based TCEs is the aluminium doped zinc oxide (ZnO: Al) often regarded as AZO. It is previously reported that AZO has a similar conductivity as that of ITO, with resistivity ranging from 10^{-4} ohm-cm to 10^{-3} ohm-cm and 80 to 90% optical transparency. However chemical instability and brittle nature with surface roughness properties limit its use as an effective candidate for an alternative of ITO based TCEs [167]. Further, recently a group of researchers reported ultrathin MoO_3/Ag transparent electrode with the sheet resistance value of 20 ohm/sq with the optical transmittance of nearly 74.22% for the application of OSCs. OSCs fabricated with this transparent electrode exhibits approximately 2.76 % PCE with a transmittance of 38% in the visible region. Apart from the aforesaid materials, some other materials were also investigated as a candidate of TCEs such as SnO_2, Cd_2SnO_4, $CdSnO_3$, $CdInO_4$, Zn_2SnO_4, $MgIn_2O_4$, $CdSb_2O_6$ and $In_4Sn_3O_{12}$ and generally passing through early-stage experiments [168-169].

Materials for Solar Cell Technologies I
Materials Research Foundations **88** (2021) 86-128

Materials Research Forum LLC
https://doi.org/10.21741/9781644901090-4

Conclusion and remarks

The search for cost-effective and sustainable TCEs for solar cell applications has withdrawn the remarkable attraction of the researchers to discover the scalable and feasible process of fabrication with an appropriate selection of materials. Much effort has already been given to the development of various transparent conducting materials as the alternatives of ITO based TCEs. As stated above, several materials have depicted different and optical transmittance, which contingent upon the selection of the materials and fabrication process. Yet ITO based TCEs still dominated the market because of high electrical conductivity and optical transparency, but cost-benefit barriers and difficulty in fabrication process impose the popularity of the ITO based TCEs. However, researcher also describes the methods in which fabrication of ITO based TCEs is possible via the solution-processable technique for porous ITO nanoparticles, but the process becomes limited due to high demand of temperature for the sintering process. Further, carbon nanomaterials such as graphene nanosheets and carbon nanotubes have emerged as the substitute for TCEs for futuristic solar cell technology. The scalable and solution-processable methods available for the fabrication of graphene and CNT based materials opens a new door for large scale solar cells. However, still, the cost-benefit analysis has become a debatable question for the implementation of these materials as TCEs for solar cells. Other TCEs that have shown significant interest for the researchers are the metal nanowires, which show high conductivity along with permissible optical transmittance for the application of solar cells. Especially, Ag nanowires are found to be the most deserving candidate among other metal nanowires due to their ease of synthesis and fabrication routes. Also, the cost-benefit analysis showed very low competitive remarks as compared to the other existing TCEs. So, while searching for the alternative transparent conducting materials with cost-benefit analysis, metal mesh-based transparent conducting electrode showed a promising window for scalable and affordable TCEs for solar cell applications. However, still, some of the parameters are needed to be filtered in to completely eradicate ITO based TCEs. Although, conducting polymers such as PEDOT: PSS is also showed a competitive environment to both metal mesh and ITO based TCEs in terms of electrical, optical and cost-benefit analysis. PEDOT: PSS based conducting electrodes are widely accepted and has become of the choice of hole conducting materials in OSCs. Other transparent conducting materials are also investigated in the last decades, but still, the performance of these conducting materials has not come to the point, from where these materials could be used as an alternative of ITO based TCEs.

Materials for Solar Cell Technologies I
Materials Research Foundations **88** (2021) 86-128

Materials Research Forum LLC
https://doi.org/10.21741/9781644901090-4

Acknowledgements

Authors acknowledge the financial support from Department of Science and Technology, INSPIRE Divison (IF150750), New Delhi, India and Ministry of Environment, Forest and Climate Change, NMHS, Kosi-Katarmal, Almora, Uttarakhand, India.

References

[1] G.A. Brian, A. Zaban, S. Ferrere, Dye sensitized solar cells: energetic considerations and applications, Z. Phys. Chem. 212 (1999) 11-22. http://doi.org/10.1524/zpch.1999.212.Part_1.011

[2] V.V Tyagi, N.A.A. Rahim, N.A. Rahim, J.A.L. Selvaraj, Progress in solar PV technology: research and achievement, Renew Sustain Energy Rev. 20 (2013) 443–61. http://doi.org/10.1016/j.rser.2012.09.028.

[3] M. Green, K. Emery, Y. Hishikawa, W. Warta, Solar cell efficiency tables (version 36), Prog. Photovolt: Res. Appl.18 (2010) 346-352. http://doi.org/10.1002/pip.1021

[4] A. Kumar, Z. Chongwu, The race to replace tin-doped indium oxide: which material will win?, ACS Nano. 4.1 (2010) 11-14. http://doi.org/10.1021/nn901903b

[5] H.S. Jung, K. Eun, Y.T. Kim, E.K. Lee, S.H. Choa, Experimental and numerical investigation of flexibility of ITO electrode for application in flexible electronic devices. Microsyst Technol. 23 (2017) 1961-1970. Http://doi.org/10.1007/s00542-016-2959-3

[6] K. Sakamoto, H. Kuwae, N. Kobayashi, A. Nobori, S. Shoji, J. Mizuno, Highly flexible transparent electrodes based on mesh-patterned rigid indium tin oxide, Sci. Rep. 8 (2018) 2825. Http://doi.org/10.1038/s41598-018-20978-x.

[7] Y.B. Park, L. Hu, G. Gruner, G. Irvin, P. Drzaic, Late-News Paper: Integration of carbon nanotube transparent electrodes into display applications, SID Symp. Dig. Tech. Pap. 39 (2008) 537. https://doi.org/10.1889/1.3069721

[8] http://www.fep.fraunhofer.de/content/dam/fep/en/documents/Produktflyer/Advanced transparent conductive coatings on flat and flexible substrates_EN_V2.0_net.pdf

[9] A.K. Geim, K.S. Novoselov, The rise of graphene, Nat. Mater. 6 (2007)183–191. http://doi.org/10.1142/9789814287005_0002

[10]K.S. Novoselov, A.K. Geim, S.V. Morozov, Electric field effect in atomically thin carbon films, Science. 306 (2004) 666–669. http://doi.org/10.1126/science.1102896

[11] P. Avouris, Z. Chen, V. Perebeinos, Carbon-based electronics, Nat. Nanotechnol. 2 (2007) 605–615. http://doi.org/10.1142/9789814287005_0018

[12] C. Cai, F. Jia, A. Li, Crackless transfer of large-area graphene films for superior-performance transparent electrodes, Carbon. 98 (2016) 457–462. http://doi.org/10.1016/j.carbon.2015.11.041

[13] W. Aloui, A. Ltaief, A. Bouazizi, Transparent and conductive multi walled carbon nanotubes flexible electrodes for optoelectronic applications, Supperlattice. Microst. 64 (2013)581–589. http://doi.org/10.1016/j.spmi.2013.10.027

[14] J.Y. Lee, Solution-processed metal nanowire mesh transparent electrodes, Nano Lett. 8 (2008) 689-692. http://doi.org/10.1021/nl073296g

[15] S. De, T. M. Higgins, P.E. Lyons, E.M. Doherty, P.N. Nirmalraj, W.J. Blau, J. N. Coleman, Silver nanowire networks as flexible, transparent, conducting films: extremely high DC to optical conductivity ratios, ACS Nano. 3 (2009) 1767-1774. http://doi.org/10.1021/nn900348c

[16] H. Park, P.R. Brown, V. Bulovi, J. Kong, Graphene as transparent conducting electrodes inorganic photovoltaics: studies in graphene morphology, hole transporting layers, and counter electrodes, Nano Lett. 12 (2011) 133–140. http://doi.org/10.1021/nl2029859

[17] Q. Bao, H. Zhang, Y. Wang, Z. Ni, Y.Yan, Z.X. Shen, Atomic-layer graphene as asaturable absorber for ultra fast pulsed lasers, Adv. Funct. Mater. 19 (2009) 3077–3083. http://doi.org/10.1002/adfm.200901007

[18] K.S. Novoselov, V. Fal, L. Colombo, P. Gellert, M. Schwab, K. Kim, A road map for graphene, Nature. 490 (2012)192–200. http://doi.org/10.1038/nature11458

[19] J.H. Chen, C. Jang, S. Xiao, M. Ishigami, M.S. Fuhrer, Intrinsic and extrinsic performance limits of graphene devices on Si02, Nature Nanotech. 4 (2008) 206. http://doi.org/10.1038/nnano.2008.58

[20] A.A. Balandin, S. Ghosh, W. Bao, I. Calizo, D. Teweldebrhan, F. Miao, Superior thermal conductivity of single-layer graphene, Nano Lett. 3 (2008) 902. http://doi.org/10.1021/nl0731872

[21] S. De, J.N. Coleman, Are there fundamental limitations on the sheet resistance and transmittance of thin graphene films?, ACS Nano. 5 (2010) 2713. http://doi.org/10.1021/nn100343f

[22] J. Wu, M. Agrawal, H.A. Becerril, Z. Bao, Z. Liu, Y. Chen, P. Peumans, Organic light-emitting diodes on solution-processed graphene transparent electrodes, ACS Nano. 1 (2009) 43. http://doi.org/10.1021/nn900728d

[23] W. Cao, J. Li, H. Chen, J. Xue, Transparent electrodes for organic optoelectronic devices: a review. J. Photon. Energ. 4 (2014) 040990. Http://doi.org/10.1117/1.JPE.4.040990

[24] C. Linda, J.C. Bernède, M. Morsli, Toward indium-free optoelectronic devices: Dielectric/metal/dielectric alternative transparent conductive electrode in organic photovoltaic cells, physica status solidi (a). 210 (2013)1047-1061. http://doi.org/10.1002/pssa.201228089

[25] S. Nam, M. Song, D. H. Kim, B. Cho, H. M. Lee, J.D. Kwon, Y.C. Park, Ultrasmooth, extremely deformable and shape recoverable Ag nanowire embedded transparent electrode, Sci. Rep. 4 (2014) 4788. http://doi.org/10.1038/srep04788

[26] M. Song, D.S. You, K. Lim, S. Park, S. Jung, C.S. Kim, Y.C. Kang, Highly efficient and bendable organic solar cells with solution-processed silver nanowire electrodes. Adv. Funct. Mater. 23 (2013) 4177-4184. http://doi.org/10.1002/adfm.201202646

[27] M. Song, J.H. Park, C.S. Kim, D.H. Kim, Y.C. Kang, S.H. Jin, J.W. Kang, Highly flexible and transparent conducting silver nanowire/ZnO composite film for organic solar cells. Nano Res. 7 (2014) 1370-1379. http://doi.org/10.1007/s12274-014-0502-3

[28] J. Yang, C. Bao, K. Zhu, T. Yu, Q. Xu, High-performance transparent conducting metal network electrodes for perovksite photodetectors. ACS Appl. Mater. Interfaces. 10 (2018) 1996-2003. http://doi.org/10.1021/acsami.7b15205

[29] H.D. Um, D. Choi, A. Choi, J. H. Seo, K. Seo, Embedded metal electrode for organic–inorganic hybrid nanowire solar cells. ACS Nano. 11 (2017) 6218-6224. http://doi.org/10.1021/acsnano.7b02322

[30] M. Layani, A. Kamyshny, S. Magdassi, Transparent conductors composed of nanomaterials. Nanoscale. 6 (2014) 5581-5591. http://doi.org/10.1039/C4NR00102H

[31] G. Haacke, New figure of merit for transparent conductors. J Appl. Phys. 47 (1976) 4086-4089. http://doi.org/10.1063/1.323240

[32] M. Dressel, G. Gruner, Electrodynamics of solids optical properties of electron in matter, Cam. Uni. Press, Cambridge, (2002) 159-165. http://doi.org/10.1017/CBO9780511606168

[33] R. Gupta, G.U. Kulkarni, Holistic method for evaluating large area transparent conducting electrodes, ACS Appl. Mater. Interfaces. 5 (2013) 730–736. https://doi.org/10.1021/am302264a

[34] Y.H. Kim, C. Sachse, M.L. MacHala, C. May, L. Müller-Meskamp, K. Leo, Highly conductive PEDOT:PSS electrode with optimized solvent and thermal post-treatment for ITO-free organic solar cells, Adv. Funct. Mater. 21 (2011) 1076–1081. https://doi.org/10.1002/adfm.201002290

[35] E.N. Dattoli, W. Lu, ITO nanowires and nanoparticles for transparent films, MRS Bulletin, 36 (2011) 782-788. http://doi.org/10.1557/mrs.2011.212

[36] J. Pütz,, N. Al-Dahoudi, M. A. Aegerter, Processing of transparent conducting coatings made with redispersible crystalline nanoparticles. Adv. Eng. Mater. 6 (2004) 733-737. http://doi.org/10.1002/adem.200400078

[37] J. Puetz, M. A. Aegerter, Direct gravure printing of indium tin oxide nanoparticle patterns on polymer foils, Thin Solid Films. 516 (2008) 4495-4501. http://doi.org/10.1016/j.tsf.2007.05.086

[38] S. Heusing, P.W. De Oliveira, E. Kraker, A. Haase, C. Palfinger, M. Veith, Wet chemical deposited ITO coatings on flexible substrates for organic photodiodes, Thin Solid Films. 518 (2009) 1164-1169. http://doi.org/10.1016/j.tsf.2009.06.056

[39] J.Yun, Y. H. Park, T.S. Bae, S. Lee, G. H. Lee, Fabrication of a completely transparent and highly flexible ITO nanoparticle electrode at room temperature, ACS Appl. Mater. Interfaces. 5 (2013)164−172. http://doi.org/10.1021/am302341p

[40] B. Azzopardi, C.J. Emmott, A. Urbina, F.C. Krebs, J. Mutale, J. Nelson, Economic assessment of solar electricity production from organic-based photovoltaic modules in a domestic environment, Energy Environ. Sci. 4 (2011) 3741-3753. http://doi.org/10.1039/C1EE01766G

[41] M. Jørgensen, K. Norrman, F.C. Krebs, Stability/degradation of polymer solar cells, Sol Energ Mat Sol C. 92 (2008) 686-714. http://doi.org/10.1016/j.solmat.2008.01.005.

[42] P. Cheng, X. Zhan, Stability of organic solar cells: challenges and strategies, Chem. Soc. Rev. 45 (2016) 2544-2582. Http://doi.org/10.1039/C5CS00593K

[43] https://www.pilkington.com/en/global/products/productcategories/specialapplications/nsg-tec-for-technical-applications, 2019

Materials Research Forum LLC
https://doi.org/10.21741/9781644901090-4

[44] Y. Ma, L. Zhi, Graphene-based transparent conductive films: Material systems, preparation and applications, Small Methods. 3 (2019) 1–32. https://doi.org/10.1002/smtd.201800199

[45] K.S. Novoselov, A.K. Geim, S.V. Morozov, D. Jiang, Y. Zhang, S.V. Dubonos, A.A. Firsov, Electric field effect in atomically thin carbon films. Science. 306 (2004). 666-669. http://doi.org/10.1126/science.1102896

[46] P. Avouris, Z. Chen, V. Perebeinos, Carbon-based electronics, Nat. Nanotechnol. 2 (2007) 605–615. http://doi.org/10.1142/9789814287005_0018

[47] L.J. Cote, F. Kim, J. Huang, Langmuir− Blodgett assembly of graphite oxide single layers, J. Am. Chem. Soc. 131 (2008) 1043-1049. http://doi.org/https://doi.org/10.1021/ja806262m

[48] X. Li, G. Zhang, X. Bai, X. Sun, X. Wang, E. Wang, H. Dai, Highly conducting graphene sheets and Langmuir–Blodgett films, Nat. Nanotechnol. 3 (2008) 538. http://doi.org/10.1038/nnano.2008.210

[49] W. Junbo, M. Agrawal, H. A. Becerril, Z. Bao, Z. Liu, Y. Chen, P. Peumans, Organic light-emitting diodes on solution-processed graphene transparent electrodes, ACS Nano. 4 (2009) 43-48. http://doi.org/10.1021/nn900728d

[50] B.A. Héctor, J. Mao, Z. Liu, R.M. Stoltenberg, Z. Bao, Y. Chen, Evaluation of solution-processed reduced graphene oxide films as transparent conductors, ACS Nano. 2 (2008) 463-470. http://doi.org/10.1021/nn700375n

[51] W. Junbo, H.A. Becerril, Z. Bao, Z. Liu, Y. Chen, P. Peumans, Organic solar cells with solution-processed graphene transparent electrodes, Appl. Phys. Lett. 92 (2008) 237. Http://doi.org/10.1063/1.2924771

[52] G. Eda, G. Fanchini, M. Chhowalla, Large-area ultrathin films of reduced graphene oxide as a transparent and flexible electronic material, Nat. Nanotechnol. 3 (2008) 270-274. http://doi.org/10.1038/nnano.2008.83

[53] J. Wang, M. Liang, Y. Fang, T. Qiu, J. Zhang, L. Zhi, Rod-coating: towards large-area fabrication of uniform reduced graphene oxide films for flexible touch screens, Adv. Mater. 24 (2012) 2874-2878. http://doi.org/10.1002/adma.201200055

[54] J. Zhao, S. Pei, W. Ren, L. Gao, H.M. Cheng, Efficient preparation of large-area graphene oxide sheets for transparent conductive films, ACS Nano. 4 (2010) 5245-5252. http://doi.org/10.1021/nn1015506

[55] J. Kang, S. Hwang, J.H. Kim, M.H. Kim, J. Ryu, S.J. Seo, B.H. Hong, M.K. Kim, J.B. Choi, Efficient transfer of large-area graphene films onto rigid substrates by hot pressing, ACS Nano. 6 (2012) 5360–5365. http://doi.org/10.1021/nn301207d

[56] X. Huang, F. Liu, P. Jiang, T. Tanaka, Is graphene oxide an insulating material?, Proc. IEEE Int. Conf. Solid Dielectr. ICSD. (2013) 904–907. http://doi.org/10.1109/ICSD.2013.6619690.

[57] M. Cecilia, G. Eda, S. Agnoli, S. Miller, K.A. Mkhoyan, O. Celik, D. Mastrogiovanni, G. Granozzi, E. Garfunkel, M. Chhowalla. Evolution of electrical, chemical, and structural properties of transparent and conducting chemically derived graphene thin films, Advanced Adv. Funct. Mater. 19 (2009) 2577-2583. http://doi.org/10.1002/adfm.200900166

[58] L. Yanyu, J. Frisch, L. Zhi, H. N. Arasi, Xi. Feng, J. P. Rabe, N. Koch, K. Müllen, Transparent, highly conductive graphene electrodes from acetylene-assisted thermolysis of graphite oxide sheets and nanographene molecules, Nanotechnology. 20 (2009) 434007. http://doi.org/10.1088/0957-4484/20/43/434007

[59] W. Xuan, L. Zhi, K. Müllen, Transparent, conductive graphene electrodes for dye-sensitized solar cells, Nano Lett. 8 (2008) 323-327. http://doi.org/10.1021/nl072838r

[60] Z. Lu, L. Zhao, Y. Xu, T. Qiu, L. Zhi, G. Shi, Polyaniline electrochromic devices with transparent graphene electrodes, Electrochim. Acta. 55 (2009) 491-497. http://doi.org/10.1016/j.electacta.2009.08.063

[61] L. Yangqiao, L. Gao, J. Sun, Y. Wang, J. Zhang, Stable Nafion-functionalized graphene dispersions for transparent conducting films, Nanotechnology. 20 (2009) 465605. http://doi.org/10.1088/0957-4484/20/46/465605

[62] E. Goki, Y.Y. Lin, S. Miller, C.W. Chen, W.F. Su, M. Chhowalla, Transparent and conducting electrodes for organic electronics from reduced graphene oxide, Appl. Phys. Lett. 92 (2008) 209. http://doi.org/10.1063/1.2937846

[63] M. Shun, S. Cui, G. Lu, K. Yu, Z. Wen, J. Chen, Tuning gas-sensing properties of reduced graphene oxide using tin oxide nanocrystals, J. Mater. Chem. 22 (2012) 11009-11013. http://doi.org/10.1039/C2JM30378G

[64] Z. Jinping, S. Pei, W. Ren, L. Gao, H.M. Cheng, Efficient preparation of large-area graphene oxide sheets for transparent conductive films, ACS Nano. 4 (2010) 5245-5252. http://doi.org/10.1021/nn1015506

[65] E. Goki, G. Fanchini, M. Chhowalla, Large-area ultrathin films of reduced graphene oxide as a transparent and flexible electronic material, Nat. Nanotechnol. 3 (2008) 270-274. http://doi.org/10.1038/nnano.2008.83

[66] G.A. Alexander, M.C. Hersam, Solution phase production of graphene with controlled thickness via density differentiation, Nano Lett. 9 (2009) 4031-4036. http://doi.org/10.1021/nl902200b

[67] H. Yenny, V. Nicolosi, M. Lotya, F.M. Blighe, Z. Sun, S. De, I.T. McGovern, High-yield production of graphene by liquid-phase exfoliation of graphite, Nat. Nanotechnol. 3 (2008) 563. http://doi.org/10.1038/nnano.2008.215

[68] B. Peter, P.D. Brimicombe, R.R. Nair, T. J. Booth, D. Jiang, F. Schedin, L.A. Ponomarenko, Graphene-based liquid crystal device, Nano Lett. 8 (2008) 1704-1708. http://doi.org/10.1021/nl080649i

[69] C.M. Gee, C.C. Tseng, F.Y. Wu, Flexible transparent Electrodes made of electro chemically exfoliated graphene sheets from low-cost graphite pieces, Displays. 34 (2013) 315– 319. http://doi.org/10.1016/j.displa.2012.11.002

[70] K.K. Soo, Y. Zhao, H. Jang, S.Y. Lee, J.M. Kim, K.S. Kim, J.H Ahn, P. Kim, J.Y. Choi, B.H. Hong, Large-scale pattern growth of graphene films for stretchable transparent electrodes, Nature. 457 (2009) 706-710. http://doi.org/10.1038/nature07719

[71] L. Xuesong, Y. Zhu, W. Cai, M. Borysiak, B. Han, D. Chen, R.D. Piner, L. Colombo, R.S. Ruoff, Transfer of large-area graphene films for high-performance transparent conductive electrodes, Nano Lett. 9 (2009) 4359-4363. http://doi.org/10.1021/nl902623y

[72] M. Cecilia, H. Kim, M. Chhowalla, A review of chemical vapour deposition of graphene on copper, J. Mater. Chem. 21 (2011) 3324-3334. http://doi.org/10.1039/C0JM02126A

[73] L. Xuesong, W. Cai, J. An, S. Kim, J. Nah, D. Yang, R. Piner, Large-area synthesis of high-quality and uniform graphene films on copper foils, Science. 324 (2009) 1312-1314. http://doi.org/10.1126/science.1171245

[74] Y. Zhang, L. Zhang, C. Zhou, Review of chemical vapor deposition of graphene and related applications, Acc. Chem. Res. 46 (2013) 2329–2339. http://doi.org/10.1021/ar300203n

[75] S. Chen, W Cai, R. D. Piner, J. W. Suk, Y. Wu, Y. Ren, J. Kang, R. S. Ruoff, Synthesis and characterization of large-area graphene and graphite Films on commercial Cu–Ni alloy foils. Nano Lett. 11 (2011) 3519–3525. http://doi.org/10.1021/nl201699j.

[76] B. Sukang, H. Kim, Y. Lee, X. Xu, J.S. Park, Yi Zheng, J. Balakrishnan, "Roll-to-roll production of 30-inch graphene films for transparent electrodes, Nat. Nanotechnol. 5 (2010) 574. http://doi.org/10.1038/nnano.2010.132

[77] L. Xuesong, C.W. Magnuson, A. Venugopal, J. An, J.W. Suk, B. Han, M. Borysiak, Graphene films with large domain size by a two-step chemical vapor deposition process, Nano Lett. 10 (2010) 4328-4334. http://doi.org/10.1021/nl101629g

[78] T.C. Vincent, L.M. Chen, M.J. Allen, J.K. Wassei, K. Nelson, R.B. Kaner,Y. Yang, Low-temperature solution processing of graphene-carbon nanotube hybrid materials for high-performance transparent conductors, Nano Lett. 9 (2009) 1949-1955. http://doi.org/10.1021/nl9001525

[79] L. Chunyan, Z. Li, H. Zhu, K. Wang, J. Wei, X. Li, P. Sun, H. Zhang, D. Wu, Graphene nano-"patches" on a carbon nanotube network for highly transparent/conductive thin film applications, J. Phys. Chem. C. 114 (2010) 14008-14012. http://doi.org/10.1021/jp1041487

[80] L.M. Sun, K. Lee, S.Y. Kim, H. Lee, J. Park, K.H. Choi, H.K. Kim, High-performance, transparent, and stretchable electrodes using graphene–metal nanowire hybrid structures, Nano Lett. 13 (2013) 2814-2821. http://doi.org/10.1021/nl401070p

[81] H. Liangbing, D.S. Hecht, G. Gruner, Carbon nanotube thin films: fabrication, properties, and applications, Chem. Rev. 110 (2010) 5790-5844. http://doi.org/10.1021/cr9002962

[82] S.H. Han, B.J. Kim, J.S. Park, Effects of the corona pretreatment of PET substrates on the properties of flexible transparent CNT electrodes, Thin Solid Films. 572 (2014) 73-78. http://doi.org/10.1016/j.tsf.2014.09.066

[83] S.H. Han, B.J. Kim, J.S. Park, Surface modification of plastic substrates via corona-pretreatment and its effects on the properties of carbon nanotubes for use of flexible transparent electrodes, Surf. Coat. Technol. 271 (2015) 100. http://doi.org/10.1016/j.surfcoat.2014.12.077

[84] T.M. Barnes, J.D. Bergeson, R.C. Tenent, Carbon nanotube network electrodes enabling efficient organic solar cells without a hole transport layer, Appl. Phys. Lett. 96 (2010) 243309. http://doi.org/10.1063/1.3453445

[85] V. Scardaci, R. Coull, J.N. Coleman, Very thin transparent,conductive carbon nanotube films on flexible substrates, Appl. Phys. Lett. 97 (2010) 23114. http://doi.org/doi.org/10.1063/1.3462317

[86] A. Schindler, J. Brill, N. Fruehauf, J.P. Novak, Z. Yaniv, Solution-deposited carbon nanotube layers for flexible display applications, Physica E Low Dimens. Syst. Nanostruct. 37 (2007) 119–123. http://doi.org/10.1016/j.physe.2006.07.016

[87] J.G. Ruiz, S. Palmero, D. Iba~nez, A. Heras, A. Colina, Press-transfer optically transparent electrodes fabricated from commercial single-walled carbon nanotubes, Electrochem. commun. 25 (2012) 1–4. http://doi.org/10.1016/j.elecom.2012.09.004

[88] A. Heras, A. Colina, J.L´opez-Palacios, Flexible optically transparent single-walled carbon nanotube electrodes for UV-Vis absorption spectroelectrochemistry, Electrochem. commun. 11 (2009) 442–445. http://doi.org/10.1016/j.elecom.2008.12.016.

[89] J.W. Jo, J.W. Jung, J.U. Lee, W.H. Jo, Fabrication of highly conductive and transparent thin films from single-walled carbon nanotubes using a new non-ionic surfactant via spin coating, ACS Nano. 4 (2010) 5382–5388. http://doi.org/10.1021/nn1009837

[90] J. Zhang, L. Gao, J. Sun, Dispersion of single-walled carbon nanotubes by nafion in water/ethanol for preparing transparent conducting films, J. Phys. Chem. C. 112 (2008) 16370–16376. http://doi.org/10.1021/jp8053839

[91] N.F. Anglada, J.P. Puigdemont, J. Figueras, M.Z. Iqbal, S. Roth, Flexible, transparent electrodes using carbon nanotubes, Nanoscale Res. Lett. 7 (2012) 571–578. http://doi.org/10.1186/1556-276X-7-571

[92] M.W. Rowell, M.A. Topinka, M.D. McGehee, Organic solar cells with carbon nanotube network electrodes, Appl. Phys. Lett. 88 (2006) 233506. http://doi.org/10.1063/1.2209887

[93] W. Aloui, A. Ltaief, A. Bouazizi, Transparent and conductive multi walled carbon nanotubes flexible electrodes for optoelectronic applications, Superlattices Microstruct. 64 (2013) 581–589. http://doi.org/10.1016/j.spmi.2013.10.027

[94] Y. H. Kim, L.M. Meskamp, A.A. Zakhidov, Semitransparent small molecule organic solar cells with laminated free-standing carbon nanotube top electrodes, Sol. Energy Mater. Sol. Cells. 96 (2012) 244–250. http://doi.org/10.1016/j.solmat.2011.10.001

[95] S.I. Na, J.S. Lee, Y.J. Noh, Efficient ITO-free polymer solar cells with pitch-converted carbon nanosheets as novel solution-processable transparent electrodes, Sol. Energy Mater. Sol. Cells. 115 (2013) 1–6. http://doi.org/10.1016/j.solmat.2013.03.019.

[96] T. Andrea, F. Kim, C. Hess, J. Goldberger, R. He, Y. Sun, Y. Xia, P. Yang, Langmuir- Blodgett silver nanowire monolayers for molecular sensing using surface-enhanced Raman spectroscopy, Nano Lett. 3 (2003) 1229-1233. http://doi.org/10.1021/nl0344209

[97] H. Liangbing, H.S. Kim, J.Y. Lee, P. Peumans, Yi Cui, Scalable coating and properties of transparent, flexible, silver nanowire electrodes, ACS Nano. 4 (2010) 2955-2963. http://doi.org/10.1021/nn1005232

[98] T. Takehiro, M. Nogi, M. Karakawa, J. Jiu, T.T. Nge, Yoshio Aso, K. Suganuma, Fabrication of silver nanowire transparent electrodes at room temperature, Nano Res. 4 (2011) 1215-1222. http://doi.org/10.1007/s12274-011-0172-3

[99] L. Jaemin, I. Lee, T.S. Kim, J.Y. Lee, Efficient welding of silver nanowire networks without post-processing, Small. 9 (2013) 2887-2894. http://doi.org/10.1002/smll.201203142

[100] L. Jinhwan, P. Lee, H. Lee, D. Lee, S.S. Lee, S.H. Ko, Very long Ag nanowire synthesis and its application in a highly transparent, conductive and flexible metal electrode touch panel, Nanoscale. 4 (2012) 6408-6414. http://doi.org/10.1039/C2NR31254A

[101] R.R. Aaron, S.M. Bergin, Y.L. Hua, Z.Y. Li, B.J. Wiley, The growth mechanism of copper nanowires and their properties in flexible, transparent conducting films, Adv. Mater. 22 (2010) 3558-3563. http://doi.org/10.1002/adma.201000775

[102] Y. Shengrong, A.R. Rathmell, Y.C. Ha, A.R. Wilson, B.J. Wiley, The role of cuprous oxide seeds in the one-pot and seeded syntheses of copper nanowires, Small. 10 (2014) 1771-1778. http://doi.org/10.1002/smll.201303005

[103] R.R. Aaron, B.J. Wiley, The synthesis and coating of long, thin copper nanowires to make flexible, transparent conducting films on plastic substrates, Adv. Mater. 23 (2011) 4798-4803. http://doi.org/10.1002/adma.201102284

[104] G. Huizhang, N. Lin, Y. Chen, Z. Wang, Q. Xie, T. Zheng, N. Gao, Copper nanowires as fully transparent conductive electrodes, Sci. Rep. 3 (2013) 1-8. http://doi.org/10.1038/srep02323

[105] H. Ziyang, C.K. Tsung, W. Huang, X. Zhang, P. Yang, Sub-two nanometer single crystal Au nanowires, Nano Lett. 8 (2008): 2041-2044. http://doi.org/10.1021/nl8013549

[106] S.I. Ana, B.R. Murias, M. Grzelczak, J.P. Juste, L.M. Liz-Marzán, F. Rivadulla, M.A. Correa-Duarte, Highly transparent and conductive films of densely aligned ultrathin Au nanowire monolayers, Nano Lett. 12 (2012) 6066–6070. http://doi.org/10.1021/nl3021522

[107] L.J. Yong, S.T. Connor, Y. Cui, P. Peumans, Solution-processed metal nanowire mesh transparent electrodes, Nano Lett. 8 (2008) 689–692. http://doi.org/10.1021/nl073296g

[108] K.C. Lin, K.C. Hwang, Nitrate ion promoted formation of Ag nanowires in polyol processes: a new nanowire growth mechanism, Langmuir. 28 (2012) 3722-3729. http://doi.org/10.1021/la204002b

[109] B. Bushra, J. Lee, T. Jang, P. Won, S. H. Ko, K. Alamgir, M. Arshad, L.J. Guo, Simple hydrothermal synthesis of very-long and thin silver nanowires and their application in high quality transparent electrodes, J. Mater. Chem. A. 4 (2016) 11365-11371. http://doi.org/10.1039/C6TA03308C

[110] W. Hui, D. Kong, Z. Ruan, P.C. Hsu, S. Wang, Z. Yu, T.J. Carney, L. Hu, S. Fan, Y. Cui, A transparent electrode based on a metal nanotrough network, Nat. Nanotechnol. 8 (2013) 421. http://doi.org/10.1038/nnano.2013.84

[111] H. Bing, K. Pei, Y. Huang, X. Zhang, Q. Rong, Q. Lin, Y. Guo, Uniform self-forming metallic network as a high-performance transparent conductive electrode, Adv. Mater. 26 (2014) 873-877. http://doi.org/10.1002/adma.201302950

[112] L.S. James, K.Y. Shin, O.J. Cheong, J.H. Kim, J. Jang, Highly sensitive and multifunctional tactile sensor using free-standing ZnO/PVDF thin film with graphene electrodes for pressure and temperature monitoring, Sci. Rep. 5 (2015) 7887. http://doi.org/10.1038/srep07887

[113] J.H. Young, S.K. Lee, S.H. Cho, J.H. Ahn, S. Park, Fabrication of metallic nanomesh: Pt nano-mesh as a proof of concept for stretchable and transparent electrodes, Chem. Mater. 25 (2013) 3535-3538. http://doi.org/10.1021/cm402085k

[114] L.Y. Hua, J.L. Xu, X. Gao, Y.L. Sun, J.J. Lv, S. Shen, L.S. Chen, S.D. Wang, Freestanding transparent metallic network based ultrathin, foldable and designable supercapacitors, Energy Environ. Sci. 10 (2017) 2534-2543. http://doi.org/10.1039/C7EE02390A

[115] G. Chengqun, X. Ding, S. Zhou, Y. Gao, X. Liu, S. Liu, Nanoscale Ni/Au wire grids as transparent conductive electrodes in ultraviolet light-emitting diodes by laser direct writing, Opt. Laser Technol. 104 (2018) 112-117. http://doi.org/10.1016/j.optlastec.2018.02.030

[116] Y.H. Liu, J.L. Xu, S. Shen, X.L. Cai, L.S. Chen, S.D. Wang, High-performance, ultra-flexible and transparent embedded metallic mesh electrodes by selective electrodeposition for all-solid-state supercapacitor applications, J. Mater. Chem. A. 5 (2017) 9032-9041. http://doi.org/10.1039/C7TA01947E.

[117] L. Li, B. Zhang, B. Zou, R. Xie, T. Zhang, S. Li, B. Zheng, J. Wu, J. Weng, W. Zhang, W. Huang, F. Huo, Fabrication of flexible transparent electrode with enhanced conductivity from hierarchical metal grids, ACS Appl. Mater. Interfaces. 9 (2017) 39110–39115. http://doi.org/10.1021/acsami.7b12298

[118] T. Iwahashi, R. Yang, N. Okabe, J. Sakurai, J. Lin, D. Matsunaga, Nanoimprint-assisted fabrication of high haze metal mesh electrode for solar cells, Appl. Phys. Lett. 105 (2014) 223901. http://doi.org/10.1063/1.4903061.

[119] T. Gao, B. Wang, B. Ding, J. Lee, P.W. Leu, Uniform and ordered copper nanomeshes by microsphere lithography for transparent electrodes, Nano Lett. 14 (2014) 2105-2110. http://doi.org/10.1021/nl5003075

[120] M. Layani, P. Darmawan, W. L. Foo, L. Liu, A. Kamyshny, D. Mandler, S.Magdassi, P. S. Lee, Nanostructured electrochromic films by inkjet printing on large area and flexible transparent silver electrodes, Nanoscale. 6 (2014): 4572-4576. http://doi.org/10.1039/C3NR06890K

[121] X. Meng, X. Hu, X. Yang, J. Yin, Q. Wang, L. Huang, Z. Yu, T. Hu, L. Tan, W. Zhou, Y. Chen, Roll-to-roll printing of meter-scale composite transparent electrodes with optimized mechanical and optical properties for photoelectronics, ACS Appl. Mater. Interfaces. 10 (2018): 8917-8925. http://doi.org/10.1021/acsami.8b00093

[122] J. M. Cho, D. H. Kim, M. S. Yoo, S. Bae, D. S. Kim, The fabrication of flexible Ag grid mesh electrode by a new thermal roll imprinting process and its application, J. Nanosci. Nanotechnol. 17 (2017) 3304-3309. http://doi.org/doi.org/10.1166/jnn.2017.14072

[123] N. Kwon, K. Kim, S. Sung, I. Yi, I. Chung, Highly conductive and transparent Ag honeycomb mesh fabricated using a monolayer of polystyrene spheres, Nanotechnology. 24 (2013) 235205. http://doi.org/10.1088/0957-4484/24/23/235205

[124] S. Han, Y. Chae, J. Y. Kim, Y. Jo, S. S. Lee, S.-H. Kim, K. Woo, S. Jeong, Y. Choi, S. Y. Lee, High-performance solution-processable flexible and transparent conducting electrodes with embedded Cu mesh, J. Mater. Chem. C. 6 (2018) 4389-4395. http://doi.org/10.1039/C8TC00307F

[125] L. Chen, X. Wei, X. Zhou, Z. Xie, K. Li, Q. Ruan, C. Chen, J. Wang, C.A. Mirkin, Z. Zheng, Large area patterning of metal nanostructures by dip-pen nanodisplacement lithography for optical applications, Small. 13 (2017) 1–6. http://doi.org/10.1002/smll.201702003

[126] S. Jang, W.-B. Jung, C. Kim, P. Won, S.-G. Lee, K. M. Cho, M. L. Jin, C. J. An, H.J. Jeon, S. H. Ko, T.-S. Kim, H.T. Jung, A three-dimensional metal grid mesh as a practical alternative to ITO, Nanoscale. 8 (2016) 14257-14263. http://doi.org/10.1039/C6NR03060B

[127] D. Bryant, P. Greenwood, J. Troughton, M. Wijdekop, M. Carnie, M. Davies, K. Wojciechowski, H. J. Snaith, T. Watson, D.Worsley, A transparent conductive adhesive laminate electrode for high-efficiency organic-inorganic lead halide perovskite solar cells, Adv. Mater. 26 (2014) 7499-7504. http://doi.org/10.1002/adma.201403939

[128] N. Gupta, K.D. M. Rao, R. Gupta, F.C. Krebs, G.U. Kulkarni, Highly conformal Ni micromesh as a current collecting front electrode for reduced cost Si solar cell, ACS Appl. Mater. Interfaces. 9 (2017) 8634-8640. http://doi.org/10.1021/acsami.6b12588

[129] F. M. Wisser, K. Eckhardt, W. Nickel, W. Bo"hlmann, S. Kaskel, J. Grothe, Highly transparent metal electrodes via direct printing processes, Mater. Res. Bull. 98 (2018) 231-234. http://doi.org/10.1016/j.materresbull.2017.10.021

[130] H. Y. Jang, S.-K. Lee, S. H. Cho, J.H. Ahn, S. Park, Fabrication of metallic nanomesh: Pt nano-mesh as a proof of concept for stretchable and transparent electrodes, Chem. Mater. 25 (2013) 3535-3538. http://doi.org/10.1021/cm402085k

[131] B.J. Kim, J.-S. Park, Y.-J. Hwang, J.S. Park, Characteristics of silver meshes coated with carbon nanotubes via spray-coating and electrophoretic deposition for touch screen panels, Thin Solid Films. 596 (2015) 68-71. http://doi.org/10.1016/j.tsf.2015.07.084

[132] Y. Han, Y. Liu, L. Han, J. Lin, P. Jin, High-performance hierarchical graphene/metal-mesh film for optically transparent electromagnetic interference shielding, Carbon. 115 (2017) 34-42. http://doi.org/10.1016/j.carbon.2016.12.092

[133] Ho, H. Lu, W. Liu, J. N. Tey, C. K. Cheng, E. Kok, J. Wei, Electrical and optical properties of hybrid transparent electrodes that use metal grids and graphene films, J. Mater. Res. Technol. 28 (2013): 620-626. http://doi.org/10.1016/j.carbon.2016.12.092

[134] I. Burgue's-ceballos, N. Kehagias, C. M. Sotomayor-torres, M. Campoy-quiles, P. D. Lacharmoise, Embedded inkjet printed silver grids for ITO-free organic solar cells with high fill factor, Sol. Energy Mater. Sol. Cells. 127 (2014) 50-57. http://doi.org/10.1016/j.solmat.2014.03.024

[135] J. W. Lim, Y. T. Lee, R. Pandey, T. Yoo, B. Sang, B. Ju, D. K. Hwang, W. K. Choi, Effect of geometric lattice design on optical/electrical properties of transparent silver grid for organic solar cells, Opt. Express. 22 (2014): 26891-26899. http://doi.org/10.1364/OE.22.026891.

[136] W. Kim, S. Kim, I. Kang, M. S. Jung, S. J. Kim, J. K. Kim, S. M. Cho, J.H. Kim, J. H. Park, Hybrid silver mesh electrode for ITO-free flexible polymer solar cells with good mechanical stability. Chem. Sus. Chem. 9 (2016) 1042-1049. http://doi.org/10.1002/cssc.201600070

[137] I. Mondal, A. Kumar, K.D.M. Rao, G.U. Kulkarni, Parallel cracks from a desiccating colloidal layer under gravity flow and their use in fabricating metal micro-patterns, J. Phys. Chem. Solids. 118 (2018) 232-237. http://doi.org/10.1016/j.jpcs.2018.03.020

[138] Y. Liu, S. Shen, J. Hu, L. Chen, Embedded Ag mesh electrodes for polymer dispersed liquid crystal devices on flexible substrate, Opt. Express. 24 (2016) 25774-25784. 10.1364/OE.24.025774

[139] A. J. Morfa, E. M. Akinoglu, J. Subbiah, M. Giersig, P. Mulvaney, Transparent metal electrodes from ordered nanosphere arrays, J. Appl. Phys. 114 (2013): 054502. http://doi.org/10.1063/1.4816790

[140] S.-H. Kwak, M.-G. Kwak, B.K. Ju, S.J. Hong, Enhancement of characteristics of a touch sensor by controlling the multi-layer architecture of a low-cost metal mesh pattern, J. Nanosci. Nanotechnol. 15 (2015) 7645-7651. http://doi.org/10.1166/jnn.2015.11213

[141] J. Park, K. Lee, H. Um, K. Kim, K. Seo, Flexible and transparent metallic grid electrodes prepared by evaporative assembly, ACS Appl. Mater. Interfaces. 6 (2014) 12380-12387. http://doi.org/10.1021/am502233y

[142] X. Chen, W. Guo, L. Xie, C. Wei, J. Zhuang, W. Su, Z. Cui, Embedded Ag/Ni metal-mesh with low surface roughness as transparent conductive electrode for optoelectronic applications, ACS Appl. Mater. Interfaces. 9 (2017) 37048-37054. http://doi.org/10.1021/acsami.7b11779

[143] H.-G. Im, B. W. An, J. Jin, J. Jang, Y.G. Park, J.-U. Park, B.S. Bae, A high-performance, flexible and robust metal nanotrough-embedded transparent conducting film for wearable touch screen panels, Nanoscale. 8 (2016) 3916-3922. http://doi.org/10.1039/C5NR07657A

[144] H.B. Lee, W.Y. Jin, M.M. Ovhal, N. Kumar, J.W. Kang, Flexible transparent conducting electrodes based on metal meshes for organic optoelectronic device applications: a review, J. Mater. Chem. C. 7 (2019) 1087-1110. http://doi.org/10.1039/C8TC04423F

[145] M.G. Kang, M.S. Kim, J. Kim, L.J. Guo, Organic solar cells using nano imprinted transparent metal electrodes, Adv. Mater. 20 (2008) 4408–4413. https://doi.org/10.1002/adma.200800750

[146] D.S. Ghosh, T.L. Chen, V. Pruneri, High figure-of-merit ultrathin metal transparent electrodes incorporating a conductive grid, Appl. Phys. Lett. 96 (2010) 2008–2011. https://doi.org/10.1063/1.3299259.

[147] K. Tvingstedt, O. Inganäs, Electrode grids for ITO free organic photovoltaic devices, Adv. Mater. 19 (2007) 2893-2897. http://doi.org/10.1002/adma.200602561

[148] Y. Galagan, J.E.J. M. Rubingh, R. Andriessen, C.C. Fan, P.W. M. Blom, S.C. Veenstra, J.M. Kroon, ITO-free flexible organic solar cells with printed current collecting grids, Sol. Energy Mater. Sol. Cells. 95 (2011) 1339-1343. http://doi.org/10.1016/j.solmat.2010.08.011

[149] L. Mao, Q. Chen, Y. Li, Y. Li, J. Cai, W. Su, S. Bai, Y. Jin, C.-Q. Ma, Z. Cui, L. Chen, Flexible silver grid/PEDOT: PSS hybrid electrodes for large area inverted polymer solar cells, Nano Energy. 10 (2014) 259-267. http://doi.org/10.1016/j.nanoen.2014.09.007

[150] Y. H. Kim, L. Mu"ller-Meskamp, K. Leo, Ultratransparent polymer/semitransparent silver grid hybrid electrodes for small-molecule organic solar cells, Adv. Energy Mater. 5 (2015) 1401822. http://doi.org/10.1002/aenm.201401822

[151] Y. H. Kahng, M.-K. Kim, J.-H. Lee, Y. J. Kim, N. Kim, D.W. Park, K. Lee, Highly conductive flexible transparent electrodes fabricated by combining graphene

films and inkjet-printed silver grids, Sol. Energy Mater. Sol. Cells. 124 (2014) 86-91. http://doi.org/10.1016/j.solmat.2014.01.040

[152] D. J. Lipomi, M. Vosgueritchian, B. C. Tee, S. L. Hellstrom,J. A. Lee, C. H. Fox, Z. Bao, Skin-like pressure and strain sensors based on transparent elastic films of carbon nanotubes, Nat. Nanotechnol. 6 (2011) 788. http://doi.org/10.1038/nnano.2011.184

[153] L. Caizhi, M. Zhang, L. Niu, Z. Zheng, F. Yan, Highly selective and sensitive glucose sensors based on organic electrochemical transistors with graphene-modified gate electrodes, J. Mater. Chem. B. 1 (2013) 3820-3829. http://doi.org/10.1039/C3TB20451K

[154] C. Z. Liao, C. H. Mak, M. Zhang, H. L. W. Chan, F. Yan, Flexible organic electrochemical transistors for highly selective enzyme biosensors and used for saliva testing, Adv. Mater. 27 (2015) 676-681. http://doi.org/10.1002/adma.201404378

[155] S. Kim, J. Yim, X. Wang, D.D.C. Bradley, S. Lee, J.C. deMello, Spin-and spray-deposited single-walled carbon-nanotube electrodes for organic solar cells, Adv. Funct. Mater. 20 (2010): 2310-2316. http://doi.org/10.1002/adfm.200902369

[156] A. W. Diah, J. P. Quirino, W. Belcher, C. I. Holdsworth, Investigation of the doping efficiency of poly (styrene sulfonic acid) in poly (3, 4-ethylenedioxythiophene)/poly (styrene sulfonic acid) dispersions by capillary electrophoresis, Electrophoresis. 35 (2014) 1976-1983. http://doi.org/10.1002/elps.201400056

[157] F. Zhang, M. Johansson, M. R. Andersson, J. C. Hummelen, O. Inganas, Polymer photovoltaic cells with conducting polymer anodes, Adv. Mater. 14 (2002) 662-665. http://doi.org/10.1002/1521-4095(20020503)14:9<662::AID-ADMA662>3.0.CO;2-N

[158] J. Ouyang, C.-W. Chu, F.-C. Chen, Q. Xu, Y. Yang, High-conductivity poly (3, 4-ethylenedioxythiophene): poly (styrene sulfonate) film and its application in polymer optoelectronic devices, Adv. Funct. Mater. 15 (2005) 203-208. http://doi.org/10.1002/adfm.200400016

[159] S.I. Na, S.S. Kim, J.Jo, D.Y. Kim, Efficient and flexible ITO-free organic solar cells using highly conductive polymer anodes, Adv. Mater. 20 (2008) 4061-4067. http://doi.org/10.1002/adma.200800338

[160] M. Vosgueritchian, D. J. Lipomi, Z. N. Bao, Highly conductive and transparent PEDOT: PSS films with a fluorosurfactant for stretchable and flexible transparent

Materials Research Forum LLC
https://doi.org/10.21741/9781644901090-4

electrodes, Adv. Funct. Mater. 22 (2012) 421-428. http://doi.org/10.1002/adfm.201101775

[161] M. Kaltenbrunner, M.S. White, E.D. Glowacki, T. Sekitani, T. Someya, N.S. Sariciftci, S. Bauer, Ultrathin and lightweight organic solar cells with high flexibility, Nat. Commun. 3 (2012) 1-7. http://doi.org/10.1038/ncomms1772

[162] Y. Xia, K. Sun, J. Ouyang, Solution-processed metallic conducting polymer films as transparent electrode of optoelectronic devices, Adv. Mater. 24 (2012) 2436-2440. http://doi.org/10.1002/adma.201104795

[163] N. Kim, S. Kee, S. H. Lee, B. H. Lee, Y. H. Kahng, Y.R. Jo, B.J. Kim, K. Lee, Highly conductive PEDOT: PSS nanofibrils induced by solution-processed crystallization, Adv. Mater. 26 (2014) 2268-2272. http://doi.org/10.1002/adma.201304611

[164] X. Fan, B. Xu, S. Liu, C. Cui, J. Wang, F. Yan, Transfer-printed PEDOT: PSS electrodes using mild acids for high conductivity and improved stability with application to flexible organic solar cells, ACS Appl. Mater. Interfaces. 8 (2016) 14029-14036. http://doi.org/10.1021/acsami.6b01389

[165] W. Song, X. Fan, B. Xu, F. Yan, H. Cui, Q. Wei, R. Peng, L. Hong, J. Huang, Z. Ge, All-solution-processed metal-oxide-free flexible organic solar cells with over 10% efficiency, Adv. Mater. 30 (2018) 1800075. http://doi.org/10.1002/adma.201800075

[166] N. Kim, H. Kang, J.-H. Lee, S. Kee, S. H. Lee, K. Lee, Highly conductive all-plastic electrodes fabricated using a novel chemically controlled transfer-printing method, Adv. Mater. 27 (2015) 2317-2323. http://doi.org/10.1002/adma.201500078

[167] K. Schulze, C. Uhrich, R. Schueppel, K. Leo, M. Pfeiffer, E. Brier, E. Reinhold, P. Baeuerle, Organic solar cells on indium tin oxide and aluminium doped zinc oxide anodes, Appl. Phys. Lett. 91 (2007) S073521-073523. http://doi.org/10.1063/1.2771050.

[168] L. Shi , Y. Cui, Y. Gao , W. Wang , Y. Zhang , F. Zhu, Y. Hao; High performance ultrathin MoO_3/Ag transparent electrode and its application in semitransparent organic solar cells; Nanomaterials, 8 (2018) 473. http://doi.org/10.3390/nano8070473

[169] J.Meiss, C.L. Uhrich, K. Fehse, S. Pfuetzner, M.K. Riede, K. Leo, Transparent electrode materials for solar cells, Proc. SPIE 7002, Photonics for solar energy systems II, 700210, (2008). http://doi.org/10.1117/12.781275

Materials Research Foundations **88** (2021) 129-147

https://doi.org/10.21741/9781644901090-5

Chapter 5

Hollow Nanostructures for Application in Solar Cells

Peetam Mandal[1], Abha bhargava[2] and Mitali Saha[1*]

[1]Department of Chemistry, National Institute of Technology, Agartala– 799046, Tripura, India

[2]Department of Textile Chemistry, Uttar Pradesh Textile Technology Institute – 208001, Kanpur, India

mitalichem71@gmail.com

Abstract

Hollow nanostructures are nanoscale materials with interior cavities, high volumetric load capacity ratio and high porosity. This new generation structure has gained huge momentum in the field of energy storage and photovoltaics due to such promising physical and chemical features. This chapter highlights contributions of various works where hollow nanostructures of metals and carbonaceous materials had been used in solar cell over the last few years. The harnessing of efficiency with structural modifications in the hollow structures over the years was shown in various works. The effect of structure engineering on the performance of solar cell has been explained in detail where voids in metallic hollow nanostructure enhance light scattering and high charge recombination. Simultaneously, carbonaceous hollow nanostructured materials are considered to be the latest photoelectrode materials and designated to be alternatives for metallic hollow nanostructures counterpart due to their high feedstock availability and fabrication charges.

Keywords

Solar Cells, Hollow Nanostructures, Carbonaceous Hollow Nanostructures, Metallic Electrodes, Photovoltaics

Contents

Hollow Nanostructures for Application in Solar Cells...................................129

1.   Introduction..130

2.   Nanostructured materials in solar cells...131

3.   **Current scenario of hollow nanostructure based solar cells** **132**

    3.1   Metallic hollow nanostructurebased solar cells..............................134

    3.2   Carbonaceous hollow nanostructure based solar cells 139

Conclusion...**140**

References **140**

1.   Introduction

In contemporary time, it has been found that use of renewable energy is outplaying the demand for natural energy which is due to natural resources of energy are gradually getting extinguished for excessive usage [1]. Amongst all the renewable energy resources, solar energy harvesting is considered as one of the holistic as well as clean energy source. The total amount of sunlight hitting the earth surface is 164 watts per square meter in twenty four hours a day due to which, abundance of sunlight hitting earth's surface has much immense impact on renewable energy resources and its production [2]. The huge quantity of sunlight hitting earth hastens the procedures of photosynthesis, radiation, convection and hydrological cycle like evaporation and precipitation as portrayed in figure 1.

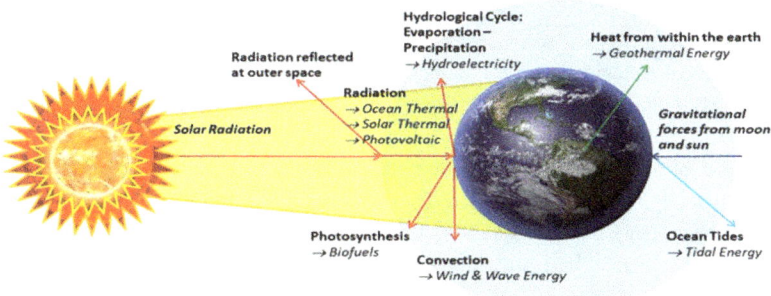

Figure 1 Impact of sunlight on various renewable energy productions

This solar energy is harvested in various solar farms ubiquitously using simple solar cell devices and its performance is based on solar cell efficiency i.e. the amount of energy that gets collected from sunlight and converted into electricity using photovoltaics. Leading photovoltaic industries and research institutes are making the era of 3rd

Materials for Solar Cell Technologies I   Materials Research Forum LLC
Materials Research Foundations **88** (2021) 129-147   https://doi.org/10.21741/9781644901090-5

generation solar cells compatible with the implementation of minimal cost raw materials and high power conversion efficiency. In 2016, tesla motors entrepreneur and business magnate Elon Musk gave proposal on manufacturing solar driven cars that will be using nanomaterials as raw materials for solar cell devices than the conventional mono-crystalline silicon solar chipset [3]. Hence a worldwide rush to fabricate solar devices composed of nanomaterials has gain importance amongst all scientific communities dealing with photovoltaics.

2. Nanostructured materials in solar cells

Commercialized available terrestrial solar cells are consisted of bulk materials: either single or multi crystalline silicon solar cells which have minimum cell thickness of 1 to 300 μm, stacked on p-type substrates. The resultant charge separation occurred due to long diffusion length of holes and electrons in silicon crystal [4]. These conventional solar cell modules were engineered to obtain high efficiency with low manufacturing cost, however, the feedstock shortage of ultra-pure silicon in the near future can't be overlooked due to colossal demand of silicon in the electronics industry. To compensate these problems, various nanomaterials and nano-alloys had been used in trace amount to escalate high efficiency (as high as 40%) with minimal budget fabrication charges [5-7]. These nanomaterials have become a key finding to the long awaiting problem of solar cell fabrication where it could achieve high power conversion efficiency than the limiting theoretical values in conventional p-n junction photovoltaic devices. Over a decade of devoted research on solar cells, various kinds of metallic nanomaterials and nanocomposites had been used immensely and various carbonaceous nanomaterials like carbon nanotubes, graphene derivatives, activated carbon, graphene quantum dots, etc. have been incorporated to prepare composite materials [8-16]. Nanostructured based solar cells are mainly restricted to dye sensitized solar cells and multi-junction solar cells. This is mainly due to their thin structures, requiring small amount of raw materials but have high potential with cost effective fabrication charges. These are also known as Grätzel cells which basically imitates the natural photosynthesis process to harvest electric energy from solar cell and it is composed of four different components viz. photoelectrode, counter electrode, photosensitizer and electrolyte as represented in figure 2a. Similarly, bulk junction solar cells are composed of anode (nano metal oxides), active layer of acceptor-donor blend, active layer comprised of conducting polymers and metallic cathode as illustrated in figure 2b. The high magnitude of research on nanomaterials in photovoltaics prompted the research fraternities to synthesize various kinds of morphologies such as nanorods, nano flowers, nano cubes, nano whiskers etc. Recently, hollow nanostructured proved to be a game-changer with diverse applications

Materials Research Forum LLC

https://doi.org/10.21741/9781644901090-5

ranging from energy storage to solar cells [17]. The interior cavities in hollow nanomaterials consist of low density, high volumetric loading capacity, large surface area, short charge-to-mass transport lengths make it potential charge carriers and the low weight makes it a favorable choice as a solar cell component.

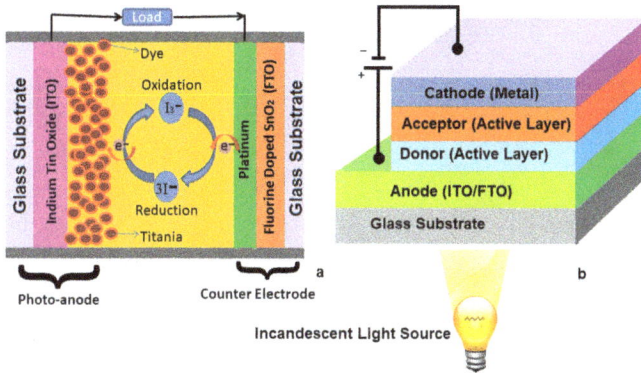

Figure 2 Illustrated schematics of (a) dye sensitized solar cell and (b) multi-junction solar cell

3. Current scenario of hollow nanostructure based solar cells

Hollow nanostructured materials in solar cells had propelled the reflections and diffraction simultaneously, thus addressing it as an efficient photon harvesting component. On contrast to simple hollow structure, complex hollow shape was found to possess greater efficiency due to multiple reflection and diffraction patterns within its porous cavities. The incorporation of metallic hollow nanostructured materials inhibited charge recombination and curtailed the charge transport length. It was also observed that multi chambers/pores in graphene derivatives and carbonaceous hollow nanostructured nanomaterials had enhanced the performance of solar cells with high efficiency. Table 1[18-53] represents different types of hollow nanostructured solar cell that had been used in the literature.

Table 1   *Tabular representation of different hollow nanostructured solar cell.*

S. No.	Counter Electrode/ Solar cell device	J_{sc} (mA/cm^2)	Voc (V)	FF (%)	PCE (η, %)	Ref.
1.	TiO$_2$ nanotubes, Pt reference	15	0.73	0.57	6.2	[18]
2.	SnO$_2$, Pt sputtered FTO	14.59	0.765	0.54	6.02	[19]
3.	TiO$_2$, Pt reference	17.26	0.76	0.53	6.91	[20]
4.	SnO$_2$ hollow sphere-TiO$_2$ nanosheet	18.2	0.76	0.6	8.2	[21]
5.	Ag@hollow SiO$_2$-TiO$_2$, Pt reference	17.5	0.74	0.69	9	[22]
6.	Multi-shell hollow TiO$_2$, Pt reference	16.52	0.77	0.73	9.4	[23]
7.	Ni$_3$Se$_4$ (same counter electrode Ni$_3$Se$_4$)	16.27	0.746	0.69	8.31	[24]
8.	10% hollow TiO$_2$ on commercial ZnO	13.15	0.59	0.58	4.59	[25]
9.	Double film of TiO$_2$ hollow boxes	15	0.69	0.61	6.4	[26]
10.	Hyper-branched Titania nanowires	15.29	0.822	0.65	8.11	[27]
11.	Hollow nanoparticle of TiO$_2$ with 1D spindle TiO$_2$	19.07	0.675	0.67	8.65	[28]
12.	Hollow SnO$_2$ nanostructures	21.3	0.71	0.5	7.5	[29]
13.	Co$_{0.85}$Se hollow nanoparticles	13.44	0.66	0.68	6.03	[30]
14.	Yb doped TiO$_2$ hollow nanoparticles	7.63	0.741	0.647	3.66	[31]
15.	Co$_9$S$_8$ nanoneedle array	16.8	0.48	0.461	3.72	[32]
16.	Irregular 1D ZnO nanorods and hollow ZnO nanostructure composite	18.58	0.62	0.62	7.08	[33]
17.	Bunchy shape TiO$_2$ nanospheres	14.3	0.76	0.69	7.5	[34]
18.	Hollow ZnO nanosphere	7.8	0.68	0.62	3.28	[35]
19.	PtNi hollow tubular alloy	17.852	0.709	0.666	8.43	[36]
20.	Al doped SiO$_2$/TiO$_2$ hollow nanostructure	18.7	1.05	0.747	14.7	[37]
21.	NiSe-Ni$_3$Se$_2$/RGO- hollow hybrid nanostructure	16.31	0.75	0.64	7.83	[50]
22.	Ta-PEDOT- hollow PPy	14.1	0.72	0.551	5.61	[51]
23.	ZnO nano-umbrella, hollow PtNi$_{1.07}$ nano-alloy CE	18.24	0.739	0.673	9.08	[38]
24.	Tri-layer anatase photo-anode of hollow nanoparticle, Sub-micro mesosphere and hierarchical microspheres	17.92	0.799	0.65	9.24	[39]
25.	Hollow TiO$_2$ nanotube and nanoparticles	16.15	0.63	0.482	4.9	[40]
26.	NiCo$_2$S$_4$ hollow nanospheres	17.4	0.843	0.647	9.49	[41]
27.	Ag nanowire on 3D hollow TiO$_2$	13.02	0.68	0.652	5.74	[42]

28.	CdSe$_{0.2}$S$_{0.8}$/CdS (7th cycle)	18.22	0.52	0.51	4.83	[43]
29.	Hyper-branched TiO$_2$ on carbon spheres grown after 2 h	19.65	0.771	0.65	9.84	[52]
30.	Co$_9$S$_8$-CuS	19.75	0.51	0.45	4.5	[44]
31.	CoS$_2$- carbon on CoS$_2$	17.8	0.791	0.67	9.32	[53]
32.	0D hollow TiO$_2$ nanoparticles	17.18	0.924	0.76	12.16	[45]
33.	Hollow rice grain shape TiO$_2$	21.6	1.07	0.61	14.2	[46]
34.	Hollow TiO$_2$ nanorods	18.43	0.765	0.68	9.58	[47]
35.	Co-Cu-WS$_x$	17.7	0.798	0.68	9.61	[48]
36.	Hollow SiO$_2$ nanofibers	13.72	0.624	0.51	4.36	[49]

3.1 Metallic hollow nanostructurebased solar cells

Metallic hollow nanostructured solar cells were first used, prior to their counterpart carbonaceous hollow nanostructured in 2012 when Qiu et al. [18] synthesized co-axial multi shelled hollow titania nanorods for dye sensitized solar cell. The quintuple titania hollow nanorods as shown in figure 3a [18], produced an open circuit voltage (V_{oc}) of 0.73 V, short-circuit current density (J_{sc})of 15 mA cm^{-2} and fill factor (FF) around ~ 0.57 leading to an efficiency of 6.2%[18]. Similarly, Wang et Al. [19] achieved almost similar value of 6.02% using hollow nanospheres of SnO$_2$ of average particle size ~ 200 nm, prepared in a green hydrothermal process without the usage of hard template (figure 3b) [19]. A year later, Huo and co-workers [20] produced TiO$_2$ assisted solar cell from spray pyrolysis technique, with a slight increase in the efficiency value (~6.91%) causing large pore size as demonstrated in figure 3c. The J$_{sc}$, V$_{oc}$ and FF values were 17.26 mA/cm^2,0.76 V and 0.53 V respectively [20].

In 2014, Ahnetal. [21] reported the fabrication of TiO$_2$-SnO$_2$ hollow nanocomposites and its usage in dye sensitized solar cell, where the current efficiency reached a value of 8.2% with V$_{oc}$ and J$_{sc}$ values of0.76 V and 18.2 mA/cm^2, respectively [21]. In figure 3d, the field emission scanning electron microscope (FESEM) images had portrayed the hollow nanospheres of SnO$_2$ having much smaller shape than the previous reported works. Thus, decrease in size played a crucial role in the elevation of power efficiency. Hwang and co-workers [22-23] produced composites of hollow SiO$_2$ and TiO$_2$ decorated silver nanoparticles that exhibited a J$_{sc}$ and V$_{oc}$ values of around ~ 17.5 mA cm^{-2} and 0.74 V, correspondingly with power efficiency of 9% [22]. They also replaced the nanocomposite of SiO$_2$-TiO$_2$ by multi shell TiO$_2$ hollow nanoparticles and got the light-to-electricity efficiency of 9.4% [23]. Urchin shape hollow Ni$_3$Se$_4$ as exhibited in figure 3e was designed by Lee et al. [24] as an alternative to replace the expensive platinum counter electrode leading to efficiency of 8.31% [24]. It was observed in Tafel plot, that nickel

Materials for Solar Cell Technologies I   Materials Research Forum LLC
Materials Research Foundations **88** (2021) 129-147   https://doi.org/10.21741/9781644901090-5

selenide blended with methanol had enhanced electro-catalytic ability to perform as a solar cell counter electrode. On the other hand, when 10% TiO_2 hollow nanofibers was amalgamated with commercial ZnO, a gradual decline in the value of efficiency (4.59%) was observed by Li and co-workers (figure 3f) [25]. This work gave an indication that even though hollow TiO_2 gave better efficiency value as a sole component but mixing ZnO decreased its efficiency in dye sensitized solar cell, which might be due to the fact that the electron recombination process in electrolyte of cell got hampered due to the presence of ZnO. Shi et al. [26] found increased efficiency up to 6.4% by replacing the previously used amalgam of TiO_2-ZnO with pure TiO_2 hollow boxes of particle size of 200 – 400 nm [26]. Interestingly, Wu et al. [27] engineered various hierarchy of titania nanowires and found that hyper-branched nanowires with diameters ranging from 3-5 nm gave 8.11 % efficiency. This improved efficiency was due to superior light scattering by hyper-branched TiO_2 wires, resulting in rapid charge transport.

In 2015, Wang et al. [28] extended the aforementioned work and designed more sophisticated one-dimensional spindle TiO_2, blended with hollow TiO_2 nanoparticles to produce a synergistic effect on the efficiency. This resultant composite with different morphology not only gave a yield of ~ 8.65 % but also required comparatively less amount of dye than previous works for sensitization. A team of Malaysian researchers replaced titania with pristine hollow SnO_2 which gave efficiency of 7.5 %, where the particle distribution was quite larger than anatase nanoparticles so resulted in the decline of the efficiency [29]. On a similar trend, pristine hollow morphology of cobalt chalcogenides ($Co_{0.85}Se$) was used for the first time in dye sensitized solar cell by Jiang and co-workers [30] to exhibit its electronic character. Although it had a high surface area, yet showed power efficiency of ~ 6.03 due to larger size of the spherical nanoparticles (>100 nm). This indicated that high surface area is interlinked with small size (< 80 nm) to attain high efficiency. To overcome the size barrier and acquire high efficiency, Cheng et al. [31] fabricated 20 nm sized 0.5 % ytterbium doped titania hollow nanostructures but found a gradual decline in the efficiency (~ 3.66 %) due to less surface area (76 m^2 /g) [31]. Co_9S_8, another cobalt chalcogenide of nanoneedle arrangement was fabricated by Chen et al. [32] gave efficiency value of 3.72 % when used as a counter electrode in quantum dot sensitized solar cell. Bai et al. [33] synthesized a nanocomposite, composed of two different morphology of zinc oxide i.e. one dimensional irregular ZnO nanorods and partially hollow ZnO. Although the composite gave a massive boost in the power efficiency (7.08 %) than other predecessor materials but the dye loaded in the cell in large quantity can't be underestimated. In 2016, Song and co-workers [34] produced various hierarchies of TiO_2 of bunchy shaped hollow nanoparticles with high surface area (> 200 m^2/g) and established efficiency above 7.5 %.

Materials for Solar Cell Technologies I Materials Research Forum LLC
Materials Research Foundations **88** (2021) 129-147 https://doi.org/10.21741/9781644901090-5

This work differentiated the hollow shapes with other nanostructures like nano-belt, nanosheets etc. where hollow shapes provided better efficiency than other nanostructures. Interestingly, the hollow shapes also exhibited long electron diffusion length which was mainly observed in cylindrical or nanorods structures. In a similar manner, the efficiency of ZnO hollow nanoshape was compared with pomegranate shape of ZnO and found that hollow shape had lower efficiency (3.28%) than the pomegranate ones [35]. It was due to the fact that pomegranate shape acquired rod like structure which had better photon trapping centers, facilitating light scattering while hollow ZnO had larger surface area but photon trapping centers were absent. Wang et al. [36] amalgamated platinum nickel tubular hollow alloy that demonstrated a phenomenal efficiency of ~ 8.43 % even though the particles were beyond the nanoscale range. This was largely due to the combined effect of Pt and Ni as well as the band gap energy of Pt in the conducting range. In the same year, Yun and co-workers [37] fabricated multi-junction perovskite solar cell composed of aluminum doped SiO_2/TiO_2 hollow nanostructure which exhibited a tremendous boost in the efficiency of ~ 14.7 %. This work canvassed that even though the solar cell was based on insulating scaffold material i.e. alumina yet the V_{oc} values were high due to interface resistance and the coverage of perovskite material. Commercializing platinum counter electrode had been an economic burden henceforth a lot of alternative had been used to replace it as well as boost the efficiency, so Li et al. [38] constructed $PtNi_{1.07}$ hollow nano-alloy based solar cell in 2017. The counter electrode boosted the electrocatlaytic ability of the umbrella shape ZnO nanoparticles photo-anode that resulted into efficiency ~ 9.08 %. This work emphasized that apart from the hollow nanostructure and morphology of the photo-anode, the conversion efficiency could be enhanced if hollow nanostructures are introduced in the counter electrode portion. Khan et al. [39] assembled a tri-layer dye sensitized solar cell comprised of various range of hollow anatase TiO_2 in nano, sub-micro and micro range that yielded 9.24% efficiency. The efficiency was ascribed due to the spherical hollow as shown in figure 3g, where less grain boundary led to high light scattering and electron transport. Wu et al. [40] reported the fabrication of hollow TiO_2 nanotubes aligned with TiO_2 nanoparticles as shown in figure 3h, had 4.9 % of efficiency. Since the surface area of the photo-anode was as low as ~ 71 m^2g^{-1} there was a drastic decline in the efficiency than previous work.

Figure 3 FESEM image of (a) a quintuple hollow shelled TiO₂, (inset) high magnification of a broken tip of hollow TiO₂ nanotube, (b) SnO₂ hollow nanospheres, (c) macro-pore of TiO₂ prepared from spray pyrolysis, (d) SnO₂ hollow spheres, (e) urchin shape Ni₃Se₄ dispersed on CH₃OH, (f) TiO₂ hollow nanofibers, (g) hollow TiO₂ nanoparticles ~ 25 nm size, (h) hollow TiO₂ nanotubes with 0.7 ml HF dosage and (i) rGO/Au doped ZnO honeycomb film. (a) Reproduced with permission. Copyright 2011, Royal Society of Chemistry. (b) Reproduced with permission. Copyright 2012, Royal Society of Chemistry. (c) Reproduced with permission. Copyright 2013, American Chemical Society. (d) Reproduced with permission. Copyright 2014, Wiley-VCH. (e) Reproduced with permission. Copyright 2014, Elsevier. (f) Reproduced with permission. Copyright 2014, Elsevier. (g) Reproduced with permission. Copyright 2017, Royal Society of Chemistry. (h) Reproduced with permission. Copyright 2017, Elsevier. (i) Reproduced with permission. Copyright 2017, Wiley-VCH

Materials for Solar Cell Technologies I   Materials Research Forum LLC
Materials Research Foundations **88** (2021) 129-147   https://doi.org/10.21741/9781644901090-5

Prior to 2018, titania had been a dormant active material in perovskite as well as in dye sensitized solar cell due to their dual role as opaque and transparent solar cell module and promoted high Schottky barrier that facilitated high electron capture. However incorporating double metal chalcogenide in solar cell was completely untouched, therefore, Jiang group [41] assembled hollow $NiCo_2S_4$ nanospheres in Grätzel cell. The efficiency rocketed up to 9.49 % due to high diffusion coefficient of redox species and low diffusion impedance imparted by the ball-in-ball hollow shape of the bi-functional metal sulfide. Lien et al. [54] synthesized ZnO nanoshell that underwent optoelectronic characterizations where omnidirectional detection of light at various incidence angles and polarization occurred aided by theoretical modeling. Various nano hierarchies of silver and gold were doped on three dimensional hollow titania by Ran and co-workers [42] yielding an efficiency of ~5.74 %. Silver nanowires on TiO_2 were better active materials than their counterpart gold nanowires due to improved electron movement in the rim of silver nanowires and the occurrence of intensified light scattering due to the large size of silver. However, the photoluminescence spectra revealed that gold was having higher e^-/h^+ recombination rate than silver. Semiconducting CdSe quantum dots proved to be an excellent sensitized layer material when Lan and co-workers [43] incorporated it with three dimensional hierarchical hollow TiO_2 and got the efficiency value of ~ 4.83 % after 7th cycle [43]. Furthermore, the absorption wavelength range of the sensitized photoelectrode was found to be broadened with proper adjustment of the Se^{-2}/S^{-2} molar ratio. Hong et al. [44] fabricated colloidal Co_9S_8-CuS architecture where hollow nanoneedles of Co_9S_8 were deposited on CuS nanosheet. This mixture produced power efficiency of ~ 4.5 % and the efficiency retained by 80 % after 500 cycles which justified the superior stability of polysulfide electrolyte and increased reaction sites on the hetero junction of the solar cell.

A remarkable change was observed in 2019 when Khan et al. [45] constructed zero dimensional hollow TiO_2 nanoparticles for perovskite solar cell where the efficiency drastically increased to 12.16 %. This efficiency was possible due to the homogeneous hollow grain size of titania and unlike other conventional junction solar cell, the upper capping agent was perovskite crystals itself. This densely packed structure of titania enabled minimal infiltration of perovskite compound so light scattering propelled rapidly. Based on similar concept, another team of researchers led by Ma and co-workers [46] constructed perovskite solar cell with hollow rice grain shape titania which was coated via spin coating technique. It was found that the thinnest layer deposited on fluorine tin oxide glass at 4000 rpm achieved a power efficiency of ~ 14.2 % and it suggests that efficiency depends not only on the homogeneous growth but must also retain an uniform thin layer for penetration of photons. Marandi and co-workers [47] improved their synthesis procedure by growing hollow titania rods and fetched an efficiency of ~ 9.58 % without the support of any template. A unique tri metal chalcogenide was assembled by Qian and co-workers [48] in dye sensitized solar cell, comprising of hollow nanospheric Co-Cu-WS$_x$ active photoelectrode, attaining efficiency of ~ 9.61 %. This tri metallic sulfide had a typical hollow surface topography, where it could accumulate incoming photons in a trap set up mechanism due to the multiple voids present on its surface. Recently, Iranian researchers modified conventional silica based solar cell with hollow silica nanofibers as active material in quantum dot solar cell [49]. However, the fabrication charge was cost effective due to the availability of precursor but the efficiency was poor (4.36 %) due to the uneven growth of silica containing rods from nano to sub-micron level.

3.2 Carbonaceous hollow nanostructure based solar cells

Most carbonaceous hollow nanostructures are quite facile to prepare and have a simplistic synthetic route compared to metallic nanostructures. This makes the fabrication cost-effective and also a worthy replacement for metallic nanoparticles due to the availability of raw feedstock. Most carbon based hollow nanostructures like carbon black, graphene derivatives, graphene quantum dots, etc. are prepared from low cost precursors as well as they have a high durability under high temperature and pressure [55-59]. Similarly, highly conducting organic polymers like poly-3,4-ethylenedioxythiophene, poly(p-phenylene sulfide), etc. demonstrated high electron mobility and electrical conductivity due to the presence of sp^2 hybridization and σ-bonded conjugated system. Hence, carbonaceous hollow structures opened a gateway for solar cell materials apart from their usage in energy storage devices and diodes. Zhang et al. [50] prepared a hybrid hollow nanocomposite consisted of NiSe-Ni$_3$Se$_2$/ reduced graphene oxide in 2016 and the material achieved power conversion efficiency of 7.87 %. The enhanced efficiency was achieved due to the self- tuning charge transfer ability of reduced graphene oxide and the electro-catalytic ability of nickel selenide. This hollow nanocomposite also demonstrated remarkable electrochemical ability as a counter electrode over conventional platinum electrode. In the same way, 3,4-ethylenedioxythiophene was blended with hollow micro/nano-horn shaped polypyrrole with a tantalum cathode by Bai and co-workers [51]. The composite produced a low efficiency of 5.61% only which might be due to the random mixture of micro and nano polymer materials. Sun et al. [60] designed a honeycomb array graphene oxide film and uniformly distributed gold coupled ZnO nanoparticles as depicted in figure 3i, to investigate the current-voltage (I-V) relationship. Another team of researchers from University of York investigated that applying polystyrene based template on aluminium doped ZnO diminished the boundary cracks which ultimately gave high current density (7.26 mA cm^{-2}) [61].

Materials for Solar Cell Technologies I   Materials Research Forum LLC
Materials Research Foundations **88** (2021) 129-147   https://doi.org/10.21741/9781644901090-5

Excited with this concept, in 2018 Robbiano group [62] applied polystyrene on hollow titania nanosphere and obtained efficiency as high as 9 % in their newly assembled planar perovskite solar cell. The enhancement in the efficiency was attributed to the feedback structure of perovskite that increased rapid light-trapping and improved electrical contact between the titania electrodes and active perovskite materials. However, Marandi et al. [52] reported an efficiency value of 9.84 after combining carbon spheres as template on hyper-branched TiO_2 hollow spheres via hydrothermal method. On further optimization in the synthesis technique, carbon spheres growth was varied at different time intervals (1-3 h) and it was observed that ~ 25 nm sized spheres gave the highest efficiency. It was evident that when the size of carbon sphere was reduced to nanoscale level, the light scattering excelled more resulting in greater light to electricity conversion. Niu and co-workers [53] assembled yolk shell CoS_2 wrapped by a hollow porous carbon nanocage which showed an efficiency of 9.32 %. Thus, engineering the hollow nanostructure on the performance of the solar cells will show great promise towards the power efficiency related energy problems and future directions of design of hollow nanostructure to solve emerging challenges and improvement in that direction.

Conclusion

The chapter demonstrated the up-to-date research efforts on hollow nanostructured materials in various types of solar cells. Apart from implementation as photoelectrode, hollow nanomaterials boosted the efficiency when used as an alternate for counter electrode materials. Despite of multiple challenges in the synthesis of metallic hollow nanostructures, carbonaceous hollow nanostructures could be produced in facile way with an economic fabrication charge. The chapter describes the state of the art on the developments concerning the synthesis of hollow nanostructures and their photovoltaic performance.

References

[1] N. Panwar, S. Kaushik, S. Kothari, Role of renewable energy sources in environmental protection: A review, Renew. Sustain.Energ.15 (2011) 1513-1524. http://doi.org/10.1016/j.rser.2010.11.037

[2] E.M. Rocco, Evaluation of the terrestrial albedo applied to some scientific missions, Space Sci. Rev. 151 (2010) 135-147. http://doi.org/10.1007/s11214-009-9622-6

[3] K. Laurischkat, D. Jandt, Business model prototyping for electric mobility and solar

power solutions, Procedia CIRP 48 (2016) 307-312.http://doi.org/10.1016/j.procir.2016.03.026

[4] T. Soga, Nanostructured materials for solar energy conversion, first ed., Elsevier, Amsterdam, 2006

[5] R. King, D. Law, K. Edmondson, C. Fetzer, G. Kinsey, H. Yoon, R. Sherif, N. Karam, 40% efficient metamorphic GaInP/GaInAs/Ge multijunction solar cells, Appl. Phys. Lett. 90 (2007) 183516.http://doi.org/10.1063/1.2734507

[6] K.T. VanSant, J. Simon, J.F. Geisz, E.L. Warren, K.L. Schulte, A.J. Ptak, M.S. Young, M. Rienäcker, H. Schulte-Huxel, R. Peibst, Toward low-cost 4-terminal GaAs//Si tandem solar cells, ACS Appl. Energy Mater. 2 (2019) 2375-2380. http://doi.org/10.1021/acsaem.9b00018

[7] M. Wiemer, V. Sabnis, H. Yuen, 43.5% efficient lattice matched solar cells, High and low concentrator systems for solar electric applications VI, Proc.SPIE 8108(2011) 810804.http://doi.org/10.1117/12.897769

[8] P. Sutradhar, M. Saha, Silver nanoparticles: synthesis and its nanocomposites for heterojunction polymer solar cells, J. Phys. Chem. C 120 (2016) 8941-8949. http://doi.org/10.1021/acs.jpcc.6b00075

[9] P. Sutradhar, M. Saha, Green synthesis of zinc oxide nanoparticles using tomato (Lycopersiconesculentum) extract and its photovoltaic application, J. Exp. Nanosci. 11 (2016) 314-327. http://doi.org/10.1080/17458080.2015.1059504

[10] D. Wongratanaphisan, K. Kaewyai, S. Choopun, A. Gardchareon, P. Ruankham, S. Phadungdhitidhada, CuO-Cu_2O nanocomposite layer for light-harvesting enhancement in ZnO dye-sensitized solar cells, Appl. Surf. Sci. 474 (2019) 85-90. http://doi.org/10.1016/j.apsusc.2018.05.037

[11] A. Hegazy, N. Kinadjian, B. Sadeghimakki, S. Sivoththaman, N.K. Allam, E. Prouzet, TiO_2 nanoparticles optimized for photoanodes tested in large area dye-sensitized solar cells (DSSC), Sol. Energy Mater. Sol. Cells 153 (2016) 108-116. http://doi.org/10.1016/j.solmat.2016.04.004

[12] S. Rafique, S.M. Abdullah, W.E. Mahmoud, A.A. Al-Ghamdi, K. Sulaiman, Stability enhancement in organic solar cells by incorporating V_2O_5 nanoparticles in the hole transport layer, RSC Adv. 6 (2016) 50043-50052. http://doi.org/10.1039/C6RA07210K

[13] K. Aitola, K. Domanski, J.P. CorreaBaena, K. Sveinbjörnsson, M. Saliba, A. Abate, M. Grätzel, E. Kauppinen, E.M. Johansson, W. Tress, High temperaturestable perovskite solar cell based on lowcost carbon nanotube hole contact, Adv. Mater. 29 (2017) 1606398. http://doi.org/10.1002/adma.201606398

[14] P. O'Keeffe, D. Catone, A. Paladini, F. Toschi, S. Turchini, L. Avaldi, F. Martelli, A. Agresti, S. Pescetelli, A. Del Rio Castillo, Graphene-induced improvements of perovskite solar cell stability: effects on hot-carriers, Nano Lett. 19 (2019) 684-691. http://doi.org/10.1021/acs.nanolett.8b03685

[15] K. Kumarasinghe, G. Kumara, R. Rajapakse, D. Liyanage, K. Tennakone, Activated coconut shell charcoal based counter electrode for dye-sensitized solar cells, Org. Electron. 71 (2019) 93-97. http://doi.org/10.1016/j.orgel.2019.05.009

[16] S. Diao, X. Zhang, Z. Shao, K. Ding, J. Jie, X. Zhang, 12.35% efficient graphene quantum dots/silicon heterojunction solar cells using graphene transparent electrode, Nano Energy 31 (2017) 359-366. http://doi.org/10.1016/j.nanoen.2016.11.051

[17] J. Wang, Y. Cui, D. Wang, Design of hollow nanostructures for energy storage, conversion and production, Adv. Mater. 31 (2018) 1801993. http://doi.org/10.1002/adma.201801993

[18] J. Qiu, F. Zhuge, X. Li, X. Gao, X. Gan, L. Li, B. Weng, Z. Shi, Y.-H. Hwang, Coaxial multi-shelled TiO_2 nanotube arrays for dye sensitized solar cells, J. Mater. Chem. 22 (2012) 3549-3554. http://doi.org/10.1039/C2JM15354H

[19] H. Wang, B. Li, J. Gao, M. Tang, H. Feng, J. Li, L. Guo, SnO_2 hollow nanospheres enclosed by single crystalline nanoparticles for highly efficient dye-sensitized solar cells, Cryst. Eng. Comm. 14 (2012) 5177-5181. http://doi.org/10.1039/C2CE06531B

[20] J. Huo, Y. Hu, H. Jiang, W. Huang, Y. Li, W. Shao, C. Li, Mixed solvents assisted flame spray pyrolysis synthesis of TiO_2 hierarchically porous hollow spheres for dye-sensitized solar cells, Ind. Eng. Chem. Res. 52 (2013) 11029-11035. http://doi.org/10.1021/ie4006222

[21] S.H. Ahn, D.J. Kim, W.S. Chi, J.H. Kim, Hierarchical doubleshell nanostructures of TiO_2 nanosheets on SnO_2 hollow spheres for highefficiency, solidstate, dyesensitized solar cells, Adv. Funct. Mater. 24 (2014) 5037-5044. http://doi.org/10.1002/adfm.201400774

[22] S.H. Hwang, D.H. Shin, J. Yun, C. Kim, M. Choi, J. Jang, SiO_2/TiO_2 Hollow nanoparticles decorated with Ag nanoparticles: enhanced visible light absorption and

improved light scattering in dyesensitized solar cells, Chem. Eur. J. 20 (2014) 4439-4446. http://doi.org/10.1002/chem.201304522

[23] S.H. Hwang, J. Yun, J. Jang, Multishell porous TiO_2hollow nanoparticles for enhanced light harvesting in dye-sensitized solar cells, Adv. Funct. Mater. 24 (2014) 7619-7626. http://doi.org/10.1002/adfm.201401915

[24] C.T. Lee, J.D. Peng, C.T. Li, Y.L. Tsai, R. Vittal, K.C. Ho, Ni_3Se_4 hollow architectures as catalytic materials for the counter electrodes of dye-sensitized solar cells, Nano Energy 10 (2014) 201-211. http://doi.org/10.1016/j.nanoen.2014.09.017

[25] F. Li, G. Wang, Y. Jiao, J. Li, S. Xie, Efficiency enhancement of ZnO-based dye-sensitized solar cell by hollow TiO_2 nanofibers, J. Alloys Compd. 611 (2014) 19-23. http://doi.org/10.1016/j.jallcom.2014.05.100

[26] Y. Shi, L. Zhao, S. Wang, J. Li, B. Dong, Z. Xu, L. Wan, Double-layer composite film based on hollow TiO_2 boxes and P25 as photoanode for enhanced efficiency in dye-sensitized solar cells, Mater. Res. Bull. 59 (2014) 370-376. http://doi.org/10.1016/j.materresbull.2014.07.012

[27] W.Q. Wu, H.S. Rao, H.L. Feng, H.Y. Chen, D.B. Kuang, C.Y. Su, A family of vertically aligned nanowires with smooth, hierarchical and hyperbranched architectures for efficient energy conversion, Nano Energy 9 (2014) 15-24.http://doi.org/10.1016/j.nanoen.2014.06.019

[28] G. Wang, X. Zhu, J. Yu, Bilayer hollow/spindle-like anatase TiO_2 photoanode for high efficiency dye-sensitized solar cells, J. Power Sourc. 278 (2015) 344-351. http://doi.org/10.1016/j.jpowsour.2014.12.091

[29] Q. Wali, A. Fakharuddin, A. Yasin, M.H. Ab Rahim, J. Ismail, R. Jose, One pot synthesis of multi-functional tin oxide nanostructures for high efficiency dye-sensitized solar cells, J.Alloys Compd. 646 (2015) 32-39. http://doi.org/10.1016/j.jallcom.2015.05.120

[30] Q. Jiang, G. Hu, $Co_{0.85}$ Se hollow nanoparticles as Pt-free counter electrode materials for dye-sensitized solar cells, Mater. Lett. 153 (2015) 114-117. http://doi.org/10.1016/j.matlet.2015.04.008

[31] L. Cheng, X. Xu, Y. Fang, Y. Li, J. Wang, G. Wan, X. Ge, L. Yuan, K. Zhang, L. Liao, Triblock copolymer-assisted construction of 20 nm-sized ytterbium-doped TiO_2 hollow nanostructures for enhanced solar energy utilization efficiency, Sci. China Chem. 58 (2015) 850-857. http://doi.org/10.1007/s11426-014-5237-1

[32] C. Chen, M. Ye, N. Zhang, X. Wen, D. Zheng, C. Lin, Preparation of hollow Co_9S_8 nanoneedle arrays as effective counter electrodes for quantum dot-sensitized solar cells, J. Mater. Chem. A 3 (2015) 6311-6314. http://doi.org/10.1039/C4TA06987K

[33] T. Bai, Y. Xie, J. Hu, C. Zhang, J. Wang, Novel one-dimensional ZnO nanorods synthesized through a two-step post-treatment for efficiency enhancement of dye-sensitized solar cells, J. Alloys Compd. 644 (2015) 350-353. http://doi.org/10.1016/j.jallcom.2015.05.040

[34] D. Song, P. Cui, T. Wang, B. Xie, Y. Jiang, M. Li, Y. Li, S. Du, Y. He, Z. Liu, Bunchy TiO_2 hierarchical spheres with fast electron transport and large specific surface area for highly efficient dye-sensitized solar cells, Nano Energy 23 (2016) 122-128. http://doi.org/10.1016/j.nanoen.2016.03.006

[35] R. Chauhan, M. Shinde, A. Kumar, S. Gosavi, D.P. Amalnerkar, Hierarchical zinc oxide pomegranate and hollow sphere structures as efficient photoanodes for dye-sensitized solar cells, Micropor. Mesopor. Mater. 226 (2016) 201-208. http://doi.org/10.1016/j.micromeso.2015.11.054

[36] J. Wang, Q. Tang, B. He, P. Yang, Counter electrodes from polymorphic platinum-nickel hollow alloys for high-efficiency dye-sensitized solar cells, J. Power Sourc. 328 (2016) 185-194. http://doi.org/10.1016/j.jpowsour.2016.08.029

[37] J. Yun, J. Ryu, J. Lee, H. Yu, J. Jang, SiO_2 /TiO_2based hollow nanostructures as scaffold layers and Al-doping in electron transfer layer for efficient perovskite solar cells, J. Mater. Chem. A 4 (2016) 1306-1311. http://doi.org/10.1039/C5TA08250A

[38] P. Li, Y. Zhang, X. Yang, Y. Gao, S. Ge, Alloyed PtNi counter electrodes for high performance dye-sensitized solar cell applications, J. Alloys Compd. 725 (2017) 1272 1281. http://doi.org/10.1016/j.jallcom.2017.07.266

[39] J. Khan, J. Gu, S. He, X. Li, G. Ahmed, Z. Liu, M.N. Akhtar, W. Mai, M. Wu, Rational design of a tripartite-layered TiO_2 photoelectrode: a candidate for enhanced power conversion efficiency in dye sensitized solar cells, Nanoscale 9 (2017) 9913-9920. http://doi.org/10.1039/C7NR03134C

[40] D. Wu, X. Wang, Y. An, X. Song, N. Liu, H. Wang, Z. Gao, F. Xu, K. Jiang, Hierarchical TiO_2 structures derived from F-mediated oriented assembly as triple-functional photoanode material for improved performances in CdS/CdSe sensitized solar cells, Electrochim. Acta 248 (2017) 79-89. http://doi.org/10.1016/j.electacta.2017.06.150

[41] Y. Jiang, X. Qian, C. Zhu, H. Liu, L. Hou, Nickel cobalt sulfide double-shelled hollow nanospheres as superior bifunctional electrocatalysts for photovoltaics and alkaline hydrogen evolution, ACS Appl. Mater. Interfaces 10 (2018) 9379-9389. http://doi.org/10.1021/acsami.7b18439

[42] H. Ran, J. Fan, X. Zhang, J. Mao, G. Shao, Enhanced performances of dye-sensitized solar cells based on Au-TiO$_2$ and Ag-TiO$_2$ plasmonic hybrid nanocomposites, Appl. Surf. Sci. 430 (2018) 415-423. http://doi.org/10.1016/j.apsusc.2017.07.107

[43] Z. Lan, X. Chen, S. Zhang, J. Wu, CdSe$_x$ S$_{1-x}$/CdS-cosensitized 3D TiO$_2$ hierarchical nanostructures for efficient energy conversion, J. Solid State Electrochem. 22 (2018) 347-353. http://doi.org/10.1007/s10008-017-3748-3

[44] X. Hong, Q. Liu, X. Gao, C. He, X. You, X. Zhao, X. Liu, M. Ye, Rational design of coralloid Co$_9$S8–CuS hierarchical architectures for quantum dot-sensitized solar cells, J. Mater. Chem. C 6 (2018) 11384-11391. http://doi.org/10.1039/C8TC04274H

[45] J. Khan, N.U. Rahman, W.U. Khan, A. Hayat, Z. Yang, G. Ahmed, M.N. Akhtar, S. Tong, Z. Chi, M. Wu, Multi-dimensional anatase TiO$_2$ materials: Synthesis and their application as efficient charge transporter in perovskite solar cells, Sol. Energy 184 (2019) 323-330. http://doi.org/10.1016/j.solener.2019.04.020

[46] S. Ma, T. Ye, T. Wu, Z. Wang, Z. Wang, S. Ramakrishna, C. Vijila, L. Wei, Hollow rice grain-shaped TiO$_2$ nanostructures for high-efficiency and large-area perovskite solar cells, Sol. Energy Mater. Sol. Cells 191 (2019) 389-398. http://doi.org/10.1016/j.solmat.2018.11.028

[47] M. Marandi, S. Bayat, M.N.S. Sabet, Hydrothermal growth of a composite TiO$_2$ hollow spheres/TiO$_2$ nanorods powder and its application in high performance dye-sensitized solar cells, J. Electroanal. Chem. 833 (2019) 143-150. http://doi.org/10.1016/j.jelechem.2018.11.023

[48] X. Qian, H. Liu, J. Yang, H. Wang, J. Huang, C. Xu, Co-Cu–WS$_x$ ball-in-ball nanospheres as high-performance Pt-free bifunctional catalysts in efficient dye-sensitized solar cells and alkaline hydrogen evolution, J. Mater. Chem. A 7 (2019) 6337-6347. http://doi.org/10.1039/C8TA12558A

[49] Z. Sherafati-Tabarestani, M. Samadpour, Simply synthesized silica hollow fibers for enhancing the performance of dye/quantum dot sensitized solar cells, Sol. Energy 183 (2019) 716-724. http://doi.org/10.1016/j.solener.2019.03.078

[50] X. Zhang, M. Zhen, J. Bai, S. Jin, L. Liu, Efficient NiSe-Ni$_3$Se$_2$/graphene electrocatalyst in dye-sensitized solar cells: the role of hollow hybrid nanostructure, ACS Appl. Mater. Interfaces 8 (2016) 17187-17193. http://doi.org/10.1021/acsami.6b02350

[51] Y. Bai, Y. Xu, J. Wang, M. Gao, J. Zhu, W.U. Rehman, Electrochemically prepared poly (3, 4-ethylenedioxy-thiophene)/polypyrrole films with hollow micro/nanohorn arrays as high-efficiency counter electrodes for dye-sensitized solar cells, ChemElectroChem 3 (2016) 1376-1383. http://doi.org/10.1002/celc.201600191

[52] M. Marandi, S. Bayat, Facile fabrication of hyper-branched TiO$_2$ hollow spheres for high efficiency dye-sensitized solar cells, Sol. Energy 174 (2018) 888-896. http://doi.org/10.1016/j.solener.2018.09.065

[53] Y. Niu, X. Qian, J. Zhang, W. Wu, H. Liu, C. Xu, L. Hou, Stepwise synthesis of CoS$_2$–C@ CoS$_2$ yolk–shell nanocages with much enhanced electrocatalytic performances both in solar cells and hydrogen evolution reactions, J. Mater. Chem. A 6 (2018) 12056-12065. http://doi.org/10.1039/C8TA03591A

[54] D.H. Lien, Z. Dong, J.R.D. Retamal, H.P. Wang, T.C. Wei, D. Wang, J.H. He, Y. Cui, Resonance enhanced absorption in hollow nanoshell spheres with omnidirectional detection and high responsivity and speed, Adv. Mater. 30 (2018) 1801972. http://doi.org/10.1002/adma.201801972

[55] S. Das, M. Saha, Potato starch-derived almond-shaped carbon nanoparticles for non-enzymatic detection of sucrose, New Carbon Mater. 30 (2015) 244-251. http://doi.org/10.1016/S1872-5805(15)60189-5

[56] P. Mandal, M.J.P. Naik, M. Saha, Room temperature synthesis of graphene nanosheets, Cryst. Res. Technol. 53 (2018) 1700250.http://doi.org/10.1002/crat.201700250

[57] P. Mandal, M. Saha, Low-temperature synthesis of graphene derivatives: mechanism and characterization, Chem. Pap. 73 (2019) 1997-2006. http://doi.org/10.1007/s11696-019-00756-3.

[58] M.J.P. Naik, P. Mandal, J. Debbarma, M. Saha, Graphene quantum dots (GQDs) from organic acids, Appl. Innovative Res.1 (2019) 128-134

[59] P. Mandal, J. Debbarma, M. Saha, One step synthesis of N-containing graphene oxide from 3-Aminophenol,Cryst. Res. Technol. 55 (2020) 1900158. http://doi.org/10.1002/crat.201900158

[60] H. Sun, Q. He, S. Yin, K. Xu, Enhanced photocurrent generation of graphene/Au@ ZnO honeycomb film, Chin. J. Chem. 35 (2017) 1627-1632. http://doi.org/10.1002/cjoc.201700347

[61] M. Zhang, R.W. Mitchell, H. Huang, R.E. Douthwaite, Ordered multilayer films of hollow sphere aluminium-doped zinc oxide for photoelectrochemical solar energy conversion, J. Mater. Chem. A 5 (2017) 22193-22198. http://doi.org/10.1039/C7TA07509J

[62] V. Robbiano, G. Paternò, G. Cotella, T. Fiore, M. Dianetti, M. Scopelliti, F. Brunetti, B. Pignataro, F. Cacialli, Polystyrene nanoparticle-templated hollow titania nanosphere monolayers as ordered scaffolds, J. Mater. Chem. C 6 (2018) 2502-2508. http://doi.org/10.1039/C7TC04070A

Materials for Solar Cell Technologies I
Materials Research Foundations **88** (2021) 148-175

Materials Research Forum LLC
https://doi.org/10.21741/9781644901090-6

Chapter 6

Monocrystalline Silicon Solar Cells

M. Rizwan[1,*], Waheed S. Khan[2], S. Aleena[3]

[1]School of Physical Sciences, University of the Punjab, Lahore, Pakistan

[2]Nanobiotechnology Group,National Institute for Biotechnology and Genetic Engineering (NIBGE), Jhang Road, Faisalabad-38000, Pakistan

[3]Department of Physics, University of Gujrat, Hafiz Hayat Campus, Gujrat City, Pakistan

*dr.rizwan@uog.edu.pk

Abstract

Monocrystalline silicon based solar cells have the attributes that includes elemental semiconductor nature and balancing properties making it extensively applicable in the field of microelectronics. Silicon based solar cells make about 90% of today's photovoltaic technology. The highest experimental efficiency reported for monocrystalline solar cells so far is 26.6%. The V-I characteristics of monocrystalline silicon based solar cells have been deliberated in the contextual of silicon as substrate material. The theoretical value of Shockely-Queisser (SQ) limit for monocrystalline solar cells is 30% that invocate further efficiency developments. The typical monocrystalline structure and recent advancements in monocrystalline solar cells are emphasized with appropriate examples to understand the photovoltaic phenomenon. Power conversion efficiency (PCE) enhancement is of prime importance in photovoltaic industry (PV) and hence different techniques analyze the question of PCE in context of cost effective solar cell production. In light of the literature, the texturizing, anti-reflecting coating and metallization are proposed as the efficient methods for reduction in losses and enhancement in efficiency.

Keywords

Photovoltaic, Solar Cell, Monocrystalline, Shockely-Queisser, PCE

Materials for Solar Cell Technologies I

Materials Research Foundations **88** (2021) 148-175

Materials Research Forum LLC

https://doi.org/10.21741/9781644901090-6

Contents

Monocrystalline Silicon Solar Cells..**148**

1.   Introduction...**149**

2.   Typical structure of monocrystalline Si solar cell...............................**153**

3.   Advanced monocrystalline solar cells...**155**
       3.1   Passivated emitter rear localized (PERL) solar cell........................155
       3.2   Heterojunction with intrinsic thin layer (HIT) solar cell................156

4.   Experimental and theoretical power conversion efficiency (PCE)...**157**
       4.1   Efficiency of monocrystalline solar cells 157

5.   Monocrystalline solar cell PCE enhancement methods....................**161**
       5.1   Optical losses reduction methods 161
       5.1.1   Texturing..161
       5.1.1.1 Chemical method 162
       5.1.1.2 Physical method 162
       5.1.1.3 Pyramidal texture uniformity...163
       5.1.2   Anti-reflecting coating (ARC) 163
       5.1.3   Metallization 165
       5.2   Thickness of the monocrystalline solar 165

Conclusions..**166**

Acknowledgements...**167**

References..**167**

1.   Introduction

With the increase in population, the energy demands have also increased. Environmental degradation, and fossil fuels depletion have caused aggravated concern about the future energy system. Mankind has to look for renewable energy sources for reduction of hazardous and poisonous emission of gasses into the atmosphere. The photvoltaic technology has the potential to replace these conventional energy sources to fulfil the needs of energy. In photovoltaic technology, solar energy is considered as one of the most abundant renewable energy source on earth as sunlight takes only 8.5 minutes to reach the earth [1,2].

Solar energy is materialized in the form of solar cells, which trap the light of sun and convert it into useful amount of electricity. The light conversion phenomenon mostly arise in the semiconductor materials. Moreover, solar energy is non-exhaustable and ecofriendly which make its efficiency more significant [3, 4]. The solar energy based technology is used from small scale to the large industrial scale. As of 2019, the renewable energy sources have coontributed 26% to the global electricity production that is 26615 TWh, out of which the solar energy contribution is 1.9 %. It is predicted that in 2021, the PV technology has the capacity to contribute 900 GW to the global power generation and by 2050 the PV has the potential to become a leading power generation source with estimated contribution of 16% [5].

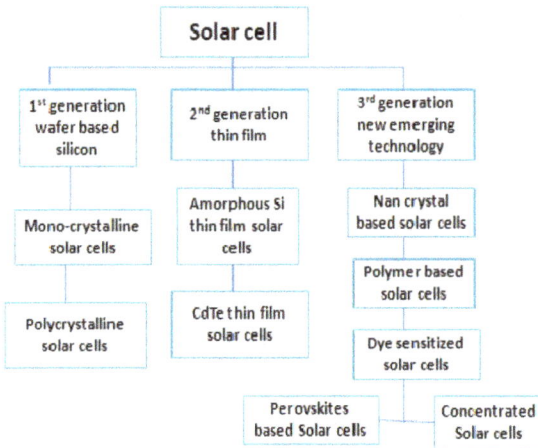

Figure 1   The fundamental classification of solar cells three generations [10].

A photovoltic cell also known as solar cell consists of various components including cell, power generator etc. New materials to enhance the efficieny of the solar cells are crucial in the PV technology [6]. For the production of photovoltaic phenomenon in solar cells different of the materials are utilized such as silicon, [7] copper indium-gallium-selenide [8, 9] and cadmium-telluride.The photovolatic industry has three generations of solar cell based on different materials utilized as shown in figure 1 [10]. One of the earliest material used for photovoltaic phenomenon is silicon. Silicon wafer based solar cells were considered as first generation solar cells which furthere categorized as the monocrystalline, polycrystalline solar cells[9, 8, 11], thin film based solar cels were considered as second generation solar cells which further categorized as amorphous Si and CdTe thin film solar cells [7, 8] and third generation solar cells are new emereninig technologies which can be categorized as the nanocrystal, polymer,dye sensitized, concentrated and perovskites solar cells.

A comparison between different types (1st, 2nd and 3rd generation) of the solar cells with respect to their efficiency, temperature, size and cost is described in detail in table 1 [11]. According to table 1, the cost is reduced on the other hand solar cell efficiency, stability increase from first generation to third.

Table 1 Comparison of different properties of all generation of solar cells [11].

Cell Type	First Generation Silicon Based		Second Generation Thin film		
	Monocrystalline	Polycrystalline	CdTe	CIGS	Amorphous Silicon
Efficiency	17%-18%	12%-14%	9%-11%	10%-12%	4%-8%
Temperature effect	Not efficient in high temperature	At high temperatures their performance is not efficient.	At low and high temperature it performance is good.	At low and high temperature it performance is good	At low and high temperature it performance is good
Cost	Two times greater cost as contrast to thin-film and considered as expensive	Two times greater cost as contrast to thin-film and considered as expensive	They are less expensive contrast to the conventional Based silicon's cells.	They are less expensive contrast to the conventional Based silicon's cells.	They are less expensive contrast to the conventional Based silicon's cells.
Size	Less volume to produce more energy	Less volume to produce more energy	Contribution to a large number of products which are flexible as well as light durable		
Additional	Oldest PV Most stable Eco-Friendly	Economical option	Toxic Cd	Some CIGS have 20% striking efficiency	Require lengthy mechanism time and large space
Cell Type	Third generation				Perovskites
	Nano crystal	Dye synthesized	Polymer	Concentrated	
Size	7%-8%	≅10%	≅3%-10%	≅ 40%	≅31%
Temperature effect	High thermal stability	Not efficient in range high temperature		High thermal stability	
Cost	As contrast to conventional Silicon , Nano crystal are 50% less expensive	As contrast conventional Silicon cells dye synthesized are 50% less expensive	As contrast conventional Silicon cells they are 50% less expensive	As contrast conventional Silicon cells they are 50% less expensive	As contrast conventional Silicon cells they are 50% less expensive Toxic
Size	Contribution to a large number of products which are flexible as well as light durable				

Materials for Solar Cell Technologies I   Materials Research Forum LLC
Materials Research Foundations **88** (2021) 148-175   https://doi.org/10.21741/9781644901090-6

Silicon is the most stable photovoltaic material. Silicon is developed as a prevailing material for the creation of power from sunlight and as of now, represents over 80% of PV productions [12]. In its unadulterated structure, silicon is a shimmering dark brilliant solid. It ranks in second abundant component in Earth's layer by mass (about 27.5%) after oxygen (about 50.5%), and accessible in adequate amount of coordinate worldwide vitality in terawatts [13]. Silica is utilized to create metallurgical evaluation silicon, which at that point experiences a few phases of cleansing and refining to deliver silicon of high excellence for industrial applications. Silicon has a non-toxic, non-destructive nature [14] unequivocally defensive silicon oxide's layer at interface between silicon & air. The recombination process at front surface diminishes due to passive layer of oxide which as electrical protecting layer and improved the silicon based technologies performance [15].

The first silicon based solar cell was discovered by the Russel Ohl in 1946 [16] which is based on the basic principle of photovoltaic effect. Photovoltaic effect was firstly observed in the 1839 by the Aleixandre-Edmond Becquerel [17]. In this process when sunlight (photon) falls on the solar cell material, an electron is generated and it is transmitted through the p-n junction from one cell to another cell. Due the production of the electron hole pair the electricity is produced by connecting it to the D.C power supply [18]. As silcon based solar cells has non-toxic, ecofriendy nature, stability, life time as well as have potential to use in various applications of reasearch and development.

Monocrystalline solar cells as the name suggests are solar cells based on a single crystal of silicon which are embedded in specific arrangement to produce electricity with the utilization of Czochralski process [18, 19]. The raw material used in single-crystalline solar cells is silicon and they are based on silicon wafer.The silicon based monocrystalline solids are unbroken, free of any grain limits, long life time, high efficincy, most stable and continuous structure. In monocrystalline silicon based solar cells, intrinsic semiconductor silicon is used or it can be doped with boron, phosphorous for making it a p or n-type silicon material [20]. Silicon based materials have most significant mechanical material of the most recent couple of years. In light of the fact that its accessibility at a reasonable expense has been basic cause for the advancement of the electronic gadgets on which the present-day hardware and information technology is based [21].

The use of the crystalline silicon based PV cells has seen much advancement in utilization and generation of electricity from sunlight [22]. The highest efficiency achieved for monocrystalline solar cells is 25% according to 2009 data [23] and now in 2020 their efficiency has exceeded to 26.6% [24]. The standard efficiency is about 15-18% for the certain mechanical solar cells, while 20% efficiency is considered as most significant for high effectiveness of solar cells. High power conversion efficiency (PCE) quest of PV cells have favorable circumstances in implementation however are frequently offensive for minimal effort generation due to large structures, the lengthy assembling procedures compulsory for manufacturing [25]. High energy efficiency and cost effectiveness must be accomplished all through the improvement of cutting edge generation innovations and gear, and the absolute most recent advances that could prompt efficiencies of more noteworthy than 25% and industrially practical creation costs are evaluated [23]. Due to all these qualities the silicon based materials are considered as the backbone of the PV technology. As for the last few decades, the production cost of the PV cell decreased but monocrystalline silicon solar cells are still costly, owing to the thicker silicon wafer which is used as a substrate. Also the wafer silicon is the most expensive step due to kerf losses in addition to grainy stainless steel wire. This leads to expansive PV technology but monocrystalline efficiency and stability is up to mark due to which it is still utilized [26]. New techniques and materials are utilized to control silicon wafer thickness.

2. Typical structure of monocrystalline Si solar cell

Monocrystalline silicon solar cells possess single crystalline wafers of silicon of superior quality, usually developed by CZ process in the form of ingot. The figure 2 represents the typical arrangement of screen printed Al-back surface field (BSF) silicon solar cell [27]. Antireflection coating (ARC) layer is placed in the form of pyramid structures on the frontal of the solar cell, which lessens the reflection of light. Silicon nitride (SiN_x) and silicon dioxide (SiO_2) has major contribution in the formation of this ARC layer. The p-doped substrate act as a substrate, upon which n+-doped layer is placed, to form a p-n junction. A profoundly doped layer of BSF is shaped at the posterior to overcome the photo-generated electrons recombination by the creation of field effect passivation. Aluminum is placed for rear contacting with the growth in Si BSF layer, which form alloy of Al-Si at eutectic temperature, where Al acts as p+ acceptor. Silver paste which contain gridlines unified by busbars are placed at the anterior to form front contacts via screen printing. The electrical contact between the Ag at the front and n+-doped layer of Si below it is established by pervading the Ag contact along the ARC layer. The 1-D current pattern that flow in the base region of the Al-BSF cell can convey the higher fill factor (FF) [5, 27].

Monocrystalline solar cells face varying difficulties during their productions such as difference in efficiency as losses due to recombination of charge carriers. The crucial technical challenges related to the Mono-Si solar cell along with their solutions are enlisted in the table 2 [5].

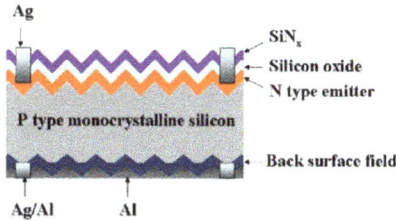

Figure 2 Structure of screen-printed Al-BSF mono-crystalline Si solar cell [27].

Table 2 Challenges associated with mono-crystalline Si solar cell and their solutions [5].

Challenges	Solutions
Theoretically, predicted efficiency according to Shockley Queisser is 30% .	Currently, the Efficiency obtained for Mono-crystalline solar cells is 26.6%.
A Significant Contrast between the module (19 – 22.5%) and cell 26.6% efficiency. The Thicker Si as a substrate leads to thread for PV growth.	a. By Efficient Developing Contact schemes. b. Minimize the optical losses, for this purpose different nanostructure, Plasmon's, optimized ARC coatings and texturization coatings are incorporated.
Loss of photons due to front side reflections and transmissions through cell.	a. Front side surface is texturized to minimize the Reflections b. Front side down shift which utilized the Core or quantum dots for Conversion of UV-light to Visible light. c. To reduce shading effects, back contact schemes are used. d. Utilization of the back surface reflectors.
Losses due to Recombination at interface and within the Bulk.	a. By employing Back side surface passivation layer and front side layers b. By optimizing the carrier's lifetime and their doping level. c. By highly doping of the BSF at rear side.
Electrical losses Occurred due to Busbars and grids as a result of their resistance.	a. By highly doping of the emitter region for reducing the resistance arisen at front side. b. By employing Narrow gridline at front side.

Materials for Solar Cell Technologies I   Materials Research Forum LLC
Materials Research Foundations **88** (2021) 148-175   https://doi.org/10.21741/9781644901090-6

3.   Advanced monocrystalline solar cells

3.1   Passivated emitter rear localized (PERL) solar cell

The figure 3(a) displays the structure of PERL solar cell which has 25% efficiency [5]. In these devices, the rear and front passivation layers are usually made up of double ARC coating layers, thermally grown oxide layers as well as textured layers with float zone (FZ) Si wafers of high quality and thickness of 450-μm [28]. The main advantage of this cell over ordinary Si cell is the possible reflection of light by thermally grown oxide layer of 20 nm along with enhanced surface passivation. Through this technique, the recombination at surface can be minimized with the increment in the lifetime of minority charge carriers. Almost 25.7 % efficiency of this cell with n-type silicon has been reported. The figure 3(b) illustrates the passivated contacts which have been formed by tunnel oxide thin layer topped with doped silicon layer at the rearmost [5]. Through this, the majority charge carriers are enabled to pass, thus result in the prevention of flow of lateral current in the bulk silicon [29]. Higher FF of 83.3% has been obtained, when joined with lowest resistive wafer. Though, it is highly efficient but much costly, attributable to the prerequisite of photolithography steps along with the edge losses controlling. Its performance can be enhanced by employing bifacial alignment and using the technique of low thermal ion-implementation [30].

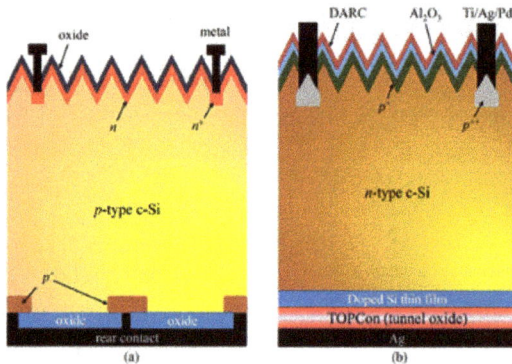

Figure 3 Schematic presentation of (a) PERL solar cell (b) n-type Si solar cell with boron doped emitter and passivating rear contact [5].

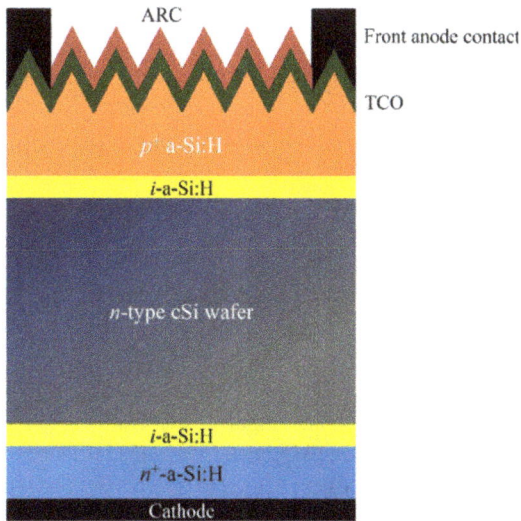

Figure 4 Heterojunction with intrinsic thin layer (HIT) solar cell technology [5].

3.2   Heterojunction with intrinsic thin layer (HIT) solar cell

Silicon heterojunction (SHJ) along with the intrinsic thin layered structure is displayed in the figure 4 [5]. It consists of an n-type CZ Si hetero-junction which is inserted between the pure and doped a-Si: H thin layers. This texture of the front surface helps in decreasing the possibility of reflection of photon incident upon it. A layer of Indium Tin Oxide (ITO) which is transparent conducting oxide (TCO) covers the surface of i- and p-type a-Si: H layers. The resistance of heavily doped a-Si: H is minimized by TCO which acts as contact resistance layer as well as transport the charge carriers through it. A -Si: H possess wide bandgap due to which hetero interface is induced which resulted in minimizing carrier's recombination and increased value of V_{oc}. The rear contact can be patterned for device formation and it also supports in reacting to the light from all directions. The appropriate managing of optical path can help in controlling the reduction in the value of current density J_{sc} which occur mainly due to the shadow loss at front electrode.

The best efficiency for HIT is considered as 24.7% [31]. The ARC layer can be substituted by nanowires and nanoparticles, which enhance the chances of light absorption. With ZnO based Si nano cell, efficiency is reported to be 20% with almost 15% addition of magnesium in it [32]. The theoretical efficiency of ZnMgO/ZnO has been reported as 37.84% while ZnO based solar cells have around 21.23%. The ZnMgO/ ZnO interface helps in increasing the value of J_{sc} along with the decrement in the defects states, with wide bandgap and low index of refraction of MgZnO window layer [33].

4. Experimental and theoretical power conversion efficiency (PCE)

William Shockley and Hans Queisser were two theoretical researchers who firstly calculated the efficiency of the solar cell in 1961 and now for an ideal solar cell it is recognized as Shockely-Queisser (SQ) limit [34].

Theoretically, the parameter to determine the solar cell efficiency is the semiconductor's band gap [34,35]. More photons absorbed by the low band gap semiconductors and for such semiconductor material the current intensity I_{sc} is higher and value of voltage V_{oc} is lowered [36]. But for wide band gap semiconductor materials the value of I_{sc} is lowered and V_{oc} is increased. For maximum output power and efficiency, a detalied study is implemented on low and high band gap materials in [35, 37] There are number of factors on which the solar cell efficincy depends such as, temprature of cell, energy conversion capabilities and maximum power point track [38].

The optimized band energy is obtained by the tradeoff between high value of V_{OC} and low I_{SC} and its obtained value is 1.1eV. p-n based solar cell, having a gap of 1.1 eV has an ideal hole mobility due to higher V_{OC} and lower I_{cs} [36]. The hole mobility of SQ limit is 30% for the silicon solar cells having band gap of 1.1eV [34]. For mono-crystalline solar cells the cell productivity is 25% and for multi crystalline silicon based solar cells it is 20%. For commercialized silicon solar cells, the best efficiency at phone level is 23% and for module level it is said to be 18-24%. The current trend in photovoltaic industry is to reduce the gap between the recorded efficiency and commercial efficiency [39].

4.1 Efficiency of monocrystalline solar cells

The efficiency of monocrystalline solar cells depends upon two types of mechanisms one is optical losses and the second is carrier losses. Optical losses are attributed to reflective surface or less absorption light. When visible light is incident on solar cell, then its interaction with material can be described using the refractive index, where n is real representing phase velocity and k is refractive index's complex part representing extinction coefficient (light intensity's loss) and equation of it is given as follows;

$$\vec{n_c} = \vec{n} + i\vec{k} \tag{1.1}$$

The reflective light is described as follows

Materials Research Forum LLC
https://doi.org/10.21741/9781644901090-6

$$(\overline{n+1})^2 + \overline{k}^2 \tag{1.2}$$

In above 1.1 and 1.2, n represents refractive index having value 1 for air.

For visible light, refractivity mainly depends on refractive index's real part, so 1.2 can be written as follows,

$$R = \frac{(\overline{n-1})^2}{(\overline{n+1})^2} \tag{1.3}$$

There are many ways to reduce the optical losses such as antireflection coating, texturizing of surface, back surface reflector which helps to trap light into cells. Yablonovitch et al. [40], described a limit length of absorption which is as a standard efficiency parameter for measuring the absorption of light by a cell, by using this engineering of optical properties, the losses can be minimized.

Efficiency of silicon based solar cell is greatly affected as a result of losses of carrier recombination due to which short circuit's current and open circuit's voltage is disturbed. Average length between carrier generation and recombination is known as diffusion length and described as

$$L_d = \sqrt{D\tau_{bulk}} \tag{1.4}$$

Here τ_{bulk} represents carrier lifetime, D diffusivity and is described as

$$D = \frac{kT}{q}\mu \tag{1.5}$$

In 1.5, k represents Boltzmann constant, q minority carrier's charge, T temperature, μ mobility.

Solar cells have non-linear relationship between the voltage and current due to temperature, force or intensity of incident light. The I-V relation for case of regular sunlight based solar cell is represented in figure 5 [41]. In ideal case, the current–voltage conduct pursues exponential link named diode condition is given as:

$$I_{dark}(V) = I_0\left(e^{\frac{qV}{kT}} - 1\right) \tag{1.6}$$

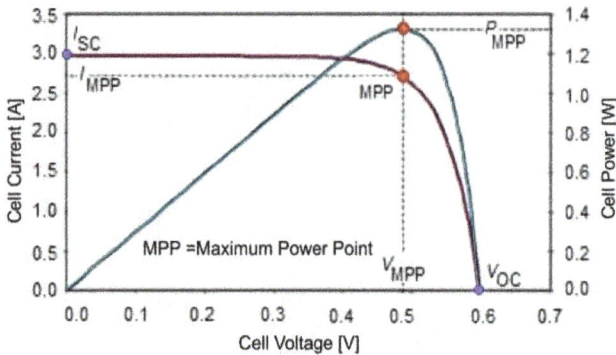

Figure 5 V-I relation of solar cells [41]

$$FF = \frac{P_{max}[W]}{I_{sc}[A] \times V_{oc}[V]} ; (P_{max}[W] = I_{mp}[A] \times V_{mp}[V]) \qquad (1.7)$$

$$\eta[\%] = \frac{P_{max}[W] \times 100}{1,000[Wm^{-2}] \times Cell\ area[m^2]} \qquad (1.8)$$

Short circuit current is represented by I_{sc} and open circuit voltage is represented by V_{oc}, FF and η both are used to represent the fill factor and efficiency respectively. η (Efficiency) is the most important factor for measuring or comparing the performance efficiency of different solar cells.

For obtaining high efficiency of solar cell, carriers which are generated by photo-excitation process must be moved from generation to collector. Due to which they have low recombination rate G, large short circuit current density (J_{SC}), and diffusion length L_d is increased [42] and given as

$$J_{SC} = qGL_d \qquad (1.9)$$

Diffusion length and short circuit current density (SCCD) both have direct relation. For a particular cell, if diffusion length [43] is less as compared to thickness, then corresponding SCCD will decrease. So, SCCD can be increased by two ways either by increasing the diffusion length or by decreasing the thickness of solar cell's absorber

Materials for Solar Cell Technologies I

Materials Research Forum LLC

Materials Research Foundations **88** (2021) 148-175

https://doi.org/10.21741/9781644901090-6

layer. If DL is much greater comparative to thickness the SCCD gains a saturation limit. The relationship between saturation SCCD represented by J_o and DL is given as

$$J_o = \frac{qDn_i^2}{L_d N}$$

(1.10)

In 1.10, n_i represents carrier concentration, N represents concentration of doping. The open circuit voltage is given as follows

$$V_{OC} = \frac{kT}{q} \ln\left(\frac{J_{sc}}{J_o} + 1\right)$$

(1.11)

By using 1.11, we get as

$$V_{OC} \propto \ln(L_d)$$

(1.12)

And further we get

$$\eta \propto J_{SC} V_{OC} \propto L_d \, ln(L_d)$$

(1.13)

Efficiency of solar cell and diffusion length both have direct relation but when diffusion length is much greater than solar cell's thickness, then this linear relation does not hold. Lifetime of charge carriers is greatly influenced via defects of solar cell (Bulk).

When light falls on the cell then external quantum efficiency (EQF) is used to measure generated carrier concentration (Numbers). For monocrystalline silicon solar cell, graph between the EQF and wavelength is given in figure 6 [44]. The EQE is given as follows

$$EQE_\lambda = \text{electrons s}^{-1}/\text{photons s}^{-1} = 1240 \times J_\lambda/(P_\lambda \times \lambda)$$

(1.14)

Here SCCD at a particular wavelength is represented by J_λ and Optical power density of incident light is represented by P_λ [45].

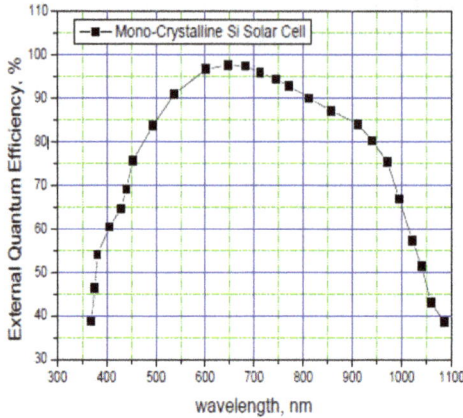

Figure 6 Graph between the quantum effiicncy and wavelength of monocrystalline silicon soalr cell [44].

Proficient PCE is a significant issue in light of the fact that the productivity impacts the whole value-chain expenditure of the PV framework [37,46], from material generation to framework fitting. The more effective a sunlight based solar cell is, the more financially savvy it is to meet the present developing vitality requests [47]. Probably the earliest PV modules utilized oblique wafers with no material between them, bringing about a low pressing thickness and, so losing force from those territories cut down the proficiency of the modules. These round wafers have been supplanted with pseudo-square wafers, which have expanded the pressing thickness and, subsequently, expanded the effectiveness of the modules [48].

5.   Monocrystalline solar cell PCE enhancement methods

5.1   Optical losses reduction methods

5.1.1 Texturing

The wafer surface is texturized which causes multiple reflections or optical path length and enhance the probability of trapping the photon of light on surface. Ideally, solar cells absorb light in two regions visible and ultraviolet before captured by rare side of cell. Solar efficacy limits as consequences of optical losses which are caused due to high reflection and low absorption in short and long wavelength regions. The silicon based solar cells efficiency can be improved by increasing the light trapping of photon of longer wavelength by surface texturizing process [2, 49-51].

The features such as high roughness and uniform surface coverage are achieved by surface texturization. Basically, the texturization of silicon solar cells is done by using two methods one is chemical another is a physical method [51].

5.1.1.1 Chemical method

In the chemical method acidic or alkaline solutions are used and isotropic wet etching is done by them which creates small micro pyramids on the surface monocrystalline solar cells (Mc − Si) 'wafer are texturized by using solution of NaOH or KOH as a consequence each orientation plane has different etching rates [51-54].

In Mc − Si texturizing process the acidic solution HCl and HF used for cleaning process and metal impurities are removed from the surface. By this process, a reduction of reflection about 35% to 12% is obtained and uniformity is also improved [52,55].

5.1.1.2 Physical method

In physical method, the dry processes are mostly used including laser ablation, mechanical grooving. In mono-crystalline and Mc − Si wafers texturization is achieved mostly by mechanical grooving due to its high texturization rate and less cost [56-60]. In this method, structural printing is achieved by using V-grooved metal blade including a diamond coating and water is used as a cooling agent. The spin speed is 150– 160 mms^{-1}, but can be varied depending upon diamond's particles grain size. After this defect free grooves are obtained.

In reactive ion etching [61] process highly reactive ion plasma are used for etching the wafer surface. It includes tetrafluoromethane or a combination of gases which acts like a reactive gas. By using RIE defined pattern(less than 1 μm), high uniform features are obtained. In mono-crystalline and Mc hi wafers texturization is done by RIE. Unlike wet etching process, RIE cannot depend upon crystallographic orientation and by RIE absorption is increased compared to other methods.

In laser ablation process, texturization is achieved by melting and evaporation of wafer surface under high pulsed laser which generates radiations. The benefit of this technique is high accuracy and defined patterns are obtained. The limitations of this technique are that during melting of silicon other material are also deposited on wafer surface [51]. Hence surface texturization is realized to remove residue which causes to increase recombination process [51, 62].

5.1.1.3 Pyramidal texture uniformity

In case of mono-crystalline silicon solar cells (MCSSC), for improving the power conversion efficiency (PCE), screen printing and chemical processes are being utilized, that plays a significant role for enhancement in the surface pyramidal texture as well as optimizing the texturization. A team of researchers attained cell processing parameters primes to reduce the light reflectivity by proper adjustment of concentration, temperature [63]. Chen at al. [64] performed experiments for uniform pyramidal texturization by alkaline solution which significantly improves the light absorption and performance of the cell. Wang et al. [63] performed experiments by utilizing a mixture of tetramethyl ammonium hydroxide (TMAH) and isopropanol (IPA) which significantly enhance the MCSSC power conversion efficiency [63]. Different studies have been conducted on the texturizing surface uniformity which depicts that, it depends upon the type of etching solution and its concertation as well as etching temperature and time [65]. Xu [67] purposed uniformity coefficient (UC) to characterize the uniformity of pyramidal texture. The uniformity coefficient (UC) is in direct relation with the PCE as well as the short circuit current and has an inverse relation with reflectivity. The maximum value of the UC for MCSSC obtained is 0.82, the reflectivity also decreased reaching minimum value of 1.2% and PCE can be enhanced up to 20.46% [68].

5.1.2 Anti-reflecting coating (ARC)

When light falls directly on the surface of silicon then about 30% of light is reflected back due to variance in refractive index of Si and air [69-71]. An ARC is a dielectric thin-film coating which is utilized in optical devices having surface reflectivity in a designated wavelength region [72, 73]. Figure 7 [74] represents the reflectance and variance in refractive index.

The basic principle of ARC is cancelation of reflected waves by destructive interference phenomenon through optical interfaces. The reflection through a surface is large, if the difference of refractive index of two materials is large. The wave reflected by the silicon surface is cancelled only if other wave is of same magnitude, wavelength but out of Phase. For this purpose, another reflected wave at different location of silicon surface is required in such a way that both waves are out of phase(half − wavelength). This can be possible only if the reflective surfaces are at quarter-wavelength far from each other. Hence the other name acquired by ARC is "quarter-wave" coatings. At a certain wavelength, minimum reflections can be obtained only be suitable choice of layer thickness and dielectric.

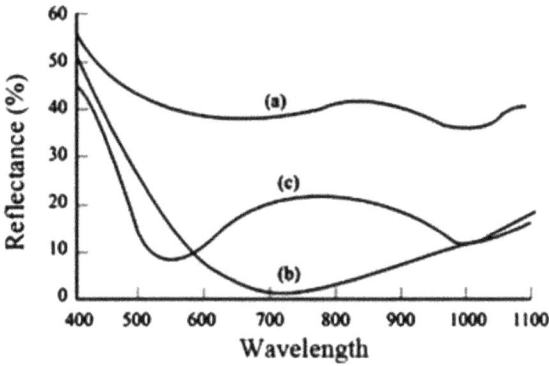

Figure 7 Graphical representation of reflectance and wavelength (a) Without ARC (b) single layer ARC (c) With double layer ARC [74].

For effective ARC, the optical thickness can be obtained by using following formula

$$n_1 d = \lambda o/4 \tag{1.15}$$

λo, Represents peak value of wavelength of photon spectrum and n_1 *refractive index of material which is being coated.*

In single layer coating, the light is reflected at low rate as compared to the reflectance of light when the cell is uncoated. Hence if we need more reflection the more coating is required. The reflected index of upper ARC layer is greater than original layer. Here n_0, n_1, n_2, n_3 are the reflective index of air, solar cells and 1st and 2nd ARC thin film layer. When multiple coating can be applied then 99% transmission of light can be obtained. By applying multiple coating, the cell works more efficiently. Hence, this coating can reduce the price of module.

Titanium dioxide and silicon dioxide are the two chemicals which have been used in the coating of cells but now silicon nitride is used as standard ARC, on industrial scale. TiO_2 lacks both surface and bulk property [75]. Using TiO_2 as coating in solar cell does not give wide range of rate of reflection. On the other hand SiO_2 doesn't have sufficient surface properties to be employed in bulk such as an ARC in Si solar cells [76-81].

Materials for Solar Cell Technologies I
Materials Research Foundations **88** (2021) 148-175

Materials Research Forum LLC
https://doi.org/10.21741/9781644901090-6

Thin film of silicon nitrite (SN) is dumped on the cell by a synthesis technique known as chemical vapor deposition (CVD). In this technique called CVD, reaction starts at low pressure. For deposition of SN thin films on the cell, plasma-enhanced (PECVD) and atmospheric pressure (APCVD) techniques are utilized. The precursors are silicon and ammonia [61, 82]. During the reaction hydrogen is produced, the most of the which transpires, while the rest hydrogen defused in bulk becomes reactive and attacks itself making the bulk long lasting.

By using silicon nitride not only optical losses are reduced but it also enhances the bulk passivation of material. SN deposition by using PECVD can act as surface as well as bulk passivation due to controlled flow of precursor gases and colossal passivation effects. It makes it highly appropriate for defect rich bulk passivation of Mc − Si.SN deposition by PECVD or sputtering thin films both have same properties of surface and bulk passivation [52, 83]. The APCVD is typically employed in microelectronic [84-86].

5.1.3 Metallization

For preventing back diffusion and extracting ions form solar cell different contacts are used. Screen printing (SP) is a method which is used for forming the back and front contacts [87]. The contacts formed by screen printing must have low contact resistance, high aspect ratio and act as a good adhesive to silicon cells.

Commercially, the net throughput of SP 1000 and 2000 wafers h^{-1} is obtained by Single and double lines. The most suitable material for front side contact is silver while for back contact aluminum is used due to its good ohmic properties in case of monocrystalline solar cells [52].

5.2 Thickness of the monocrystalline solar

Reduction in cost of the solar modules (per Wp) is done by reducing cost of manufacturing and by increasing efficiency of the solar cell. To improve solar cell efficiency, back-surface field (BSF) is used and material wafer-thickness and properties of emitter are taken under consideration [88]. According to the theoretical studies, BSD effect on thin solar cells is 100 μm. In thin silicon solar cells, BSE plays basic role in improving efficiency. Silicon wafer thickness is what makes a solar cell costly [89]. The thickness of the monocrystalline solar cell wafer is reduced because silicon wafer contributes 50-60% of total cost. The thickness of the typical wafer is 300 μm and nowadays the 200 μm thickness based solar cells are being developed [90]. At industrial level, during the manufacturing process grid metallization is used which increases the front side and backside efficiency collectively 50% and for (100-150 μm) thin solar cells and 10-15% power out gain [25]. Efficiency of monocrystalline sun based cells is increased by decreasing thickness. Utilizing determined and calculated affirmed yields, without using BSF in assembling process is about 14.5% effective. Using the BSF, by decreasing the thickness to (250 μm and 150μm) the efficiency can be increased up to 16% [36]. Additionally, the utilization of a BSF brings about an improvement of about 10% moderately in efficiency. Utilizing the basic model for the normal yields, a decrease in wafer thickness of around 50 μm brings about a cost decrease for each Wp of about 5%. When cell thickness reduces from 300 μm to 150 μm along with utilization of a BSF, a cost decrement of about 10% is acquired. With indistinguishable yields and 250 μm thick solar cells, the expense decrease of 20% was determined cost effectiveness is very significant to keep up the handling of yield for extra thin cells. Other than the investment funds in material, the decrease in wafer thickness likewise gives a higher assembling limit in the wafer generation, which is hard to take into genuine quantitative thought [88].

Conclusions

The world is running out of energy sources to meet rising energy demands. It is high time that all economies adopt sustainable and renewable energy sources. The photovoltaic industry is a major contributor in world's energy production. Solar cells have dominated the photovoltaic industry for decades now due to silcon's abundant and non-toxic nature. Solar cells have three generations, out of which the most widely commercialized solar cells are silicon based. Monocrystalline silicon solar cells (MCSSC) have dominated commercialized PV technology due to their high stability, efficiency, large life time, high carrier mobility, homogenous structure. Cost effectiveness and high power conversion efficiency is the main target of the solar industry. Silcon's wafer thickness is what makes the solar cells costly, tailoring the thickness with different methods and factors are deliberated in order to enhance the MCSSC efficiency, to break theoretical efficiency limit of 30% as purposed as Schokley–Quiesser (SQ). The experimentally obtained efficiency achieved is 26.6% which is less that SQ limit that indicated more room for improvement. The implementation of ARC and BSF are proposed to minimize reflection and electron-hole recombination process for enhancement of power conversion efficiency such as in In PERL solar cells, a double layer of ARC coating as well as rear and front side passivation surface layers are used which effectively reduced the electron-hole recombination as well as enhanced lifetime of the charge carriers. PERL solar cells have a higher efficiency of 25.7 % and fill factor of 83.3% and HIT solar cells have efficiency of 26.6% and FF is 83.8%. There are a number of parameters that are suggested to influence solar cell manufacturing includes material cost, efficiency, stability, throughput, output production yield. Production cost of solar cells can be increased by reducing optical losses through the implementation of ARC coating or texturing and

Materials for Solar Cell Technologies I | Materials Research Forum LLC
Materials Research Foundations **88** (2021) 148-175 | https://doi.org/10.21741/9781644901090-6

controlling carrier losses as well as the implementation of BSF. The monocrystalline solar cell is a vast field and still has room for further improvement, this deliberation is a motivation for new methods and materials to be employed in order to make monocrystalline solar cells economical, stable and effective for energy production.

Acknowledgements

All the authors appreciate the support from Department of Physics, University of Gujrat, Gujrat-50700, Pakistan and National Institute for Biotechnology & Genetic Engineering (NIBGE), Faisalabad-38000, and Pakistan for providing healthy and conducive environment for such work.

References

[1] W.B. Stine, R.W. Harrigan, Solar energy fundamentals and design, (1985) 536

[2] P.M. Ushasree, B. Bora, Silicon solar cells, R. Soc. Chem. (2019)1-55. https://doi.org/10.1039/9781788013512-00001

[3] N. Panwar, S. Kaushik, S. Kothari, Role of renewable energy sources in environmental protection: a review, Renew. Sust. Energ. Rev. 15 (2011) 1513-1524. https://doi.org/10.1016/j.rser.2010.11.037

[4] P.P. Barker, J.M. Bing, Advances in solar photovoltaic technology: an applications perspective, IEEE Power Engineering Society General Meeting, IEEE. (2005)1955-1960.https://doi.org/10.1109/PES.2005.1489304

[5] H. Mehmood, T. Tauqeer, S. Hussain, Recent progress in silicon based solid state solar cells, Int. J. Electron. 105 (2018) 1568-1582.https://doi.org/10.1080/00207217.2018.1477191

[6] H. Chen, T.N. Cong, W. Yang, C. Tan, Y. Li, Y. Ding,Progress in electrical energy storage system: a critical review, Prog. Nat. Sci. 19 (2009) 291-312.https://doi.org/10.1016/j.pnsc.2008.07.014

[7] A. McEvoy, T. Markvart, L. Castaner, Solar cells: materials, manufacture and operation, Solar cells: materials, manufacture and operation, Academic Press. (2012)

[8] A. Fahrenbruch, R. Bube, Fundamentals of solar cells: photovoltaic solar energy conversion, Elsevier. (2012)

[9] M. Bertolli, Solar cell materials, Course: Solid State II. Department of Physics, University of Tennessee, Knoxville. (2008)

[10] K.H. Raut, H.N. Chopde, D.W. Deshmukh, A review on comparative studies of diverse generation in solar cell, IJEE (2018).https://doi.org/2455-9771

[11] S. Sharma, K.K. Jain, A. Sharma, Solar cells: in research and applications: a review, Mater.Sci. Appl. 6 (2015) 1145-1155. https://doi.org/10.4236/msa.2015.612113

[12] M.A. Green, Polycrystalline silicon on glass for thin film solar cells, Appl. Phys. A. 96 (2009) 153-159. https://doi.org/10.1007/s00339-009-5090-9

[13] R.M. Izatt, S.R. Izatt, R.L. Bruening, N.E. Izatt, B.A. Moyer, Challenges to achievement of metal sustainability in our high-tech society, Chem. Soc. Rev. 43 (2014) 2451-2475. https://doi.org/10.1039/C3CS60440C

[14] J. Narayan, S. Kim, K. Vedam, R. Manukonda, Formation and nondestructive characterization of ion implanted silicononinsulator layers, Appl. Phys. Lett. 51 (1987) 343-345.https://doi.org/10.1063/1.98435

[15] M.A. Green, High efficiency silicon solar cells, Seventh EC Photovoltaic Solar Energy Conference, Springer, (1987), pp. 681-687. Https://doi.org/10.1007/978-94-009-3817-5-121

[16] R.N. Castellano, Solar panel processing, Archives contemporaines. (2010)

[17] A. Yadav, P. Kumar, M. RPSGOI, Enhancement in efficiency of PV cell through P&O algorithm, IJTRE. 2 (2015) 2642-2644. https://doi.org/10.4236/msa.2015.612113

[18] B. Srinivas, S. Balaji, M. Nagendra Babu, Y. Reddy, Srinivas, B., S. Balaji, Review on present and advance materials for solar cells, . Int. J. Eng. Res. 3 (2015) 178-182

[19] P. Würfel, U. Würfel, Physics of solar cells: from basic principles to advanced concepts, John Wiley &Sons.(2016) 122-288

[20] N. ElAtab, N. Qaiser, R. Bahabry, M.M. Hussain, Corrugation enabled asymmetrically ultrastretchable (95%) monocrystalline silicon solar cells with high efficiency (19%), Adv. Energy Mater.(2019) 1-7. https://doi.org/10.1002/aenm.201902883

[21] K. Wang, D. Yang, C. Wu, J. Shapter, S. Priya, Monocrystalline perovskite photovoltaics toward ultrahigh efficiency?, Joule. 3 (2019) 311-316.https://doi.org/10.1016/j.joule.2018.11.009

[22] A. Goetzberger, J. Knobloch, B. Voss, Crystalline silicon solar cells, New York (1998) 114-118

[23] M.A. Green, The path to 25% silicon solar cell efficiency: history of silicon cell evolution, Prog. Photovolt. 17 (2009) 183-189. https://doi.org/10.1002/pip.892

[24] D.I. Paul, M. Smyth, Enhancing the performance of a building integrated compound parabolic photovoltaic concentrator using a hybrid photovoltaic cell, Int. J. Renew. Energy Technol. 11 (2020) 49-69.https://doi.org/10.1504/IJRET.2020.106519

[25] L. Yang, Q. Ye, A. Ebong, W. Song, G. Zhang, J. Wang, et al., High efficiency screen printed bifacial solar cells on monocrystalline CZ silicon, Prog. Photovolt. 19 (2011) 275-279.https://doi.org/10.1002/pip.1018

[26] J. Liu, Y. Yao, S. Xiao, X. Gu, Review of status developments of high efficiency crystalline silicon solar cells,J. Phys. D appl. Phys. 51 (2018) 123001.https://doi.org/10.1088/1361-6463/aaac6d

[27] C.P. Liu, M.W. Chang, C.L. Chuang, Effect of rapid thermal oxidation on structure and photoelectronic properties of silicon oxide in monocrystalline silicon solar cells, Curr. Appl. Phys. 14 (2014) 653-658.https://doi.org/10.1016/j.cap.2014.02.017

[28] J. Zhao, A. Wang, M.A. Green, 24·5%Efficiency silicon pert cells on mcz substrates and 24·7% efficiency perl cells on fz substrates, Prog. Photovolt. 7 (1999) 471-474.https://doi.org/10.1002/(SICI)1099-159X(199911/12)7:6<471::AID-PIP298>3.0.CO;2-7

[29] A. Richter, J. Benick, F. Feldmann, A. Fell, M. Hermle, S.W. Glunz, Ntype si solar cells with passivating electron contact: identifying sources for efficiency limitations by wafer thickness and resistivity variation, Sol. Energy Mater. Sol. Cells. 173 (2017) 96-105.https://doi.org/10.1016/j.solmat.2017.05.042

[30] D.D. Smith, P.J. Cousins, A. Masad, S. Westerberg, M. Defensor, R. Ilaw, et al., Sunpower's maxeon gen iii solar cell: high efficiency and energy yield,2013 IEEE 39th Photovoltaic Specialists Conference (PVSC), IEEE, 2013, pp. 0908-0913.https://doi.org/10.1109/PVSC.2013.6744291

[31] M. Taguchi, A. Yano, S. Tohoda, K. Matsuyama, Y. Nakamura, T. Nishiwaki, et al., 24.7% record efficiency HIT solar cell on thin silicon wafer, IEEE J.Photovolt. 4 (2013) 96-99.https://doi.org/10.1109/JPHOTOV.2013.2282737

[32] S. Maity, C. Bhunia, P. Sahu, Improvement in optical and structural properties of ZnO thin film through hexagonal nanopillar formation to improve the efficiency of a Si–ZnO heterojunction solar cell, J. Phys. D.appl. Phys.49 (2016) 205104.10.1088/0022-3727/49/20/205104

[33] D. Muchahary, S. Maity, High-efficiency thin film ZnMgO/ZnO solar cell simulation approach: Temperature dependency, BSF and efficient small signal analysis, Superlattice. Microst. 109 (2017) 209-216.https://doi.org/0.1016/j.spmi.2017.05.012

[34] T. Markvart, L. Castañer, Solar cells: materials, manufacture and operation/edited by tom markvart and luis castañer, Elsevier J. Adv. Tech.(2010)

[35] S.W. Glunz, High-efficiency crystalline silicon solar cells, Adv. Opt. Photonics. (2007)15. https://doi.org/10.1155/2007/97370

[36] T. Saga, Advances in crystalline silicon solar cell technology for industrial mass production, npg asia mater. 2 (2010) 96-102.https://doi.org/10.1038/asiamat.2010.82Saga, T., Advances in crystalline silicon solar cell technology for industrial mass production. NPG Asia Mater. 2 (2010) 96-102. https://doi.org/10.1038/asiamat.2010.82

[37] A.R. Whitson, H.L. Walster, Soil fertility, Webb Publishing Company. (1912)

[38] D. Furkan, M. Mehmet Emin, Critical factors that affecting efficiency of solar cells, SGRE. 1(2010) 1-4. https://doi.org/10.4236/sgre.2010.11007

[39] M.A. Green, Y. Hishikawa, W. Warta, E.D. Dunlop, D.H. Levi, J. Hohl-Ebinger, et al., Solar cell efficiency tables (version 50), Prog. Photovolt.25 (2017) 668-676. https://doi.org/10.1002/pip.2909

[40] E. Yablonovitch, G.D. Cody, Intensity enhancement in textured optical sheets for solar cells, IEEE
Trans. Electron Devices. 29 (1982) 300-305.https://doi.org/10.1109/T-ED.1982.20700

[41] J.A. Ramos Hernanz, J. Campayo, E. Zulueta, O. Barambones, P. Eguía, I. Zamora, Obtaining the characteristics curves of a photocell by different methods, IJRET.1(2013)1-6 https://doi.org/10.24084/repqj11.455

[42] M.A. Green, Solar cells: operating principles, technology, and system applications, Englewood Cliffs NJ. Prentice Hall Inc. (1982)288

[43] T. Rendler, J. Neburkova, O. Zemek, J. Kotek, A. Zappe, Z. Chu, et al., Optical imaging of localized chemical events using programmable diamond quantum nanosensors , Nat. Commun. 8 (2017) 1-9. https://doi.org/10.1038/ncomms14701

Materials Research Forum LLC
https://doi.org/10.21741/9781644901090-6

[44] A.A. El Maksood, Performance dependence of (IV) and (CV) for solar cells on environmental conditions, (2018) 5329-5349. https://doi.org/10.24297/jap.v14i1.7315

[45] R.P. Smith, A.A.C. Hwang, T. Beetz, E. Helgren, Introduction to semiconductor processing: fabrication and characterization of pn junction silicon solar cells, Am. J. Phys. 86 (2018) 740-746. https://doi.org/10.1155/2019/7945683

[46] L. Zhao, Y. Zuo, C. Zhou, H. Li, H. Diao, W. Wang, Theoretical investigation on the absorption enhancement of the crystalline silicon solar cells by pyramid texture coated with SiNx: H layer, Elsevier Sol. energy 85 (2011) 530-537. https://doi.org/10.24084/repqj11.455

[47] N. Kannan, D. Vakeesan, Solar energy for future world: a review, renew. Sust. Energ. Rev. 62 (2016) 1092-1105.https://doi.org/10.1016/j.rser.2016.05.022

[48] L. Lalouat, H. Ding, B. Gonzalez-Acevedo, A. Harouri, R. Orobtchouk, V. Depauw, et al., Pseudo disordered structures for light trapping improvement in mono crystalline Si thin films, Sol.Energ. Mat. Sol. C. 159 (2017) 649-656. https://doi.org/10.1016/j.solmat.2016.04.031

[49] Y. Jiang, I. Almansouri, S. Huang, T. Young, Y. Li, Y. Peng, et al., Optical analysis of perovskite/silicon tandem solar cells, J. Mater. Chem. C. 4 (2016) 5679-5689. https://doi.org/10.1039/C6TC01276K

[50] K. Nguyen, D. Abi-Saab, P. Basset, E. Richalot, M. Malak, N. Pavy, et al., Study of black silicon obtained by cryogenic plasma etching: approach to achieve the hot spot of a thermoelectric energy harvester, Microsyst. Technol. 18 (2012) 1807-1814. https://doi.org/10.1007/s00542-012-1486-0

[51] M. Abdullah, M. Alghoul, H. Naser, N. Asim, S. Ahmadi, B. Yatim, et al., Research and development efforts on texturization to reduce the optical losses at front surface of silicon solar cell, Renew. Sust. Energ. Rev. 66 (2016) 380-398. https://doi.org/10.1016/j.rser.2016.07.065

[52] D.H. Neuhaus, A. Münzer, Industrial silicon wafer solar cells, Adv. Opt. Photonics. (2007)1-15. https://doi.org/10.1155/2007/24521

[53] U. Gangopadhyay, K. Kim, A. Kandol, J. Yi, H. Saha, Role of hydrazine monohydrate during texturization of large area crystalline silicon solar cell fabrication, Sol. Energ. Mat. Sol. C. 90 (2006) 3094-3101. https://doi.org/10.1016/j.solmat.2006.06.014

[54] H. Saha, S.K. Datta, K. Mukhopadhyay, S. Banerjee, M. Mukherjee, Influence of surface texturization on the light trapping and spectral response of silicon solar cells, IEEE T. Electron dev. 39 (1992) 1100-1107. https://doi.org/10.1109/16.129089

[55] P. Cousins, D. Smith, H. Luan, J. Manning, T. Dennis, A. Waldhauer, et al., 35th IEEE Photovoltaic Specialists Conference (PVSC 2010), June 20–25, 2010, Honolulu, Hawaii, USA (2010) 000275

[56] I. Zubel, M. Kramkowska, The effect of isopropyl alcohol on etching rate and roughness of (1 0 0) Si surface etched in KOH and TMAH solutions, Sensor Actuat. A.phys. 93 (2001) 138-147. https://doi.org/10.1016/S0924-4247(01)00648-3

[57] K. Araki , M.Yamaguchi,An Si concentrator cell by single photolithography process, Sol. Energy Mater. Sol. Cells.65(2001)437-443.https://doi.org/10.1016/S0927-0248(00)00124-0

[58] P. Fath, C. Borst, C. Zechner, E. Bucher, G. Willeke, S. Narayanan, Progress in a novel high throughput mechanical texturization technology for highly efficient multicrystalline silicon solar cells, Sol. Energ. Mat. Sol. C. 48 (1997) 229-236. https://doi.org/10.1016/S0927-0248(97)00105-0

[59] L.L. Kazmerski, Photovoltaics: A review of cell and module technologies, Renew. Sust. Energ. Rev 1 (1997) 71-170. https://doi.org/10.1016/S1364-0321(97)00002-6

[60] B. Terheiden, P. Fath, Highly efficient double side mechanically textured novel silicon solar cell concepts, 3rd World Conference onPhotovoltaic Energy Conversion, 2003. Proceedings of, IEEE, (2003)1443-1446

[61] W. Soppe, H. Rieffe, A. Weeber, Bulk and surface passivation of silicon solar cells accomplished by silicon nitride deposited on industrial scale by microwave PECVD, Prog. Photovolt. 13 (2005) 551-569. https://doi.org/10.1002/pip.611

[62] B. Fischer, Loss analysis of crystalline silicon solar cells using photoconductance and quantum efficiency measurements, Cuvillier Göttingen. (2003)

[63] P.K. Basu, D. Sarangi, K.D. Shetty, M.B. Boreland, Liquid silicate additive for alkaline texturing of mono Si wafers to improve process bath lifetime and reduce IPA consumption, Sol. Energy Mater. Sol. Cells 113 (2013) 37-43.https://doi.org/10.1016/j.solmat.2013.01.037

[64] K. Chen, Y. Liu, X. Wang, L. Zhang, X. Su, Novel texturing process for diamond wire sawn single crystalline silicon solar cell, Sol. Energy Mater. Sol. Cells 133 (2015) 148-155.https://doi.org/0.1016/j.solmat.2014.11.016

[65] L. Wang, F. Wang, X. Zhang, N. Wang, Y. Jiang, Q. Hao, et al., Improving efficiency of silicon heterojunction solar cells by surface texturing of silicon wafers using tetramethylammonium hydroxide, J. Power Sources. 268 (2014) 619-624.https://doi.org/10.1016/j.jpowsour.2014.06.088

[66] P.K. Basu, A. Khanna, Z. Hameiri, The effect of front pyramid heights on the efficiency of homogeneously textured inline-diffused screen-printed monocrystalline silicon wafer solar cells, Renew. Energg.78 (2015) 590-598.https://doi.org/10.1016/j.renene.2015.01.058

[67]Z. Xu, X. Xu, C. Cui, H. Huang, A new uniformity coefficient parameter for the quantitative characterization of a textured wafer surface and its relationship with the photovoltaic conversion efficiency of monocrystalline silicon cells, Sol. Energy 191 (2019) 210-218. https://doi.org/10.1016/j.solener.2019.08.028

[68] Z. Fang, Z. Xu, D. Wang, S. Huang, H. Li, The influence of the pyramidal texture uniformity and process optimization on monocrystalline silicon solar cells, J. Mater. Sci. (2020) 1-9.https://doi.org/10.1007/s10854-020-03185-1

[69] H.K. Raut, V.A. Ganesh, A.S. Nair, S. Ramakrishna, Anti reflective coatings: A critical, in depth review, Energy Environ. Sci. 4 (2011) 3779-3804.https://doi.org/10.1039/C1EE01297E

[70] M.A. Green, M.J. Keevers, Optical properties of intrinsic silicon at 300 K, Prog. Photovolt. 3 (1995) 189-192.https://doi.org/10.1002/pip.4670030303

[71] G. Shao, C. Lou, D. Xiao, Enhancing the efficiency of solar cells by down shifting YAG: Ce3+ phosphors, J. Lumin. 157 (2015) 344-348.https://doi.org/10.1016/j.jlumin.2014.08.064

[72] B. Richards, Comparison of TiO2 and other dielectric coatings for buried-contact solar cells: a review, Prog. Photovolt. 12 (2004) 253-281.https://doi.org/10.1002/pip.529

[73] N.S. Beattie, P. See, G. Zoppi, P.M. Ushasree, M. Duchamp, I. Farrer, et al., Quantum engineering of InAs/GaAs quantum dot based intermediate band solar cells, Acs Photonics 4 (2017) 2745-2750.https://doi.org/10.1021/acsphotonics.7b00673

[74] I.G. Kavakli, K. Kantarli, Single and double layer antireflection coatings on silicon, *Turk. J. Phys.*26 (2002) 349-354

[75] A. Rohatgi, P. Doshi, J. Moschner, T. Lauinger, A. Aberle, D. Ruby, Comprehensive study of rapid, low cost silicon surface passivation technologies, IEEE T. Electron dev. 47 (2000) 987-993. https://doi.org/10.1109/16.841230

[76] C. Keavney, V. Haven, S. Vernon, 21st IEEE Photovoltaic Specialists Conference Proc. (1990) 141-144 https://doi.org/10.1109/PVSC.1990.111606

[77] A. Stephens, A. Aberle, M. Green, Surface recombination velocity measurements at the silicon silicon dioxide interface by microwave-detected photoconductance decay, J. Appl. Phys. 76 (1994) 363-370.https://doi.org/10.1063/1.357082

[78] S.W. Glunz, A.B. Sproul, W. Warta, W. Wettling, Injection level dependent recombination velocities at the Si-SiO2 interface for various dopant concentrations, J. Appl. Phys. 75 (1994) 1611-1615. https://doi.org/10.1063/1.356399

[79] N. Bosco, Reliability concerns associated with PV technologies, NREL(2010)

[80] A.G. Aberle, Crystalline silicon solar cells: advanced surface passivation and analysis, Centre for Photovoltaic Engineering. University of New South wales. (1999)

[81] A.G. Aberle , Surface passivation of crystalline silicon solar cells: a review, Prog. Photovolt.8(2000)473-487. https://doi.org/10.1002/1099-159X(200009/10)8:5<473::AID-PIP337>3.0.CO;2-D

[82] V. Sharma, C. Tracy, D. Schroder, S. Herasimenka, W. Dauksher, S. Bowden, Manipulation of K center charge states in silicon nitride films to achieve excellent surface passivation for silicon solar cells, Appl. Phys. Lett. 104 (2014) 053503. https://doi.org/10.1063/1.4863829

[83] G. Bauhuis, P. Mulder, J. Schermer, E. Haverkamp, J. van Deelen, P. Larsen, Proceedings of the 20th European Photovoltaic Solar Energy Conference, (2005) 468-471

[84] A. Sassella, A. Borghesi, S. Rojas, L. Zanotti, Optical properties of CVD-deposited dielectric films for microelectronic devices, J. Phys. Colloq. IV. 5 (1995) C5-843-C845-859.https://doi.org/10.1051/jphyscol:19955100

[85] J. Hong, W. Kessels, W. Soppe, a. W. Weeber, WM Arnoldbik, and M. CM van de Sanden,Influence of the high temperature firing step on high-rate plasma deposited silicon nitride films used as bulk passivating antireflection coatings on silicon solar

Materials for Solar Cell Technologies I

Materials Research Forum LLC

Materials Research Foundations **88** (2021) 148-175

https://doi.org/10.21741/9781644901090-6

cells, J. Vac. Sci. Technol. B Microelectron. Nanom. Struct. 21 (2003) 2123.https://doi.org/10.1116/1.1609481

[86] J. Shingledecker, M. Takeyama, Joint EPRI 123HiMAT International conference on advances in hightemperature materials, ASM International. (2019)

[87] R.G. Loasby, Handbook of thick film technology, Electrochem. Publ. Ltd.(1976)

[88] K.A. Munzer, K.T. Holdermann, R.E. Schlosser, S. Sterk, Thin monocrystalline silicon solar cells, IEEE T. Electron dev. 46 (1999) 2055-2061. https://doi.org/10.1109/16.791996

[89] J.F. Nijs, J. Szlufcik, J. Poortmans, S. Sivoththaman, R.P. Mertens, Advanced manufacturing concepts for crystalline silicon solar cells, IEEE T. Electron dev. 46 (1999) 1948-1969. https://doi.org/10.1109/16.791983

[90] R.J. Knuesel, H.O. Jacobs, Self assembly of microscopic chiplets at a liquid liquid solid interface forming a flexible segmented monocrystalline solar cell, PNAS. 107 (2010) 993-998.https://doi.org/10.1073/pnas.0909482107

Materials for Solar Cell Technologies I
Materials Research Foundations **88** (2021) 176-235

Materials Research Forum LLC
https://doi.org/10.21741/9781644901090-7

Chapter 7

Low Band-Gap Materials for Solar Cells

Yadavalli Venkata Durga Nageswar[1]*, Vaidya Jayathirtha Rao[2]

[1]Retired chief scientist, CSIR - Indian Institute of Chemical Technology, Hyderabad - 500 007, India

[2]Emeritus Scientist, Advisory Consultant, Hetero Research Foundation, Balanagar, Hyderabad – 500018, India

vaidya.opv@gmail.com, dryvdnageswar@gmail.com

Abstract

Organic solar cells (OSCs) are discussed at length in terms of its performance leading to the generation of electricity. The key materials required for OSCs are the small organic molecules having donor and acceptor with suitable light absorption and electro-chemical properties of low energy band gap. Various structural scaffolds are highlighted with their structural design leading to film forming in an orderly manner and this morphology of film having a pivotal role in photo-induced charge separation, migration and collection at an electrode. Present day research informs that OSCs involving non fullerene based donors and acceptors are functioning with high photo conversion efficiency [PCE] of >17% and are promising candidates for practical applications.

Keywords

Organic Solar Cells, Low Band Gap, Energy Materials, Fullerene Acceptors, Photo Conversion Efficiency, Photovoltaic Parameters, Energy Transfer Process, Perovskite Solar Cells

Contents

Low Band-Gap Materials for Solar Cells 176

1. Introduction 177

2. Examples from recent literature 177

    2.1   Efficiency <5% - Fullerene 177

    2.2   Efficiency >5% - Fullerene 187

Materials Research Forum LLC

https://doi.org/10.21741/9781644901090-7

2.3   Efficiency,<5% - Non-Fullerene 199

2.4   Efficiency >5% - Non-Fullerene 204

Conclusions **227**

Acknowledgements **228**

References **228**

1.   Introduction

The earth is receiving solar radiation of approximately in the range of thousands of terawatt TW. It is a big challenge to convert such immense solar energy reaching earth into useful form such as electricity, making it a global issue. Dye sensitized solar cells, [1] organic solar cells [2,3,4] (Bulk Hetero Junction Solar Cells = BHJSC) and perovskite [5] type solar cells are actively pursued all over the world to convert solar energy into electricity. The present chapter deals with the design and synthesis of the materials useful for organic solar cells (BHJSC) and associated organic materials. The organic solar cell functioning leading to the higher efficiencies are highlighted. The required organic materials, their optical, thermal and electrochemical properties are considered in tuning the photovoltaic performances of organic solar cells. Information on fullerene acceptors, non-fullerene acceptors, polymer donors and small organic molecule donors applied for organic solar cell device preparation is collated. Specifically, small organic molecules having low band gap energy and acting as donors and acceptors are considered for discussion. Data compiled is organized into four categories like fullerene acceptors with photo conversion efficiency [PCE] of <5%, fullerene acceptors with PC efficiency of >5%, non-fullerene acceptors with PC efficiency of <5%, and non fullerene acceptors with PC efficiency of >5%, with examples involving low band gap energy materials.

2.   Examples from recent literature

2.1   Efficiency <5% - Fullerene

Oleg Kozlov et al. [6] studied the ultrafast electron and hole transfer phenomenon in bulk hetero junction solar cells containing polymer blends and C70-fullerene [6]. Donors employed were low band gap BTT-DPP (Poly[(5-decylbenzo[1,2-b:3,4-b':5,6-d″]trithiophene-2,8-diyl)-alt-co-(3,6-bis(2-thienyl)-2,5-dihydro-2,5-di(2-octyldodecyl)pyrrolo[3,4c]pyrrolo-1,4-dione-5,5′-diyl)]) and PCPDTBT (cyclopenta dithiophene and benzothiadiazole) polymers (Figure 1) [6] and C70 fullerene acting as acceptor. Ultrafast photo-induced spectroscopy was used to study the dynamics of

Materials for Solar Cell Technologies I   Materials Research Forum LLC
Materials Research Foundations **88** (2021) 176-235   https://doi.org/10.21741/9781644901090-7

electron and hole transfer processes. PCPDTBT was found to be very good and involved in electron and hole transfer processes very efficiently, whereas the BTT-DPP was good only in hole transfer process and it did not participate in electron transfer process. Rationale given for non electron transfer process between BTT-DPP and C70 Fullerene was due to unfavourable energy levels. Furthermore, these studies conclude that the energy levels of the blend materials have to be carefully understood before being used for fabricating BHJSC.

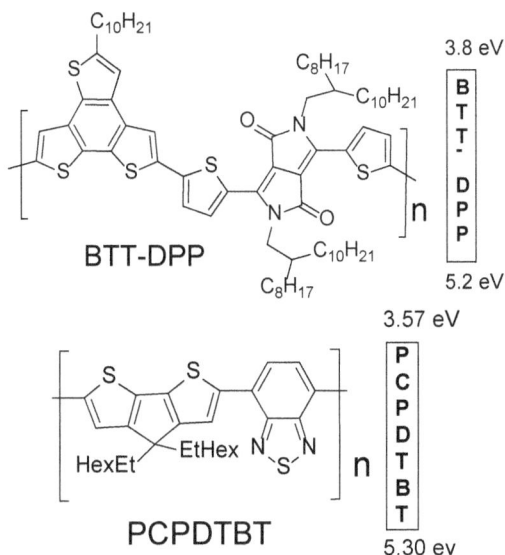

Figure 1 Polymers [6].

The utility of a low band gap squaraine dye molecule, DIB-SQ (1,3-bis(4-(diisobutylamino)-2,6-dihydroxyphenyl)cyclobutane-1,3-diylium-2,4-bis(olate)) (Figure 2) [7] in enhancing absorption of the blend and further improving the efficiency of solar cell containing P3HT (Poly-3-hexylthiophene) and PCBM71 was described by Qiaoshi An et al. [7]. The blend absorption spectra recorded covers 300 to 600 nm and along with a small band covering 600 to 700 nm was attributed to DIB-SQ. The solar cell configuration developed was: ITO/PEDOT:PSS/Blend(PCBM70 + P3HT + DIB-SQ)/LiF/Al. Authors systematically varied weight percentage of DIB-SQ in the standard

Materials for Solar Cell Technologies I   Materials Research Forum LLC
Materials Research Foundations **88** (2021) 176-235   https://doi.org/10.21741/9781644901090-7

blend of [PCBM71 + P3HT] and determined the solar cell efficiency (PCE). Best result was obtained at 1.2 wt. percent of DIB-SQ (ternary blend) showing 3.72% PCE, Voc of 0.58V and Jsc of 9.7 mA. Further authors also established energy transfer process from P3HT to DIB-SQ by fluorescence quenching and fluorescence lifetime measurements. The rationale provided for the enhancement in the efficiency by doping the solar cell with DIB-SQ was harvesting more photons in the less energy light region, increased excited state dissociation, energy transfer process, and charge mobility in the ternary blend films.

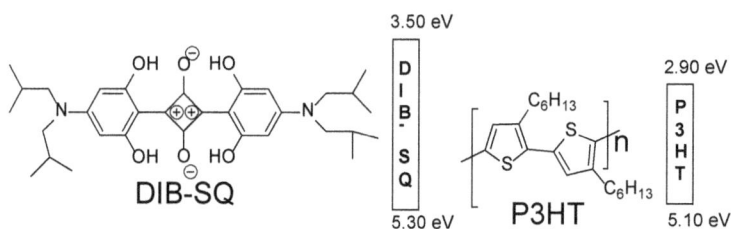

Figure 2 Squaraine dyes [7].

Yuan Fang et al. [8] synthesized a benzothiadiazole derivative YF25 (2-((7-(4,4-dipropyl-4H-silolo[3,2-b:4,5-b']dithiophen-2-yl)benzo[c][1,2,5]thiadiazol-4-yl)methylene)malononitrile) (Figure 3)[8] as a low band gap acceptor for the purpose of unravelling the mechanism of functioning of BHJ solar cells[8]. Silicon containing moiety – dithienosilol unit acts as donor and dicyanovinyl-benzothiadiazole group acts as acceptor in the YF25 molecule. YF25 has excellent absorption, suitable oxidation and reduction potentials and has band gap of ~1.70 eV. YF25 – P3HT blend and 60PCBM – P3HT blend photovoltaic parameters were compared. It was concluded that absorption by both components YF25 and P3HT led to charge generation and further it was supported by transient microwave conductivity experiments. Authors claimed in the mechanism that the light induced excitation of both acceptor and donor are very important to be considered for the efficient generation of photo current.

Jianhua Liu et al. [9] synthesized *tri* -diketopyrrolopyrrole small molecule low band gap donor materials (Figure 4) [9], differing in their alkyl chain attached at the middle core of diketopyrrolopyrrole moiety, for solar cell device fabrication [9]. Material properties like frontier energy levels, optical, film structure, charge mobilization and others were determined for two *tri* –diketopyrrolopyrrole molecules. PC71BM as acceptor and *tri* – diketopyrrolopyrrole molecules as donors were used for solar cell fabrication to observe efficiency of 4.8% for ethylhexyl alkyl chain attached molecule and 5.5% for hexyl alkyl chain attached molecule. The basic alkyl chain structural difference in these molecules was reflected in their thermal properties, film morphology and charge transport property, which impacted solar cell performance. Authors claim that this is a new design approach and will have long standing effect to achieve excellent performances for solar cell functioning.

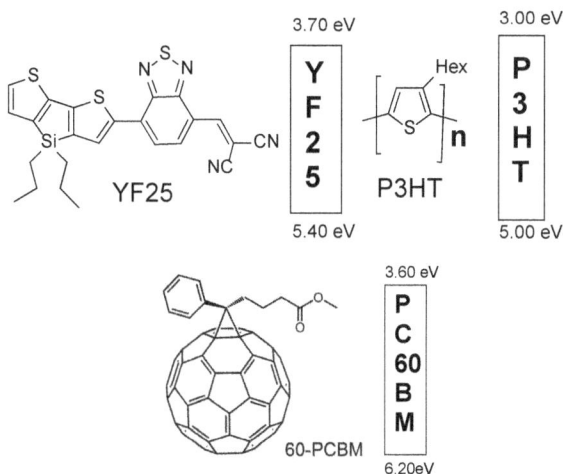

Figure 3 Benzo-thiadiazole – Dithienosilol conjugate [8].

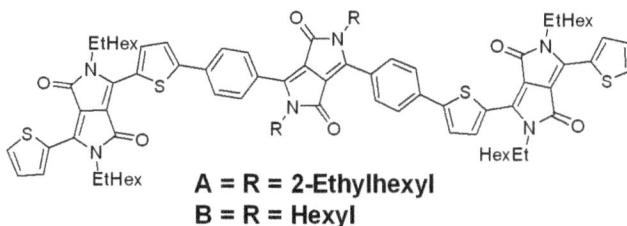

Figure 4 Diketopyrrolopyrrole based molecules [9].

Materials Research Forum LLC
https://doi.org/10.21741/9781644901090-7

Small molecule based on isoindigo acceptor, flanked with tri-phenylamine donor on both sides was synthesized (Figure 5) [10] as a low band gap donor material for the purpose of solar cell fabrication by Yingping Zou and co-workers [10]. This was a linear molecule with DAD (Donor-Acceptor-Donor) architecture, displayed band gap energy of 1.69 eV and moderate hole-mobility of 2.4×10^{-4} cm^2/V. Standard solution processed solar cell fabrication provided device parameters like, Voc = 0.78 V, Jsc = 2.94 mA, FF = 36.5% and PCE = 0.84%. Authors claimed it to be a new design and also expected it to be more efficient.

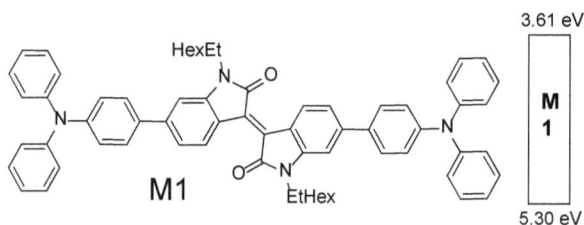

Figure 5 Indigo-Triphenylamine dyes [10].

Cira Maglione et al. [11] reported the synthesis of three low band gap small molecule donor materials, DPP1 (Diketopyrrolopyrole 1), DPP2 (Diketopyrrolopyrole 2) and DPP3 (Diketopyrrolopyrole 3) (Figure 6) [11] with DAD (Donor-Acceptor-Donor) architecture [11]. Diketopyrrolopyrrole (DPP) was the core group, which was flanked on by either side with heterocyclic spacer followed by phenothiazine ring, and these molecules differed in their heterocyclic spacer group. The effect of heterocyclic spacer was clearly visible from the properties of molecules synthesized. Solar cell devices were fabricated according to the structure given: ITO/PEDOT:PSS (40 nm)/DPP1orDPP2orDPP3:PC71BM/Al(120 nm). Solar cell device parameters were: Voc = 0.84 V; Jsc = 2.9 mA; FF = 29%; PCE = 0.71% for DPP1 and Voc = 0.73 V; Jsc = 3.1 mA; FF = 31%; PCE = 0.7% for DPP2. The low efficiency reported may be due to improper fabrication of solar cell devices.

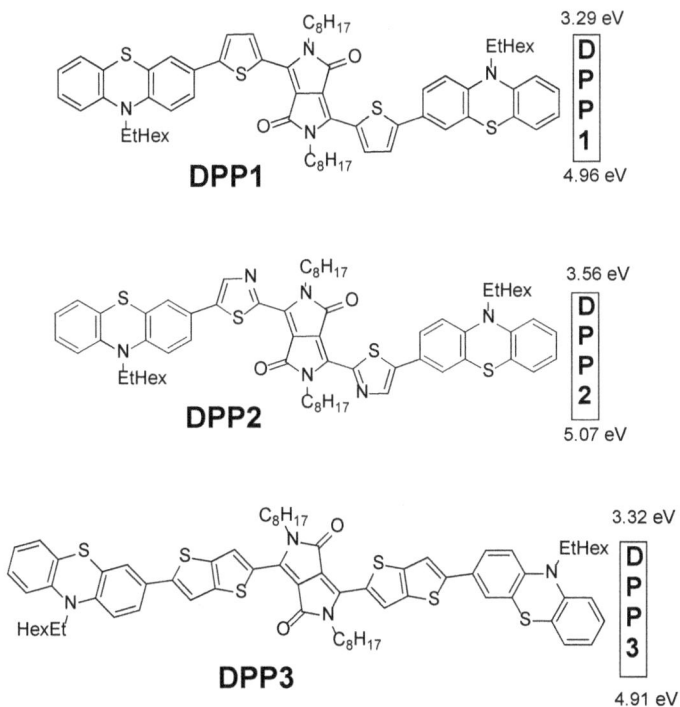

Figure 6 DPP – Thiazine molecules [11].

Prabhat Gautam et al. [12] described the synthesis of a low band gap small molecule BTD3 (Benzothiadiazole 3) (Figure 7) [12] acting as donor with D-A-D architecture [12]. BTD3 has tetra-cyano moiety coupled with benzothiadiazole as an electron withdrawing moiety, attached on one side with triphenylamine group and another side with 9-acetylenylanthracene group. PC71BM was used as acceptor in making solar cells. Photovoltaic properties were generated by adopting the ITO/PEDOT-PSS/BDT3:PC71BM/Al standard device structure. Photovoltaic parameters found were: Voc = 0.94 V, Jsc = 7.45 mA, FF = 45%, ECE = 3.15%, When the solar cell fabricated using chloro-naphthalene as additive, there was an increase in the ECE to 4.61%. Authors advocated that the improvement of efficiency of solar cell was due to chloro-naphthalene

Materials for Solar Cell Technologies I   Materials Research Forum LLC
Materials Research Foundations **88** (2021) 176-235   https://doi.org/10.21741/9781644901090-7

additive which provided good film morphology, balance charge transport and enhancement in the light harvesting ability of the active layer.

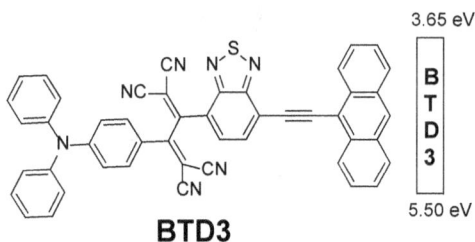

Figure 7 Tetracyano-Benzo-thiadiazole-Triphenylamine Triad [12].

Mahalingavelar Paramasivam et al. [13] synthesized four molecules (Figure 8) [13], BFBFB (Benzocarbazole-Flurene-Benzothiadiazole-Flurene-Benzocarbazole), CFBFC (Carbazole-Flurene-Benzothiadiazole-Flurene-Carbazole), BTBTB (Benzocarbazole-Thiophene-Benzothiadiazole-Thiophene-Benzocarbazole) and CTBTC (Carbazole-Thiophene-Benzothiadiazole-Thiophene-Carbazole), with systematically controlling the band gap energy and showing BTBTB as low band gap material [13]. All the compounds, BFBFB, CFBFC, BTBTB and CTBTC, displayed excellent thermal, optical properties and with a band gap range of 2.36 eV to 1.90 eV . PC60BM was employed as acceptor in these studies. Solar cell was fabricated with device structure as :ITO/PEDOT:PSS (38 nm)/active layer/Ca (20 nm)/Al (100 nm). Best photovoltaic parameters obtained for CTBTC compound are as follows: Voc = 0.96 V; Jsc = 4.63 mA; FF = 34% and PCE = 1.62%. Results obtained indicated that extending the light absorption window, charge carrier mobility and solvent selection during fabrication lead to higher efficiency. Most importantly these molecules can be used as acceptors with low band gap donors to observe higher efficiency.

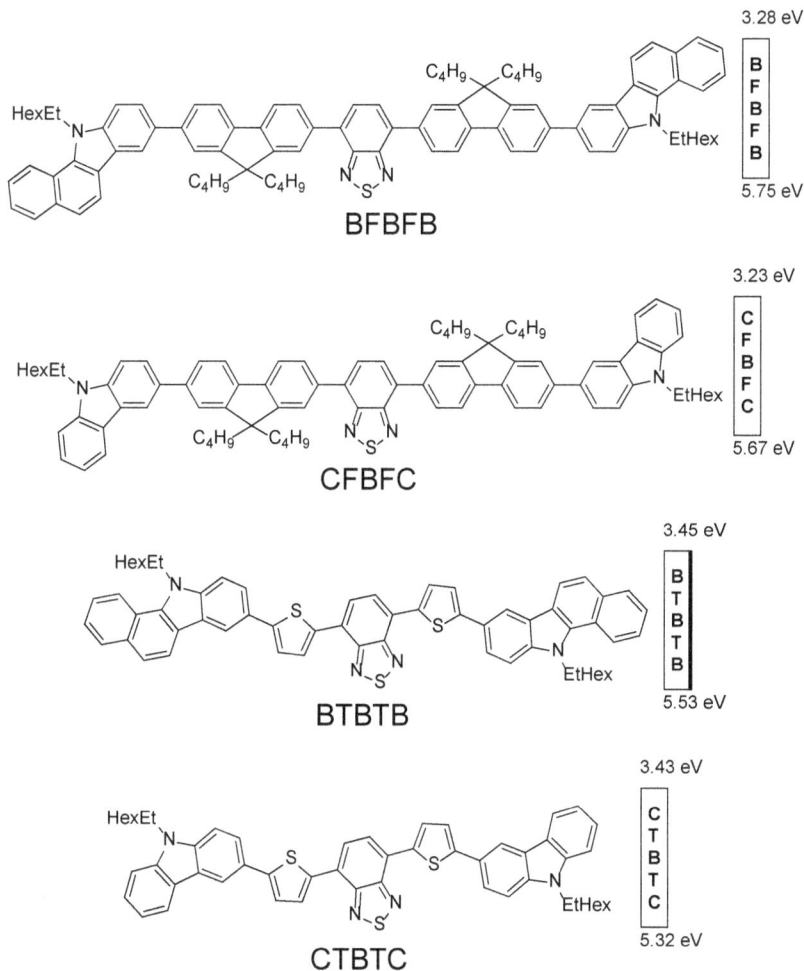

Figure 8 Benzo-thiazole linked with Flurene and Benzo-carbazole [13].

Ananda Rao et al. [14] described three novel low band gap squaraine based small donor molecules, TBU-SQ (t-Butyl-Squaraine), MeTBU-SQ (Methy-t-Butyl-Squaraine) and EtTBU-SQ (Ethyl-t-Butyl-Squaraine) (Figure 9) [14] for the purpose of solar cell device fabrications [14]. These squaraine based molecules absorbed in the region of 600 nm to 800 nm. PC70BM was employed as acceptor for device fabrications. A simple device configuration adopted was :ITO/PEDOT:PSS/SQ dyes:PC70BM (1:3)/Al. Photovoltaic parameters obtained for MeTBU-SQ were given: Jsc = 9.8 mA; Voc = 0.64 mA; FF = 50%; PCE = 3.14%. Authors claimed that the increased absorption and polar/charged nature of the molecules played a role in improving the efficiency.

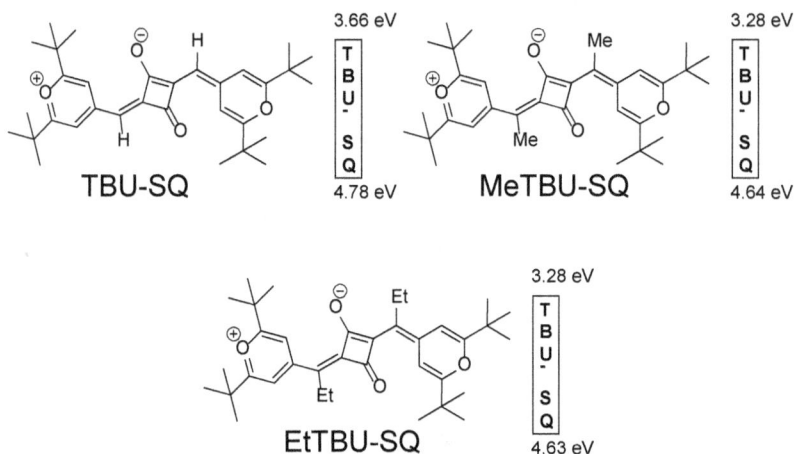

Figure 9 Pyrrilium Linked Squaraines [14].

Mahalingavelar Pramasivam et al. [15] synthesized novel funnel shaped compounds, CTDB (Carbazole-Triphenylamine-diBenzothiadiazole), CTDP (Carbazole Triphenylamine-diPyridothiadiazole), DCTB (diCarbazole-Triphenylamine-Benzothiadiazole) and DCTP (diCarbazole-Triphenylamine-Pyridothiadiazole) (Figure 10) [15] and evaluated these towards solar cells [15]. Triphenylamine was the core moiety, which was linked with di-t-butylcarbazole (one or two) donor on one side and benzo/pyrido thiadiazole (one or two) on the other side as a DDA$_2$ and D$_2$DA architecture. HOMO-LUMO levels were in the range of 2.1 to 2.4 eV. All the molecules showed very good thermal stability and also displayed fluorescence solvatochromism indicating the presence of ICT in the excited state. Solar cell devices developed have the configuration ITO/PEDOT:PSS (38nm)/active layer/Ca (20 nm)/Al (100 nm)

Materials for Solar Cell Technologies I   Materials Research Forum LLC
Materials Research Foundations **88** (2021) 176-235   https://doi.org/10.21741/9781644901090-7

architecture. Among the four compounds DCTP exhibited good photovoltaic parameters: Voc = 0.871; Jsc = 6.75 mA; FF = 38%; PCE = 2.21%. Funnel shaped molecules reported can act as very good acceptors.

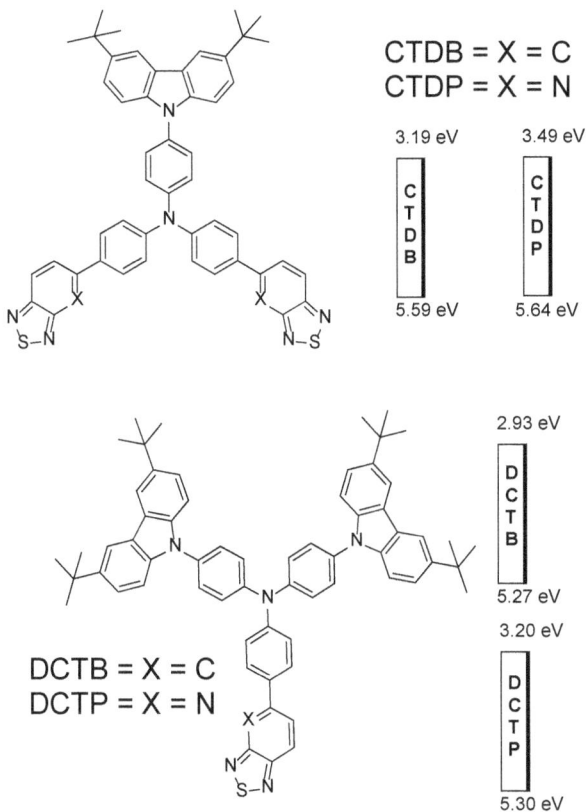

Figure 10 Funnel shaped molecules [15].

2.2   Efficiency >5% - Fullerene

Acceptor-Donor-Acceptor (ADA) small molecule D-IDTT-SQ was synthesized as a low band gap donor material for organic solar cells by Daobin Yang et al. [16]. The central moiety acts as donor and the other part acts as acceptor (Figure 11) [16]. UV-Vis-NIR absorption spectrum of D-IDTT-SQ (Bi-Squaraine) in chloroform solution with high molar absorption coefficient and in film state may be taken as an indication for low energy level optical band gap of 1.49 eV. $PC_{71}BM$ was used as acceptor material along with D-IDTT-SQ (donor material) for casting solution processed BHJSMOSCs to get 7.05% efficiency (PCE) and with high open-circuit voltage (Voc) of 0.93 V. The highlight of this work is that for the first time a low energy loss of 0.56 eV is demonstrated during the operation of BHJSMOSC. Properties of this compound like deep HOMO and LUMO energy levels, high hole mobility, good absorption wavelength coverage (300 nm to 800 nm) of casted film and low band gap energy are cited in favour of observed performance of D-IDTT-SQ.

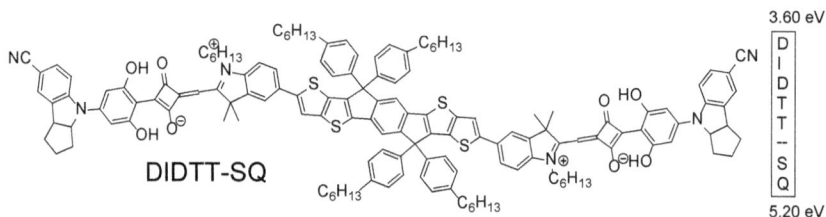

Figure 11 Bi-Squaraine derivative [16].

Phenoxazine based ADA (Acceptor-Donor-Acceptor) structured, low band gap, material M1 (Benzodithiophene linked Phenoxazine and Rhodanine) was synthesized for the purpose of understanding the solar cell behaviour towards its efficiency by Ming Cheng et al. [17]. Dialkoxybenzodithiophene (DBT) was the central core unit of M1 and it was attached either side with ethylrhodanine end group attached phenoxazine (Figure 12) [17]. The *O*-alkyl and *N*-alkyl groups help in solubility, DBT and phenoxazine act as donor and rhodanine as acceptor group. Configuration adopted for the BHJ solar cell device was: ITO/ZnO/PEIE/Blend (PCBM70 + M1) /MoO$_3$/Ag. Authors varied ZnO and PEIE layer thickness to achieve best efficiency of 6.9% (PCE), Voc of 0.897 V and Jsc of 12.5 mA. Authors employed the same M1 material as a hole transporting material in fabricating perovskite service to observe 13.5% efficiency. Authors were of the opinion

Materials Research Forum LLC
https://doi.org/10.21741/9781644901090-7

that the M1 can be synthesized from readily available starting compounds to provide a lead molecular structure useful in both BHJ solar cells and perovskite solar cells.

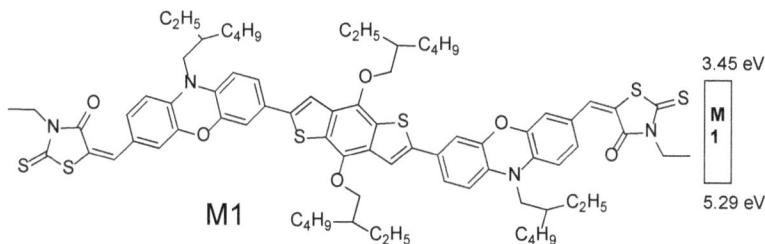

Figure 12 Benzodithiophene linked Phenoxazine and Rhodanine [17].

Kimin Lim et al. [18] synthesized two star shaped, low band gap donor small molecules for the solar cell device formation (Figure 13) [18]. The core is tri-phenylamine (TPA[DTS-PyBTTh₃]₃) or fused tri-phenylamine (DMM-TPA[DTSPyBTTh₃]₃) with three arms attached with linear linkage of dithienosilol moiety, pyridodithiazole and three thiophene units for each compound. These were used as donor materials along with PC71BM as acceptor material to fabricate solar cells. PCE was improved by fabricating the devices using 1-chloronaphthalene as additive. Observed PC efficiencies are 3.88% for TPA[DTS-PyBTTh₃]₃ and 5.81% for DMM-TPA[DTSPyBTTh₃]₃. Role of additive was ascertained by analyzing the casted films using AFM, TEM pictures, where 1-chloronaphthalene improved the morphology of the film and also hole and electron mobilities as determined for the films with and without additive.

TPA-[DTS-PyBTTh₃]₃

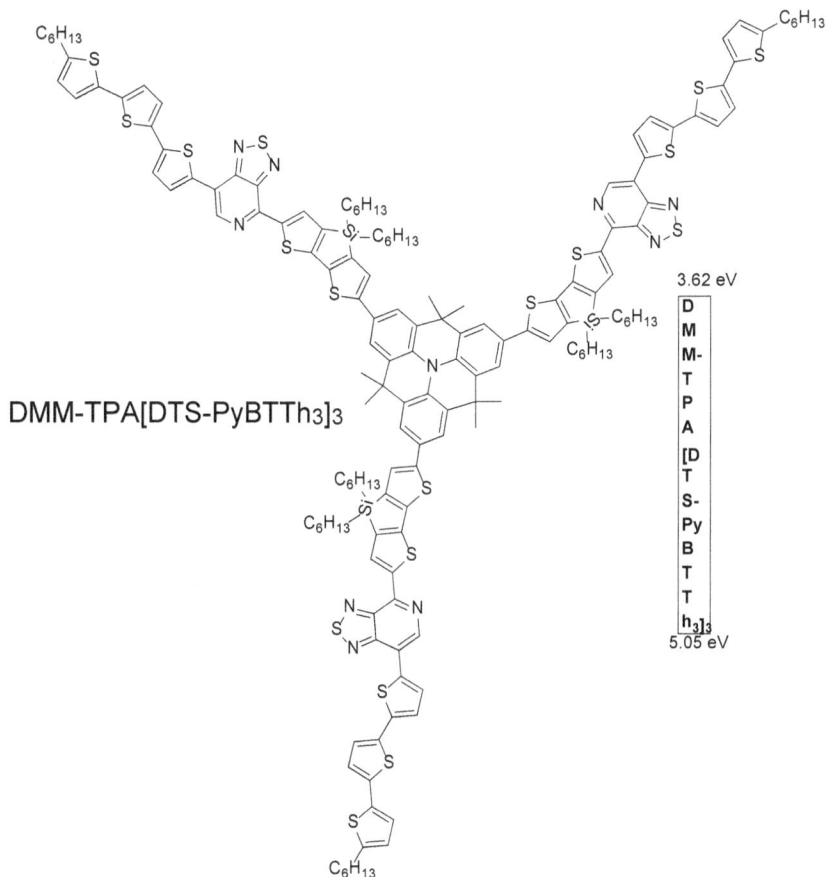

DMM-TPA[DTS-PyBTTh3]3

Figure 13 Star shaped molecules [18].

Bin Kan et al. [19] designed and synthesized several low band gap small molecule donor material - DRCN4T to DRCN9T (linearly linked thiophenes with rhodanine end groups), with a backbone of thiophene linked (4, 5, 6, 7, 8 and 9 thiophenes) linearly and flanked either side by electron withdrawing rhodanine derivative group (Figure 14) [19]. Length of conjugation and conformation/ configuration played an important role in defining their optical and electrochemical properties. Their band gap found to be ranging from 1.88 eV to 1.53 eV DRCN4T to DRCN9T (for 4 thiophene unit compound to 9 thiophene compound). Optical, electrochemical, thermal and photovoltaic properties were evaluated systematically. All these low band gap small donar molecules synthesized fall under A-D-A (Acceptor-Donor-Acceptor) type architecture. PC71BM was used as acceptor for fabrication of solar cells. Device structure adopted by authors was: ITO/PEDOT:PSS/DRCNnT:PC71BM/PFN/Al. Film morphology was monitored using AFM, TEM, charge mobility and 2D GIXD techniques. Estimated photovoltaic parameters for DRCN5T and DRCN7T were : Voc = 0.92 V; Jsc = 15.66 mA; FF = 68%; PCE = 9.8% for DRCN5T and Voc = 0.90 V; Jsc = 14.77 mA; FF = 68%; PCE = 9.05% for DRCN7T and the PCE determined was certified numbers. It may be interesting to predict that these small molecule low band gap materials may provide a higher efficiency with non-fullerene acceptors. Since these small molecule low band gap donor compounds can be designed and synthesized easily, they can provide a good handle for easy tuning of desired properties leading to higher efficiencies.

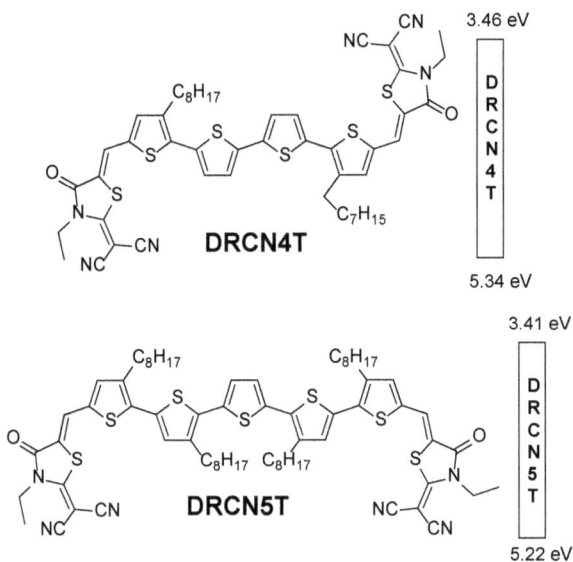

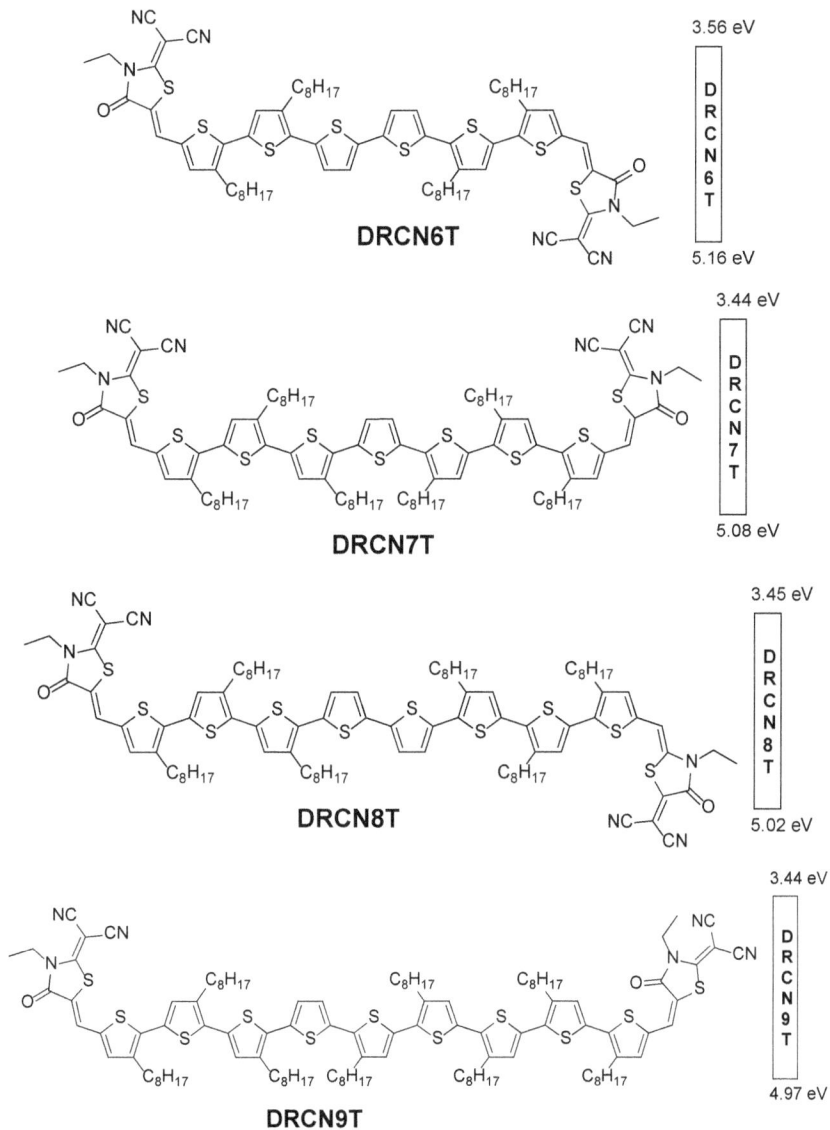

Figure 14 Thiophene oligomers linked with Dicyano rhodanine [19].

Wang Ni et al. [20] reported synthesis of two low band gap small molecule donor materials (Figure 15) [20], differing at one atom in the core structure, DR3TDTC (Dithiophene fused cyclopentene with bis-trithiophene rhodanine attached both sides) – carbon atom and DR3TDTS (Dithiophene fused Silocyclopentene with bis-trithiophene rhodanine attached to both sides) carrying Silicon atom [20]. Indeed this change in one atom reflected big change in their film absorption, molecular packing and charge transport properties, favoring silicon atom containing molecule DR3TDTS. Film morphology was investigated using X'ray Diffraction (RSoXS), GIWACS, AFM, TEM and charge mobility techniques. Solar cell devices were fabricated in a conventional style like ITO/PEDOT:PSS/donor:acceptor PC71BM/ETL-1/Al. Device parameters determined for DR3TDTS were found to be excellent as given: Voc = 0.82 V; Jsc = 13.67 mA; FF = 69%; PCE = 7.8%. For other molecule DR3TDTC exhibited very poor PCE of <1.0%. It was evident that by replacing "carbon atom" with "silicon atom" at core structure made the difference in properties and finally recorded a higher PCE value.

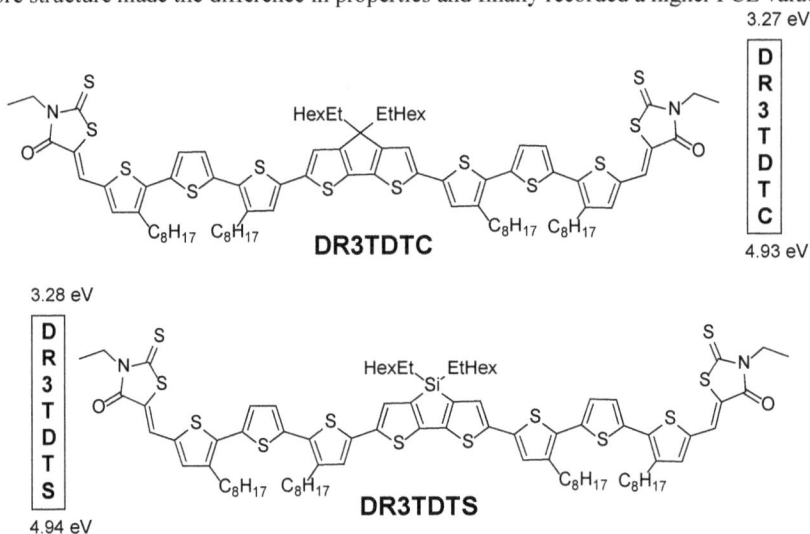

Figure 15 Fused three rings with silicon atom [20].

Manohar Reddy et al. [21] designed a novel NIR absorbing low band gap small donor molecule, ICT3 (Figure 16) [21] to determine solar cell efficiency [21]. In ICT3(Tristrithienopyrrole with rhodanine end groups)-- molecule three dithieno-pyrrole

(DTP) moieties were linked linearly with both sides carrying rhodanine end groups and it is denoted with architecture as A-D-D-D-A (Acceptor-Donor-Donor-Donor-Acceptor) small donor molecule. PC70BM was employed as acceptor in the solar cell fabrication. "ICT3 + PC70BM" blended films were analyzed by AFM and X-ray diffraction techniques to understand film morphology. Thermal annealing and solvent vapour annealing were adopted to refine the morphology of the blended film. Solar cell devices were fabricated using ITO)/PEDOT:PSS/ICT3:PC71BM/PFN/Al structure. Photovoltaic measurements were determined under three different conditions (i) as cast; (ii) thermal annealing (TA) and (iii) thermal annealing (TA) and solvent vapour annealing (SVA). Best results were obtained under TA & SVA conditions- Voc = 0.84 V; Jsc = 11.94 mA; FF = 65.6% and PCE = 6.53%. The rationale was that the active blend layer under gone morphological changes after thermal annealing and solvent vapour annealing and these changes effected the efficiency results.

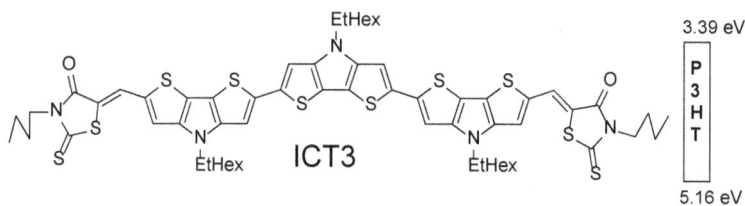

Figure 16 Dithienopyrrole trimer coupled rhodanine [21].

Manohar Reddy et al. [22] reported design and synthesis of medium band gap small donor molecules, ICT1(diAlkoxy-Benzodithiophne with dithienopyrrole rhodanine groups) and ICT2 (Alkylthiophene-benzodithiophne with dithienopyrrole rhodanine groups) (Figure 17) [22] with $AD_1D_2D_1A$ (Acceptor-Donor1-Donor2-Donor1-Acceptor) architecture [22]. Core structure for both ICT1 and ICT2 was benzodithiophene - BDT (D_2) as central moiety linked either side with dithienopyrrole - DTP (D1) and further attached with rhodanine moiety. The difference between ICT1 and ICT2 was the alkoxy versus alkylthiophene residues attached to central BDT core. Both the molecules exhibited film state light absorption, covering 500 nm to 750 nm region. Band gap values estimated were 1.96 eV and 2.02 eV for ICT1 and ICT2 respectively. PC70BM was used as acceptor in these solar cell fabrications. Solar cells with ITO/PEDOT:PSS /blend/Al structure were fabricated using these ICT1 and ICT2 materials. The "as cast" film was annealed to generate higher efficiency. ICT2 displayed photovoltaic parameters like: Voc ¼ 0.92 V Jsc = 10.68 mA; FF = 60%; PCE = 5.90%. The improvement in the efficiency was attributed to the formation of well organized film after annealing.

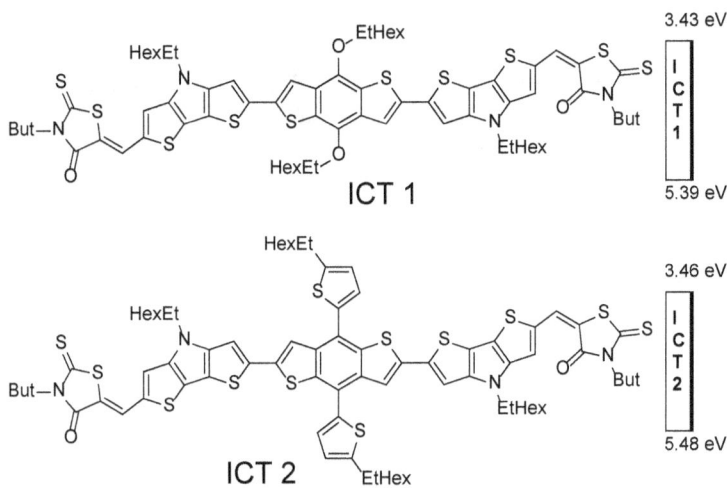

Figure 17 BDT and DTP Linked molecules [22].

Manohar Reddy et al. [23] designed and synthesized two donor small molecules with and without fluorine substitution having architecture of D_1-A-D_2-A-D_1 (Figure 18) [23] to understand the effect of fluorine on the solar cell photovoltaic parameters [23]. Bisethylhexyloxybenzodithiophene – D_2 – (OBDT) was the central core donor, attached either side with benzothiadiazole - A – (BTD) acceptor and followed by linking of ethylhexydithienopyrrole – D_1 – (EDTP) providing ICT4. ICT6 had extra four fluorine atoms attached on the benzothiadiazole acceptor skeleton (Figure 18). PC71BM was employed as acceptor in the device fabrications and the adopted device structure was: ITO/PEDOT:PSS/ICT4 or ICT6:PC71BM/PFN/Al. ICT4 and ICT6 displayed impressive performance by showing PCE 5.46 and 7.91% respectively. The improvement in the solar cell efficiency upon fluorine attachment on the acceptor benzothiadiazole ring was rationalized in terms of ICT6 forming thin film of excellent morphology leading to enhancing the dissociation of excitons, charge mobility and charge collection as well as reducing bimolecular recombination. Authors claim that the observed 7.91% efficiency is one of the best among BT based donor small molecules in photovoltaics.

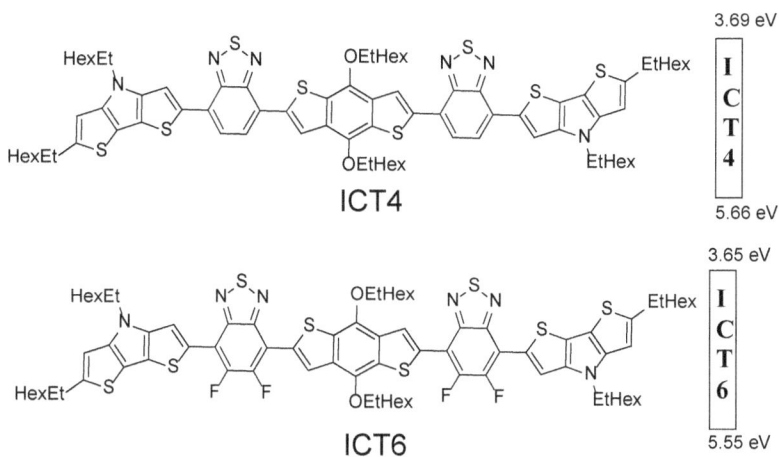

Figure 18 BDT – BTD – DTP hybrids [23].

Manohar Reddy et al. [24] synthesized small-molecule donor materials ICT7 (D1-A1-D2-A1-D1) and ICT9 (D1-A2-D2-A2-D1) (Figure 19) [24] to evaluate the effect of fluorine atom substitution on the processing conditions and photovoltaic properties [24]. Dithienopyrrole is D1, Benzothiadiazole is A1, Benzodithiophene with thiophenes is D2 and Difluorobenzothiadiazole is A2. The presence of the highly electronegative fluorine atom was expected to introduce intermolecular interactions in thin films (layers) prepared which resulted in better hole mobility (1.02V and 2.18V, respectively for ICT7:PC71BM and ICT9:PC71BM active layers) and balanced charge transport. Further, ICT9 carrying four fluorine atoms exhibited relatively more ordered nanoscale film (layer) morphology when compared to the ICT9:PC71BM active layer leading to improved exciton dissociation, charge transport, charge-collection efficiency and suppressed bimolecular recombination in the solar cells. Solar cells fabricated using ICT7 and ICT9 as donor molecules and PC71BM as acceptor displayed PCE values of 6.43% and 8.34 %, respectively. The 8.34% exhibited by the ICT9 is one among the best-performing molecules in BT-based donor materials. The present investigations also indicated the importance of the central core to generate efficient donor materials for BHJSCs.

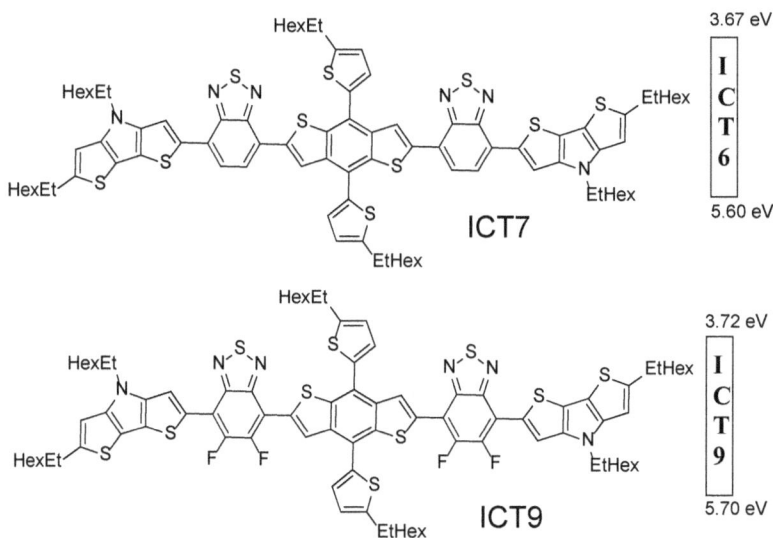

Figure 19 BDT with thiophene – BTD – DTP hybrids [24].

Manohar Reddy et al. [25] have synthesized three small molecule donor materials, ICT18 (D1-D2-A), ICT19 (A-D2-D1-D2-A) and ICT20 (A-D2-D2-D2-A) (Figure 20) [25], and investigated the effect of the donor material (Dialkoxybenzodithiophene is D1, Dithienopyrrole is D2 and Indocenone is A)architecture /backbone length and the nature of central donor units on their optical and electrochemical properties [25]. UV-Vis absorption spectrum gradually red shifted from ICT18 to ICT19 and to ICT20, leading to the NIR region in thin films and this red shift in absorption maxima was explained based on the charge transfer character in the molecules. Solar cell device structure followed was just conventional: ITO/PEDOT:PSS/SM donor:PC71BM/PFN/Al. These molecules were used as donor materials along with the PC71BM electron acceptor for the fabrication of solar cells and after the optimization (donor to acceptor weight ratio and solvent vapor annealing time), OSCs based on ICT18, ICT19 and ICT20 showed overall PCE of 5.69%, 6.92% and 8.13%, respectively. Photovoltaic parameters determined for ICT20 are: Voc = 0.87V; Jsc = 14.67 mA; FF = 64%; PCE = 8.13%. These observations indicate that the overall PCE may be related to the electron donating ability of the central core unit, leading the high hole mobility, balanced charge transport, higher exciton dissociation and charge collection probability.

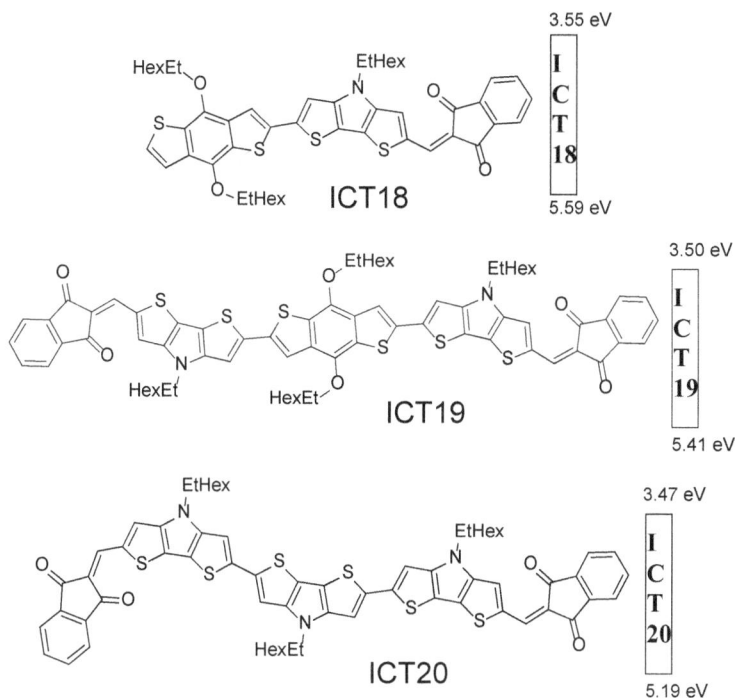

Figure 20 BDT – DTP – indenone linked molecules [25].

Ananda Rao et al. [26] came up with new arrangement of blend where P3HT was used as donor, PC70BM acted as acceptor and a low band gap squaraine compound acted as bridge or relay (Figure 21)[26] between the donor and acceptor to improve the efficiency of solar cell [26]. Light absorption, HOMO and LUMO energy levels of TBU-SQ were matched well with the P3HT and PC70BM. The device configuration adopted is: ITO/PEDOT-PSS/P3HT+TBU-SQ+PC70BM/Al. Photovoltaic parameters determined are: Voc = 0.66 V; Jsc = 12.6 mA; FF = 66%; PCE = 5.12% at 2.5 wt% of TBU-SQ. The PCE was found to be 3.4% without doping TBU-SQ. The reasons accounted for the improvement in efficiency of the solar cell device are; (a) annealing effect providing better morphology for the blend (P3HT+TBU-SQ+PC70BM), (b) increased absorption of the blend in the visible region, (c) TBU-SQ acting as sensitizer or relay of energy.

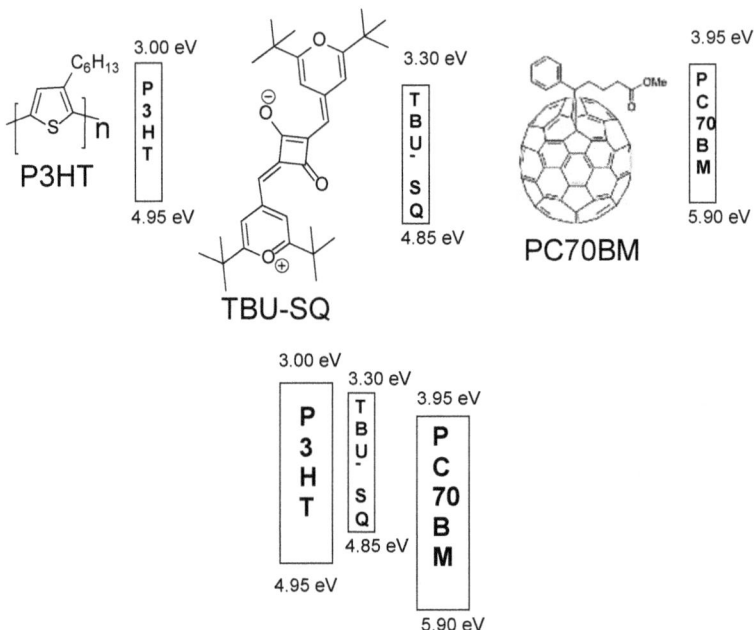

Figure 21 Squaraine – pyrrilium molecule[26].

2.3 Efficiency,<5% - Non-Fullerene

Jae Woong Jung and Won Ho Jo [27] communicated, design, synthesis (from readily available starting compounds) and solar cell fabrication studies for small acceptor molecules (Figure 22) [27]. Benzothiadiazole is the central core acceptor, which was flanked either side by two thiophene moieties, followed by DPP unit (DTBT(TDPP)2). Alkyl groups attached to the acceptor molecules provided solubility. DTBT(TDPP)2, did not carry fluorine substituent whereas DTDfBT(TDPP)2 had two fluorine atoms attached with it (Figure 22) [27]. The donor employed in these studies was PTB7 polymer. These synthesized non fullerene acceptors, DTBT(TDPP)2 and DTDfBT(TDPP)2 made them attractive for solar cell fabrication because of their optoelectronic properties like, high crystallinity, good electron mobility, well placed energy levels, low band gap energy and facile synthesis. Configuration of the solar cell fabricated was: ITO/ZnO/Donor-Acceptor Blend/MoO₃/Ag. Solar cell efficiency recorded was found to be of 5.0% with Jsc 12 mA, which indicated that small molecules with low band gap and good crystallinity had excellent scope for developing solar cells with higher efficiency.

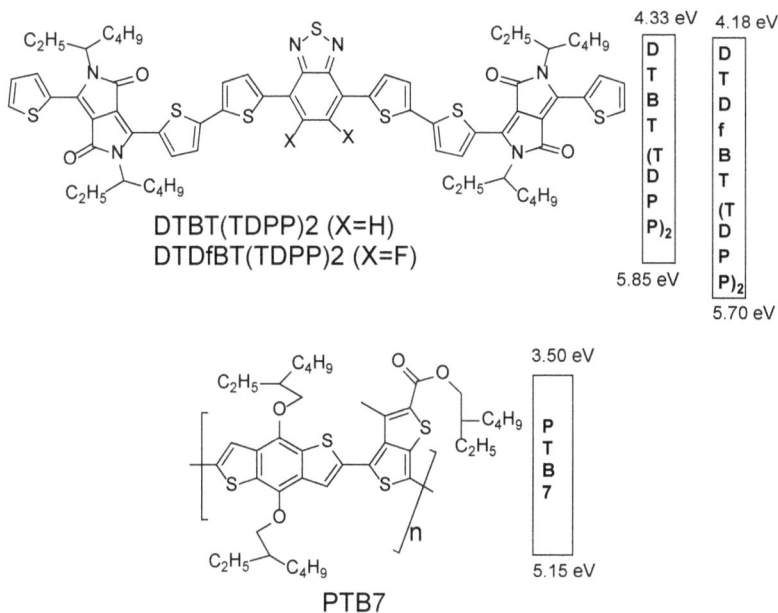

Figure 22 BTD – thiophene – DPP derivatives [27].

Bazan and co-workers [28] synthesized four (1 to 4) isomorphic low band gap donor molecules (Figure 23) [28] and also defined direction of the dipole moment in the molecule by using pyridine moiety "N" position [28]. The four compounds were described as D1-A-D2-A-D1 charge transfer molecules based on the linking of donor and acceptor groups within the molecule. D1 = Thiophene, A = Benzothiadiazole, D2 = Silicon based fused three rings. Indeed, designed direction of dipole-moment showed in forming ordered films and its morphology was analyzed by HR-TEM. Device fabrication was made using PC70BM and synthesized four low band gap materials. Molecules 1 and 2, where position of "N" atom defines the direction of dipole moment, formed more ordered films, leading to higher PC efficiency of 6.7 and 5.6% for 1 and 2 respectively. The other molecule 3 did not form ordered films (relatively to 1 and 2) and displayed low efficiency, whereas the molecule 4 lacked "N" atom to define direction of dipole-moment and its solar cell efficiency was negligible. Authors concluded that molecules with defined dipole-moment direction undergo self-assembly in forming more ordered films, which leads to higher efficiency.

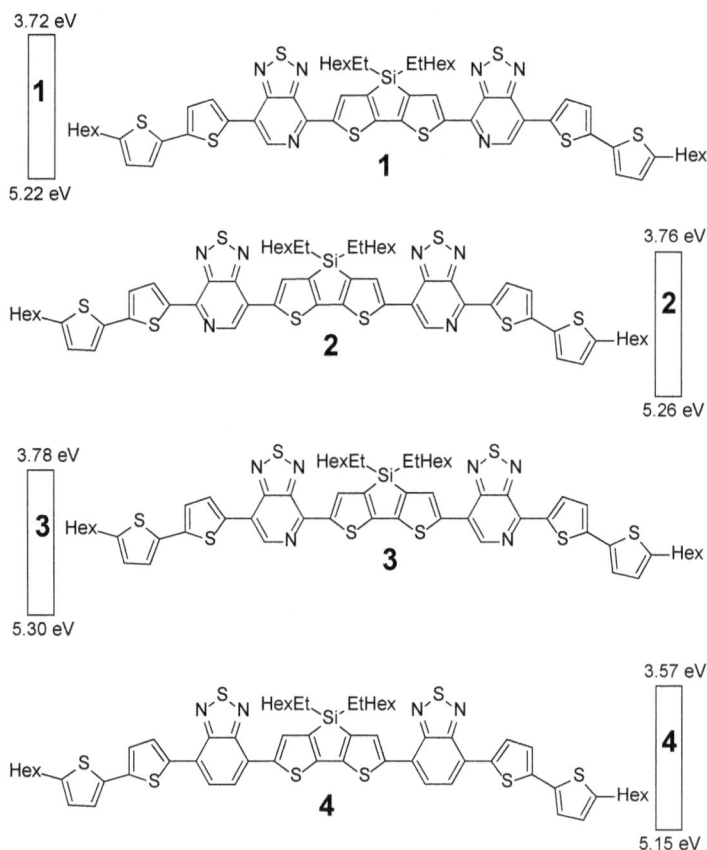

Figure 23 Thiophene-PyridoThiadiazole-Dithienosilol compounds [28].

Yuze Lin et al. [29] synthesized for the first time a dibenzosilol and diketopyrrolopyrrole based acceptor material (DBS-2DPP) for OSC. DBS-2DPP (Figure 24) [29] has dibenzosilole as central core with dithieno-diketopyrrolopyrrole attached on both sides [29]. Optical band gap determined is 1.83 eV, interestingly which is unusually smaller than band gap, 2.02 eV measured from oxidation-reduction potentials. P3HT was employed as donor along with DBS-2DPP as acceptor for device fabrications. The device configuration was: ITO/PEDOT:PSS/P3HT:DBS-2DPP/Ca/Al and the photovoltaic parameters evaluated were Voc = 0.97 V; Jsc = 4.91 mA; FF = 43%; and PC Efficiency = 2.05.

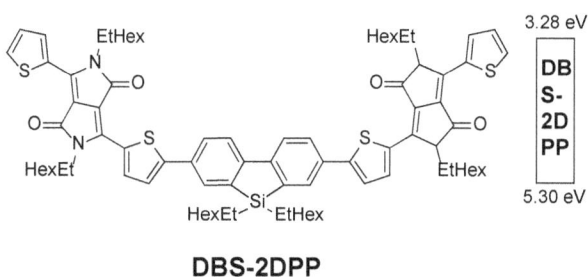

DBS-2DPP

Figure 24 Thiophene – DPP – silicon containing fluerene compound [29].

Wang Ni et al. [30] designed and reported small ADA (Acceptor-Donor-Acceptor) type acceptor molecule – DTBTF (thiophene with thiono barbiturate attached both sides of dimethylflurene), carrying terminal end groups as thiobarbituric acid and low band gap small molecule –DR3TSBDT (trithiophene-rhodanine attached both sides of dialkoxybenzodithiophene) as donor molecule differing in their middle core structure, number of thiophene residues and also terminal end groups (Figure 25) [30] for non-fullerene organic solar cells [30]. Configuration of solar cell fabricated was glass/ITO/PEDOT:PSS/ DR3TSBDT:DTBTF/PDIN/Al and energy loss in the device is estimated to be of 0.59 eV. Device parameters determined are Voc = 1.15; Jsc = 4.51 mA; FF = 34% and PCE = 1.65%, further PCE improved by annealing the spin coated layers to 3.64%. TEM images generated before and after annealing indicate improved morphology of the film leading to increase in the efficiency of solar cell to 3.64%. Authors further mentioned that the increase in the hole and electron mobilities of blended films after annealing was also cited in favor of improved morphology contributing to the increased efficiency.

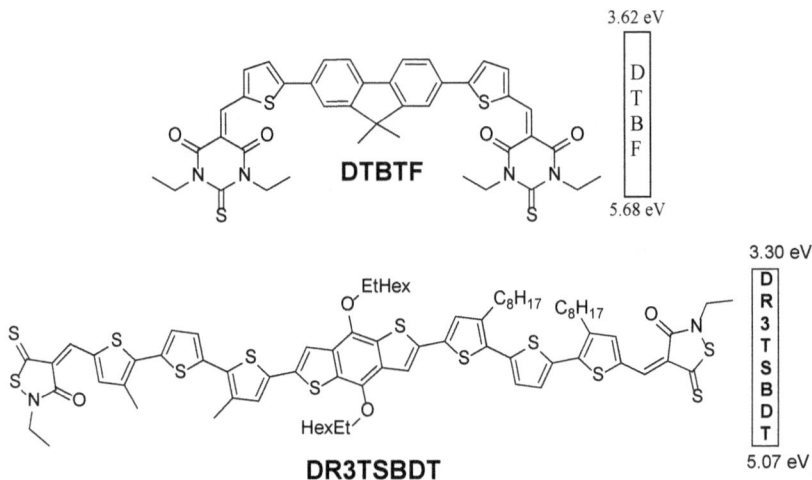

Figure 25 Thiobarbituric acid derivative [30].

Feng Liu et al. [31] reported the synthesis of a new small low band gap donor molecule STDR-TbT (Figure 26) [31] for high performance organic solar cells [31]. STDR-TbT- is a linearly attached sepithiophene with central thiophene replaced by substituted thieno[3,4-b]thiophene moiety and flanked on either side with rhodanine group as A-D-A type architecture. PC70BM was employed as acceptor in device fabrication. Photovoltaic parameters were determined by fabricating device in a simple configuration: ITO/PEDOT:PSS/activelayer/PDINO/Al. Indeed STDR-TbT displayed conversion efficiency of 5.05% and other parameters were: Voc = 0.755V, Jsc = 10.9 mA, FF = 61.4%. Authors measured photovoltaic parameters for a reference compound – STDR (seventhiophenes linked linearly and attached with rhodanine both sides) and found conversion efficiency lower like 2.31%. The rationale put forwarded by the authors for higher PCE of STDR-TbT compared to the standard reference STDR, was due to increased intramolecular charge transfer, improved morphology of the film, balanced charge transport and favourable charge collection. Authors suggested that enhancing quinonoidal resonance of D-A small molecule would improve photovoltaic performance.

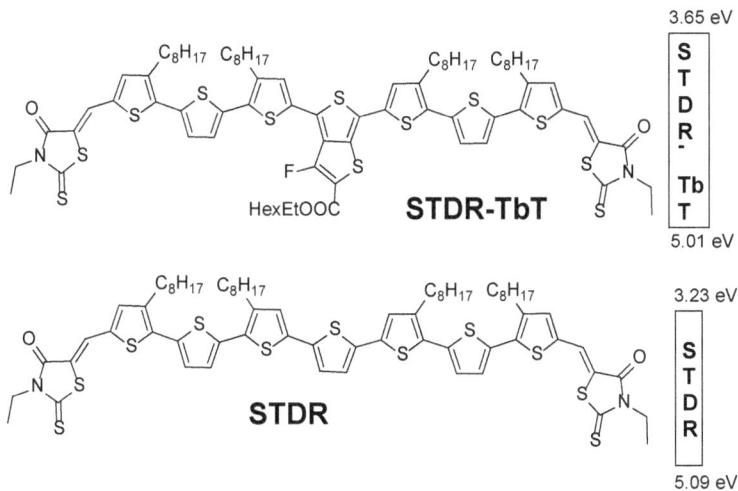

Figure 26 Thienothiophene – sepithiophene compounds [31].

2.4   Efficiency >5% - Non-Fullerene

Dongxue Liu et al. [32] reported two low band gap acceptor materials, donor as central core moiety and the remaining as acceptor (Figure 27) [32]. Thiophene and selenophene moieties are embedded as π – spacers to improve conjugation and also intra molecular charge transfer leading to red shift in the absorption spectrum. Two end groups with electron withdrawing nature, 2-(5,6-difluoro-3-oxo-2-2,3-dihydro-1H-inden-1-ylidene)malononitrile, stands for enhancing intra-molecular charge transfer (ICT). The optical band gap estimated for 4TO-T-4F (contiguously fused six rings attached with thiophene linked indenone part flanked both sides) and 4TO-Se-4F (contiguously fused six rings attached with selenophene linked indenone part flanked both sides), to be 1.3 eV and 1.27 eV for both the compounds respectively. Optical, thermal and electrochemical properties determined are found to be suitable as non-fullerene acceptors. PTB7-Th is used as donor and the other two, 4TO-T-4F and 4TO-Se-4F acted as non-fullerene acceptors in solar cell fabrications and the configuration of cell is as follows: ITO/ZnO/PTB7-Th:acceptor/MoO3/Ag. Solar cell performance was determined as 8.7% (PCE) for 4TO-T-4F with energy loss of 0.55 eV and 7.4% (PCE) for 4TO-Se-4F with energy loss of 0.57 eV.

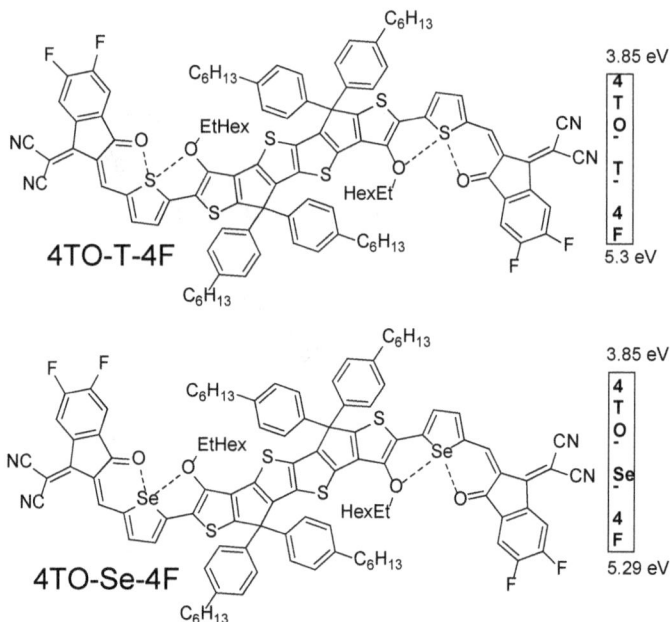

Figure 27 Vinylidinedicyano – Difluoroindanone compounds [32].

Benzothiadiazole central ring fused with pyrrole and bithiophene rings on both sides (7 rings contiguously fused together) and carrying electron withdrawing end groups 2-(5,6-difluoro-3-oxo-2-2,3-dihydro-1H-inden-1-ylidene)malononitrile and 2-(5,6-dichloro-3-oxo-2-2,3-dihydro-1H-inden-1-ylidene)malononitrile (BTP) is the main core design developed (Figure 28) [33] by Yong Cui et al. [33] for organic photovoltaic low band gap materials. Various techniques like, UV-Vis-NIR absorption, fluorescence, thermal, electrochemical, atomic force microsopy, transmission electron microscopy, and grazing incidence wide angle X-ray scattering characterizations (morphology of films), computational, X-ray diffraction (molecular packing), charge carrier mobility and measurements and highly sensitive external quantum efficiency measurements were applied to generate data. OPV fabrications were conducted using PBDB-TF as polymer donor and the BTP-4F (Fluoro) & BTP-4Cl (Chloro) as non-fullerene acceptors (Figure 28). OPV cells were fabricated with an inverted configuration of indium tin oxide (ITO)/ZnO nanoparticles/ PBDB-TF:BTP-4X/MoO₃/Al, where PBDB-TF was selected as electron donor material. Photoconversion efficiency determined from solar cell fabrications was 15.6% for BTP-4F and 16.5% for BTP-4Cl with 0.834 V and 0.867 V respectively. The high PCE was attributed to the chlorine effect by reducing the non-radiative energy (~0.26 eV) loss. Further it is expected that fine tuning the low band gap materials properties has a great potential to achieve higher conversion efficiencies.

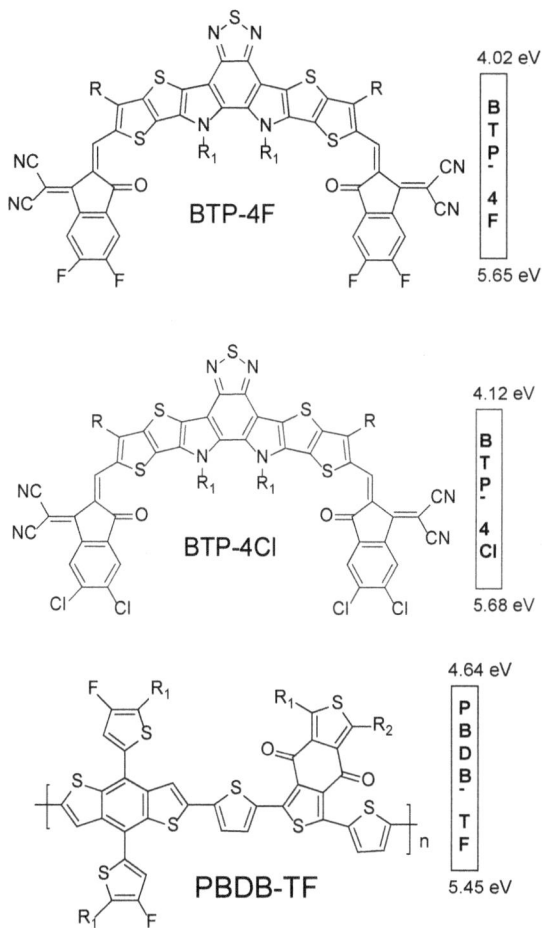

Figure 28 Fused Benzo-thiadiazole – Dicyanoindenone compounds [33].

Yangkang Yang et al. [34] in their communication highlighted the solution processable OSCs. SM1 (three thiophenes with ethyl-cyano vinylidene attached both sides of alkyldithiophene-benzodithiophene core) and DR3TBDDT (three thiophenes with rhodanine attached both sides of alkyldithiophene-benzodithiophene core) were selected as small molecule donors and PZ1 polymer as a non-fullerene acceptor material (Figure 29) [34]. SM1/DR3TBDDT donor with polymer PZ1 acceptor blend produced 400 nm to ~800 nm absorption region. Fabricated solar cells exhibited 3.97% efficiency for SM1 (donor) and PZ1a non fullerene polymer acceptor (Figure 29) and 5.86%. for DR3TBDDT/PZ1 blend. The authors were of the opinion that it is the best PCE value reported during that period.

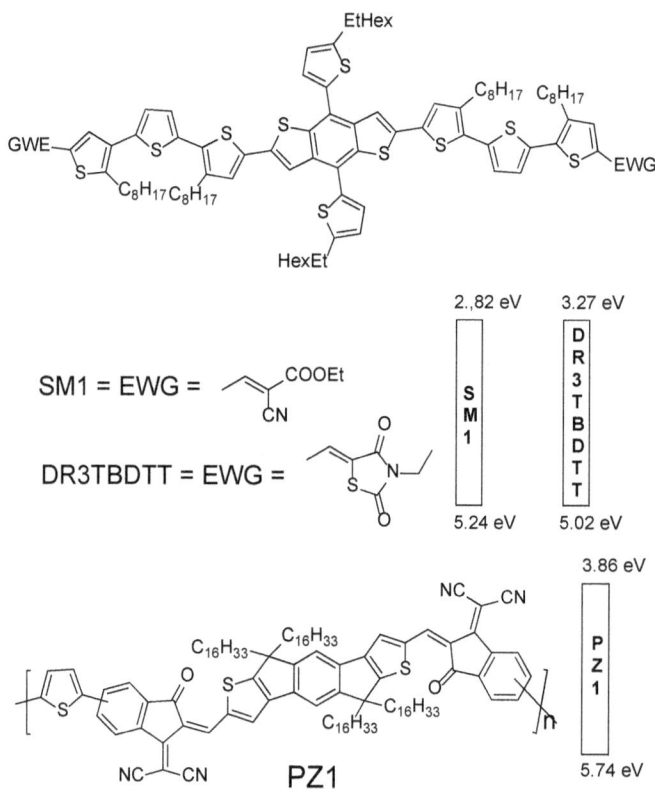

Figure 29 Thiophene BDT – thiophene compounds [34].

Bin Kan et al. [35] reported synthesis of small molecule donor NCBDT-4Cl (Dichloro-dicyanov inylidine indenone attached both sides of seven ring fused core structure: Acceptor Donor Acceptor type) with an optical band gap of ~1.40eV and used it for solar cell fabrication along with non-fullerene acceptor PBDB-T-SF (Figure 30) [35]. The NCBDT-4Cl (ADA type) donor and polymer acceptor PBDB-T-SF blend provided absorption ranging from 300 nm to 900 nm with very good molar absorption co-efficient. Solution processed solar cells were constructed with the given cell configuration: ITO/PEDOT-PSS/PBDB-T-SF:NCBDT-4Cl/PDINO/Al; PDINO acted as electron transport layer. Fabricated device as casted (without any treatment) produced efficiency of PC 13.1%. The same device after annealing and solvent additive, gave a result of PCE 14.1% with Voc of 0.85V, Jsc of 22.35 mA cm-2 and accompanied by energy loss of 0.55 eV. Authors attributed the excellent display of 14.1% PC efficiency to the blend morphology as well as to the excellent charge transport property due to attached four chlorine atoms in NCBDT-4Cl (ADA type) donor at appropriate places.

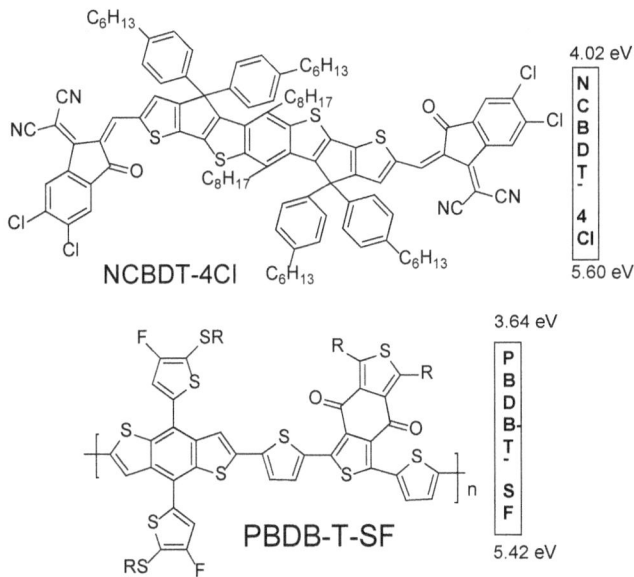

Figure 30 Fused seven ring with Dicyanoindenone compounds [35].

Materials Research Forum LLC

https://doi.org/10.21741/9781644901090-7

Jiahui Wan et al. [36] reported design and synthesis of small donor molecules suitable for solar cell fabrication studies. These molecules, BDTTNTTR (Acceptor1-Donor1-Acceptor2-Donor1-Donor2-Donor1-Acceptor2-Donor1-Acceptor1) and BDTSNTTR (Acceptor1-Donor1-Acceptor2-Donor1-Donor3-Donor1-Acceptor2-Donor1-Acceptor1) (Figure 31) [36] possessed low band gap energy, 400 nm to 800 nm light absorption range, higher carrier mobility's and low lying energy levels making them more suitable for solar cell fabrications. Fabricated devices provided the following device parameters for BDTTNTTR as: $V_{oc} = 0.89$ V; $J_{sc} = 15.70$ mA; FF = 71.70% and PCE = 10.02%, for BDTSNTTR as: $V_{oc} = 0.93$ V; $J_{sc} = 16.21$ mA; FF = 76.50% and PCE = 11.53%. UV-Visible absorption of the blend exhibited 300 nm to 800 nm absorption, there by covering entire visible range of light. Solvent carbondisulfide played a role to improve morphology of the casted film to a nanophase and thereby improving FF leading to a higher efficiency of 11.53% with low energy loss of 0.57 eV. Design of small molecule donor materials with low band gap energy and introducing tactful fabrication method afforded very good photo conversion efficiencies.

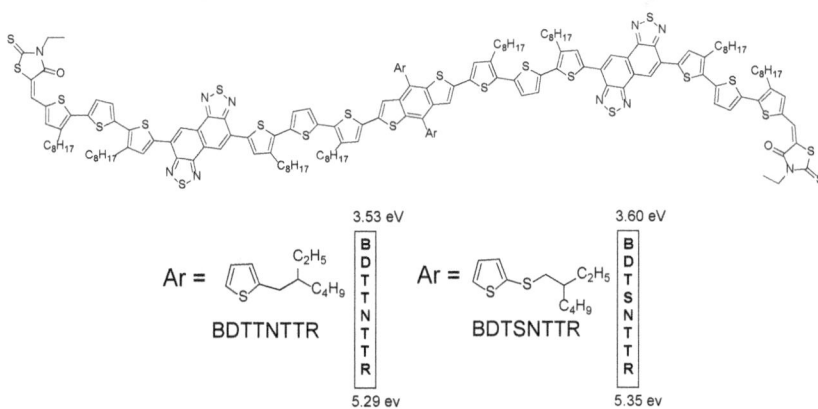

Figure 31 BDT linked dyes [36].

Central core fused five rings moiety linked with alkylthiophene spacer either side and further attached with electron withdrawing indenone group (IEIC) and central core fused five rings moiety linked with alkoxythiophene spacer, either side and further attached with electron withdrawing indenone group (IEICO) small molecules were synthesized by Huifeng Yao et al. [37] (Figure 32) [37] as low band gap acceptor materials Light absorption for IEIC and IEICO extended in to NIR (near infrared) region 600 nm to 900 nm and the optical band gap deduced is found to be 1.5 eV for IEIC and 1.34 eV for IEICO. The difference in optical band gap between IEIC and IEICO was attributed to the "alkoxy group" present in IEICO. Device configuration developed after optimization was: ITO/PEDOT:PSS/PBDTTT-E-T:IEICO or IEIC/PFN-Br/Al. PEDOT:PSS and PFN-Br were used as anode and cathode interface layers. Morphology of the casted film was studied by AFM and TEM to get further insight of device fabricated. The measured photovoltaic parameters were PCE for IEICO is 8.4 %; Voc = 0.82 V; and Jsc = 17.7 mA, further the PCE was improved to 10.7% via tandem device fabrication. The loss energy was estimated to be 0.5 eV. Photovoltaic parameters for IEIC were found to be inferior compared to IEICO. It was claimed by authors that the design leading to low band gap energy level of 1.34 eV, by introducing alkoxy group at suitable place of the small molecule helps to give higher efficiency.

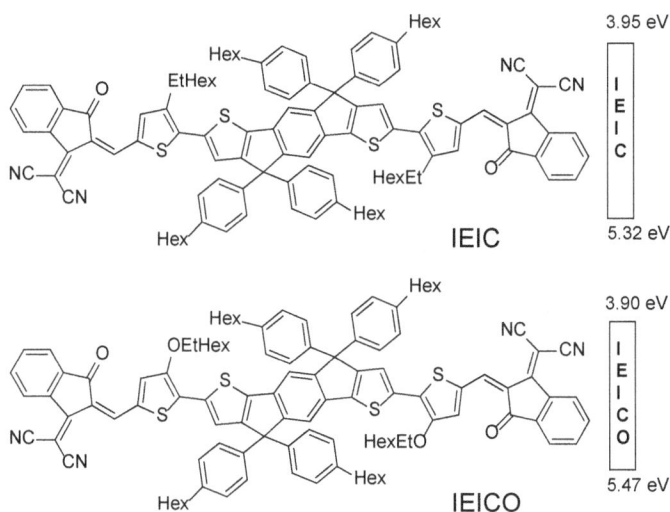

Figure 32 Fused ring linked with Thiophene – Dicyanoindenone [37].

Liyan Yang et al. [38] reported a combination of non-fullerene wide band gap donor (DRTB-T) and non-fullerene low band gap acceptor (IC-C6IDT-IC) as shown in Figure 33 [38] leading to a PCE of 9.08%. Wide band gap donor DRTB-T was designed, synthesized and material properties were evaluated. Absorption spectrum of donor DRTB-T (three alkylthiophene-benzodithiophne units linked with rhodanine groups either side) and acceptor IC-C61DT-IC (fused five rings core donor linked either side with dicyanovinylidine acceptor) indicated their complementarities and also covered ~90% of the solar spectrum in the region of 300 nm to 900 nm. Solar cell device fabricated structure was given as: ITO/MoO3/DRTB-T:IC-C6IDT-IC/Al. Solar cell structure was optimized in terms of it thickness, film morphology, donor/acceptor ratio and other aspects to get better efficiency. Indeed authors reported 9.08% PCE with Voc of 0.98V, Jsc of 14.25 mA, and FF of 65%. The efficiency produced was reported to be the highest ever reported in non-fullerene small molecule solar cells.

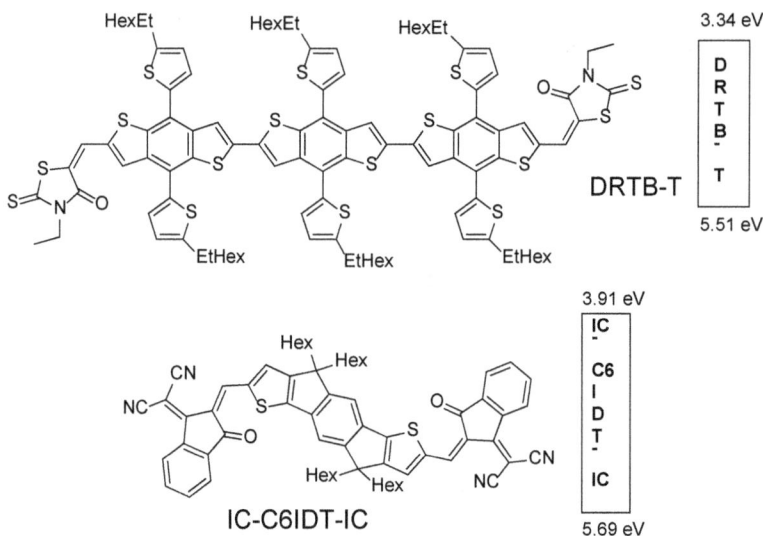

Figure 33 Trimeric BDT linked rhodanine [38].

Huifeng Yao et al. [39] synthesized fused five rings core flanked with thiophene attached electron withdrawing difluoro-dicyanovinylidine end groups having efficient intramolecular charge transfer, leading to a ultralow band gap acceptor material, IEICO-4F (Figure 34) [39]. The IEICO-4F optical band gap is found to be 1.24 eV, charge mobility calculated is 1.14×10^{-4} cm^{2}V^{-1} s^{-1} and the absorption of thin solid film extended in to NIR region with λ_{max} of ~900 nm. Polymer PBDTTT-EFT (PTB-7-Th) and Polymer

J52 (Figure 34) [39] were employed as donors for device fabrication based on their favourable absorption properties and also with suitable energy levels. Fabricated device parameters are as follows: Voc = 0.739 V; Jsc = 22.8 mA; FF = 59.4% and PCE = 10.0% using polymer donor PBDTTT-EFT (PTB-7-Th), J52 polymer donor exhibited a little smaller values compared to PBDTTT-EFT. The authors have advocated that ultralow band gap materials definitely have a role to play to improve the solar cell efficiencies.

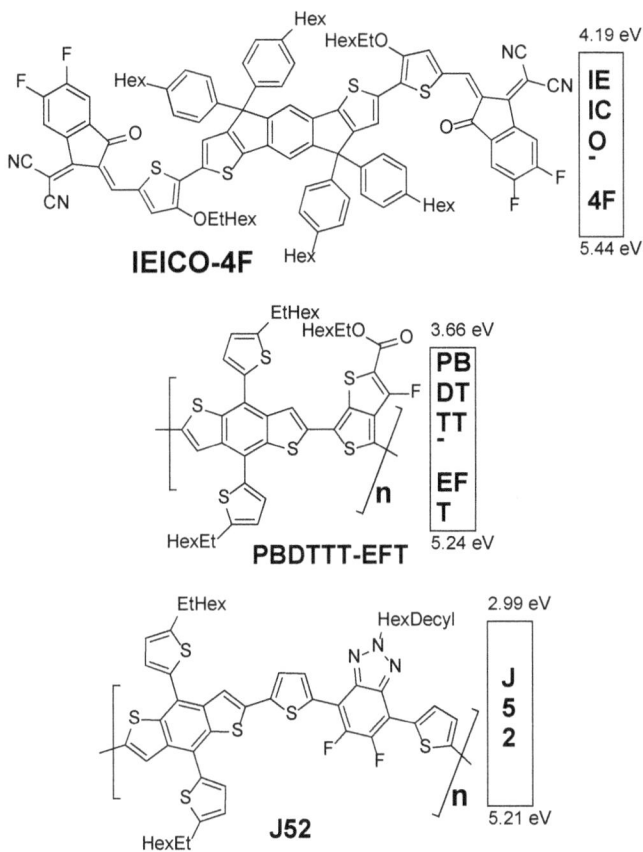

Figure 34 Fused ring coupled with Fluorodicyanoindenone [39].

Oh Kyu Kwon et al. [40] came up with a new idea of employing all small molecule non-fullerene solar cells with low band gap small molecule donor, *p*-DTS(FBTTh$_2$)$_2$ (Donor1-Acceptor-Donor2-Acceptor-Donor1) and small molecule acceptor, NIDCS–MO (Acceptor1-Donor1-Acceptor2-Donor2-Acceptor2-Donor1-Acceptor1) (Figure 35) [40]. The device structure is: ITO/PEDOT:PSS/*p* -DTS(FBTTh$_2$)$_2$:NIDCS–MO/Ca/Al and the generated device parameters, show a maximum PCE of 5.44% with a *V* oc of 0.85 V, a *J* sc of 9.68 mA cm $^{-2}$, and a FF of 0.66. UV-Visible absorption and AFM data for *p*-DTS(FBTTh$_2$)$_2$/NIDCS–MO blend, before and after annealing at suitable temperature informed that morphology of the casted film has a big role to play in improving charge mobilities, Jsc, FF and PC Efficiency.

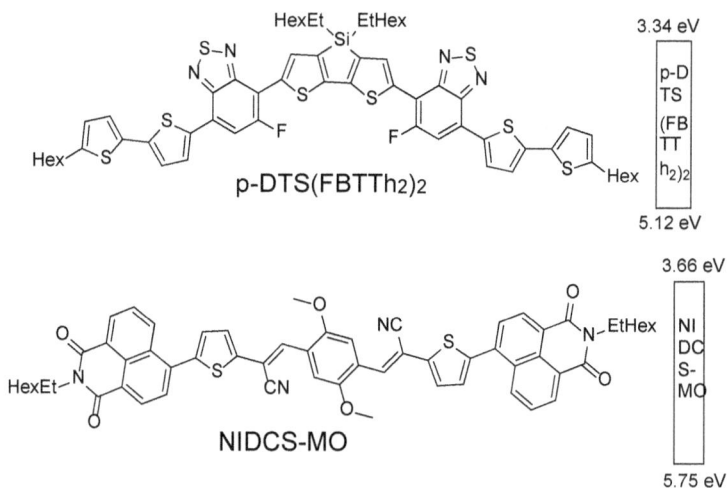

Figure 35 FluoroBTD – DTSilole based compound [40].

Dan Deng et al. [41] designed and synthesized non-fullerene small acceptor molecules – BTID-0F, BTID-1F and BTID-2F, by systematically varying fluorine and alkyl group substitution (Figure 36) [41] with a purpose of tuning HOMO (highest occupied molecular orbital) – LUMO (lowest unoccupied molecular orbital) energy levels and the blended film morphology [41]. Oxidation-Reduction potentials determined informed the variation in HOMO-LUMO energy levels upon (BTID-0F), mono (BTID-1F) and (BTID-2F) difluorinations (Figure 20) [41]. Device structure developed was: ITO/ZnO/ Blend layer/MoOx/Ag and the resultant optimized device parameters are: Voc = 0.93 V, Jsc = 14.0 mA, FF = 64% and PCE = 8.3% for BTID-0F(Acceptor1-Donor1-Donor2-Donor3-Donor2-Donor1-Acceptor1); Voc = 0.94 V, Jsc = 15.3 mA, FF = 72% and PCE = 10.4% for BTID-1F(Acceptor2-Donor1-Donor2-Donor4-Donor2-Donor1-Acceptor2); and Voc = 0.95 V, Jsc = 15.7 mA, FF = 76% and PCE = 11.08% for BTID-2F(Acceptor3-Donor1-Donor2-Donor5-Donor2-Donor1-Acceptor3). Various techniques like grazing incidence wide angle X-ray scattering, atomic force microscopy, resonant soft X-ray scattering, transmission electron microscopy and X-ray photoelectron spectroscopy were utilized to study the morphology of film, to modify and improve the vertical and lateral morphology to balance charge separation, charge transfer and charge collection associated with casted film, thereby improving the efficiency for BTID-2F to 11.08%.

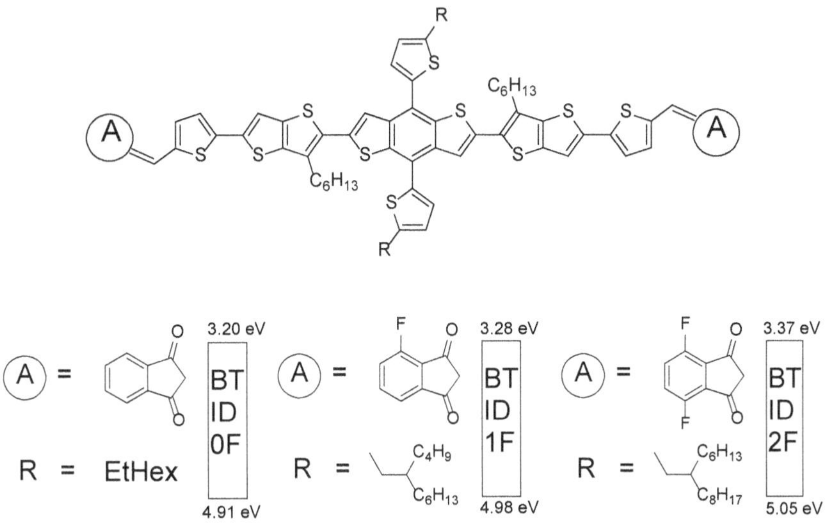

Figure 36 Thiophene BTID linked Fluoro derivatives [41].

Zhi Li et al. [42] designed and synthesized low band gap small donor molecule DERHD7T (seven thiophene rings linearly linked with rhodanine end groups), linearly linked hepta-thiophene with rhodanine terminal end groups (Figure 37) [42]. The six alkyl chains attached at specified region of DERHD7T improved solubility and also exerted influence on film morphology [42]. Solution phase cyclic voltametry measurements provided HOMO energy level as 5.00 eV and LUMO energy level as 3.28 leading to a band gap energy of 1.72 eV, which is found to be different from film state HOMO-LUMO energy levels. Intra-molecular charge transfer was indicated based on the redshift observed in DERHD7T film state UV-Visible absorption spectrum, which facilitated charge separation and mobility. Conventional solution spin-coating process was adopted to fabricate device with structure: ITO/PEDOT: PSS/DERHD7T:acceptor/LiF/Al. Device with highest PCE had these parameters: Voc = 0.92 V; Jsc = 13.98 mA; FF = 47.4% and PCE = 6.10%.

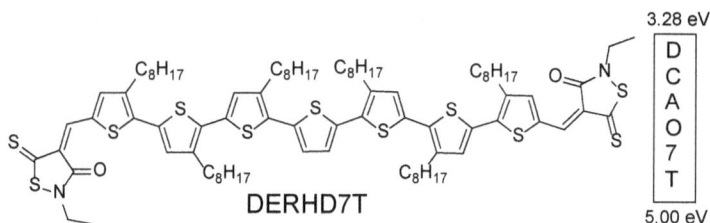

Figure 37 Heptathiophene – rhodanine derivative [42].

Huifeng Yao et al. [43], designed and synthesized a novel small molecule, ITCC (fused seven rings with dicyanithienoindenone end groups) as a non-fullerene acceptor material for organic solar cells to determine PC efficiency (Figure 38) [43]. Linearly fused 7 rings acted as donor, which was flanked both sides with acceptors (thienyl fused indanone end groups) and it was designated as ADA type architecture. Thin films were subjected to GIWAXS investigations to understand whether there existed π – π stacking to some extent and facilitated charge mobility in a better way than ITIC (which was reported earlier). Further it indicated role of thieno indanone and the superior design. Device architecture formulated was: ITO/PEDOT-PSS/ITCC or ITIC + PBDB-T/PFN-Br/Al. Solar cell photovoltaic parameters determined were: Voc = 1.01 V; Jsc = 15.9 mA; FF = 71%; PCE = 11.4% for ITCC and for ITIC (fused seven rings with dicyanoindenone end groups): Voc = 0.93; Jsc = 17.0 mA; FF = 67%; PCE = 10.6%. Authors claim that investigations have high relevance, because of 11.4% PCE noted in non-fullerene solar cell device. The results are of promising nature and suggest that the PCE can be improved to higher levels with these non-fullerene based solar cells.

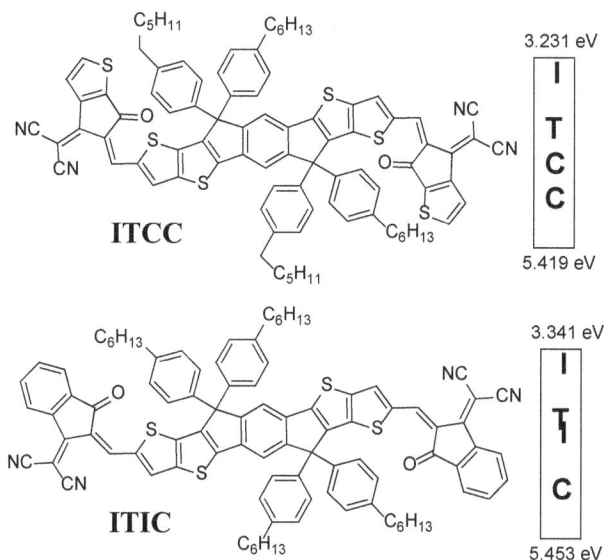

Figure 38 Fused Seven membered ring – Thienodicyanoindenones [43].

Yunlong Ma et al. [44] synthesized two novel low band gap small molecules as non-fullerene acceptors for solar cell fabrication (Figure 39) [44]. Both the molecules, DTNIC6 (six fused ringscarrying hexylchain, attached either side with dicyanovinylidine indanone) and DTNIC8 (six fused rings carrying ethylhexylchain, attached either side with dicyanovinylidine indanone), are of ladder type because of linearly fused six rings and the two molecules differ in their alkyl groups attached. Both the molecules have strong absorption in 500 nm to 720 nm region. Hole and Electron mobilities, AFM, TEM and GIWACS information obtained to analyse film morphology. Solar cell devices were fabricated using polymer PBDB-T as donor with device architecture as: ITO/TiO2:TOPD/ PBDB-T:DTNIC8/MoO3/Ag. Photovoltaic parameters generated were: for DTNIC6:- Voc = 0.96 V; Jsc = 7.71 mA; FF = 45.6%; PCE = 3.39%; for DTNIC8:- Voc = 0.96 V; Jsc = 12.92 mA; FF = 72.84%; PCE = 9.03%. DTNIC8 (which carries ethylhexyl alkyl chain) exhibited a high photo conversion efficiency compared to DTNIC6 which differed in alkyl chain from DTNIC8 showed less PCE. The alkyl chain did not influence energy levels and light absorption properties, but exerted sizable effect on the solar cell efficiency. The alkyl chain group effect was believed to have control over film morphology.

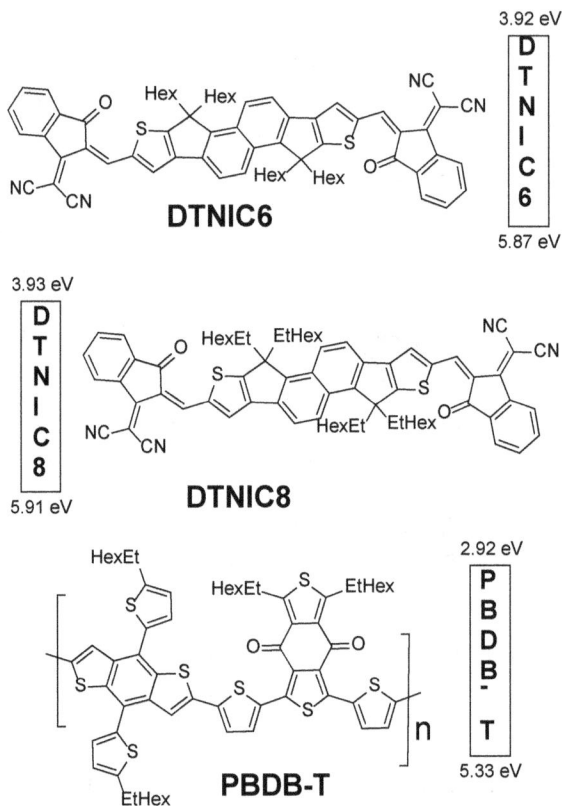

Figure 39 Fused six membered ring with Dicyanoindenones [44].

Yuze Lin et al. [45] employed (Figure 40) [45] medium band gap polymer donor – FTAZ (BG = 2.41 eV) with a nonfullerene low band gap acceptor – IDIC (BG = 1.6 eV) to process solar cells to understand the photo current efficiency [45]. FTAZ and IDIC have complementary absorption to cover 450 nm to 800 nm region, relatively have high electron and hole mobilities and well-matched energy levels. Single junction solar cell fabrication structure was: ITO/ZnO/FTAZ:IDIC/MoOx/Ag. Solar cell parameters observed were: Voc = 0.840 V; Jsc = 20.8 mA; FF = 71.8% and PCE = 12.5%, diiodoocatane was used to tune the film morphology in these fabrications. The 12.5% PCE determined for non-fullerene solar cell was very high compared to FTAZ-PCBM blend, which showed only ~6%. Femto second transient absorption studies on the casted films indicated the formation of radical cation and radical anion (charged species) and their mobilities. Authors inferred that FTAZ-PCBM combination film provided only ~40% generation of charged species compared to non-fullerene FTAZ-IDIC film combination. Authors claim that non-fullerene blends have superiority over fullerene blends in achieving higher PCE values.

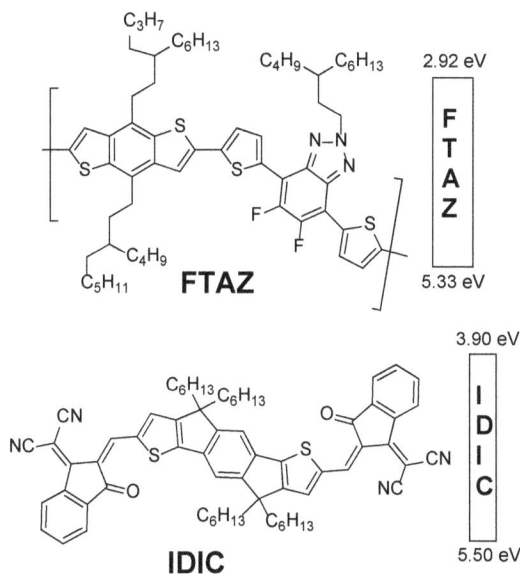

Figure 40 Fused five membered ring with Dicyanoindenone [45].

Jie Zhang et al. [46] designed and synthesized a low band gap small acceptor molecule-IFTIC (Figure 41) [46] for evaluating its solar cell efficiency. IFTIC carries fused bifluorene attached on both sides with thiophene as electron donating central core, with either side holding indenone moiety as electron acceptor. IFTIC showed absorption covering 450 nm to 700 nm and had suitable energy levels like 5.42 eV HOMO and 3.85 eV LUMO. PTB7-Th polymer was used as donor in these investigations for luminescence quenching (with IFTIC) and as donor material. AFM and TEM techniques were used for monitoring morphology of the casted film. Fabricated device parameters were: Voc = 0.92 V; Jsc = 12.71 mA; FF = 54%; PCE = 6.33%. Authors claim that non-fullerene materials with relatively simple device structure and simple method of preparation of materials make this work attractive.

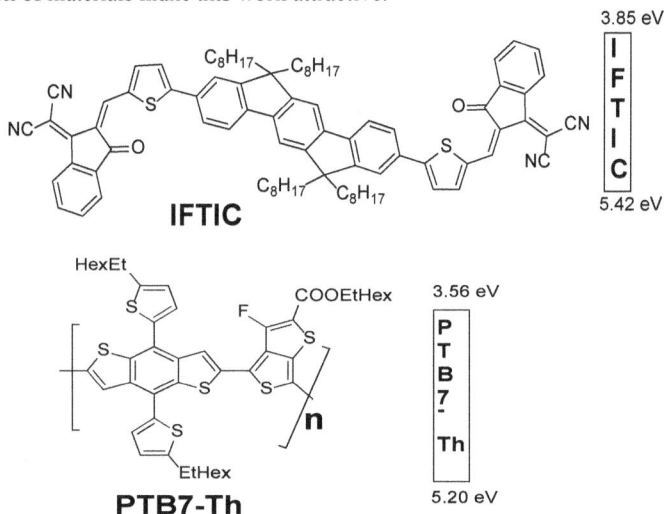

Figure 41 Bifluerene – Dicyanoindenone [46].

Oh Kyu Kwon et al. [47] described a non-fullerene base material photovoltaic device which exhibited a high percent of photo current (Figure 42) [47]. PPDT2FBT, a very well ordered polymer was used as donor and NIDCS-HO, a small molecule as an acceptor for building a solar cell device [47]. Overlayed absorption spectra for polymer donor and small molecule acceptor indicated complementary absorption, by covering 350 nm to 700 nm region. Non-fullerene based conventional single cell device adopted structure was given: ITO/PEDOT:PSS/active layer/Ca/Al. Blend film morphology was investigated systematically using charge mobility data, AFM, TEM, GIWACS, annealing temperature and other techniques. Device photovoltaic parameters determined were: Voc = 1.03 V; Jsc = 11.88 mA; FF = 63%; PCE = 7.64%. Investigations indicated complementarities in absorption, well placed energy levels along with good film morphology can improve the solar cell efficiency.

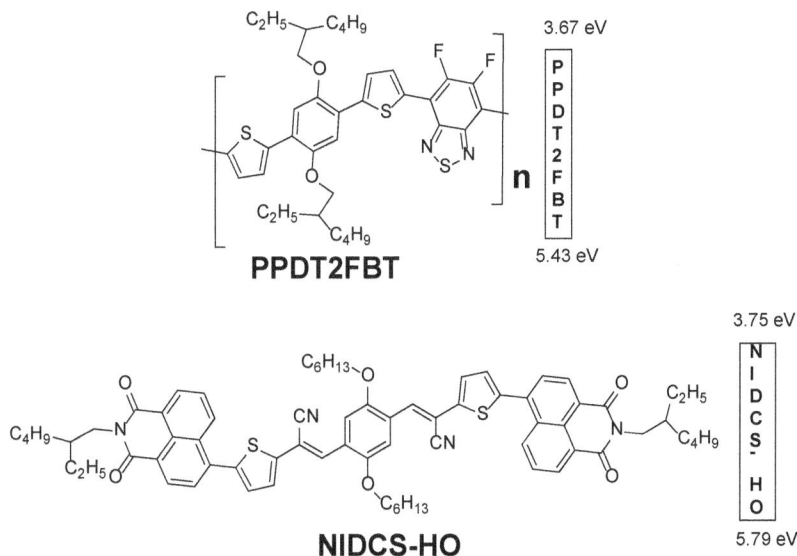

Figure 42 Dialkoxybenzene core based molecule [47].

Sunsun Li et al. [48] synthesized a series of novel methoxyl-modified dithieno[2,3-d:2′,3′-d′]-s-indaceno[1,2-b:5,6-b′]dithiophene-based (ITIC based) low band gap small-molecule acceptors, IT-OM-1, IT-OM-2, IT-OM-3, and IT-OM-4 (Figure 43) [48], with A-D-A architecture, for the purpose of developing non-fullerene based organic solar cells [48]. Position of "methoxy" substituent on terminal group was systematically varied to understand the positional effect of substitution on optical, electrochemical, charge mobility and more importantly molecular packing of these isomers. Donor molecule used in these investigations was PBDB-T polymer. PBDB-T donor blended with IT-OM acceptor film morphology was thoroughly investigated using AFM, GIWACS, TEM and charge mobility techniques. Devices were fabricated by adopting conventional cell configuration like: ITO/ZnO/active layer/MoO₃/Al to evaluate photovoltaic parameters and in particular PCE. Additive 1,8-diodooctane was employed in these fabrications and also annealed at 150 °C to make film. The Donor – Acceptor blend exhibited excellent

UV-Visible absorption covering 350 nm to 800 nm region. Among the four isomers synthesized, IT-OM-2 demonstrated excellent photovoltaic parameters like: Voc = 0.93 V; Jsc = 17.53 mA; FF = 73%; PC Efficiency of 11.9%. Further the PCE of >10% maintained when the thickness of the solar cell increased to 250 nm for IT-OM-2 blend system. The authors claimed that it is possible to modulate intrinsic molecular properties and also bulk film morphology to achieve excellent PCE in solar cells by designing molecules for fullerene free solar cells.

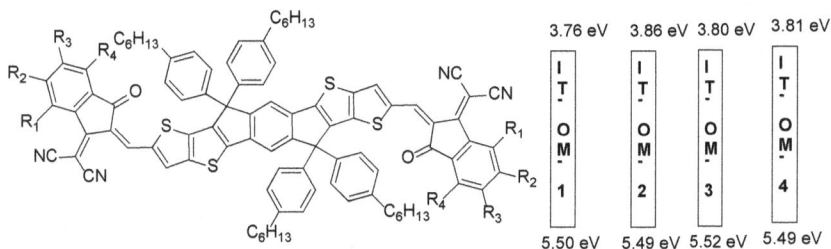

IT-OM-1 = R1 = OMe; R2=R3=R4=H
IT-OM-2 = R2 = OMe; R1=R3=R4=H
IT-OM-3 = R3 = OMe; R1=R2=R4=H
IT-OM-4 = R4 = OMe; R1=R2=R3=H

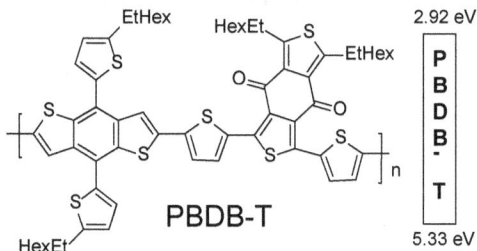

Figure 43 Fused seven membered ring – Methoxydicyanoindenones [48].

Huanran Feng et al. [49] reported synthesis of a new non-fullerene A-D-A type low band gap acceptor small molecule – FDNCTF (Figure 44) [49]. Linearly fused seven rings, three five- membered, two thiophene and two benzene rings, acting as donor and with electron withdrawing end group (NINCN) flanked on either side of the molecule – (FDNCTF) was designed. Properties of FDNCTF are (i) UV-Visible absorption reaching near infrared with high molar absorption coefficient (~3 x 10^5 L mol^{-1}), (ii) exhibited higher charge mobility, and (iii) more ordered arrangement of molecules in the film state, besides other properties. PBDB-T polymer having wide band gap acting as donor was employed for fabricating solar cells with FDNCTF. The solar cell devices were fabricated and determined photo voltaic parameters by adopting convention device architecture like ITO/PEDOT:PSS/PBDB-T:FDNCTF/PDINO/Al (PDINO is a cathode interlayer). FDNCTF as a small molecule low band gap acceptor displayed impressive efficiency of 11.2% PCE along with other parameters like Voc = 0.93V, Jsc = 16.5 mA, and FF = 72.7%. For comparison purpose they fabricated solar cell device with FDICTF as small molecule low band gap acceptor with PBDB-T polymer as donor to understand structural aspects of small molecule on the solar cell efficiency. Interestingly FDICTF with PBDB-T showed 10.06% showcasing the better design of FDNCTF in performance. FDNCTF & PBDB-T blend transient absorption studies indicated that charge mobility was very good as correlated with film morphology. These investigations highlighted that the molecular design with larger and conjugated electron withdrawing end groups improved absorption, molecular packing in the film state which in-turn improved the solar cell efficiency.

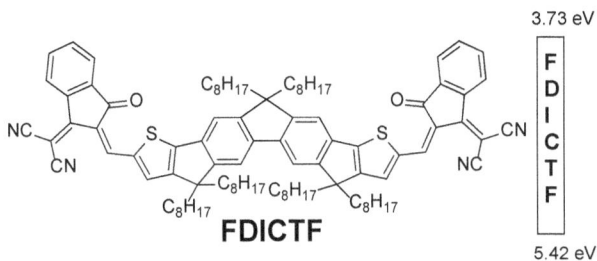

FDICTF

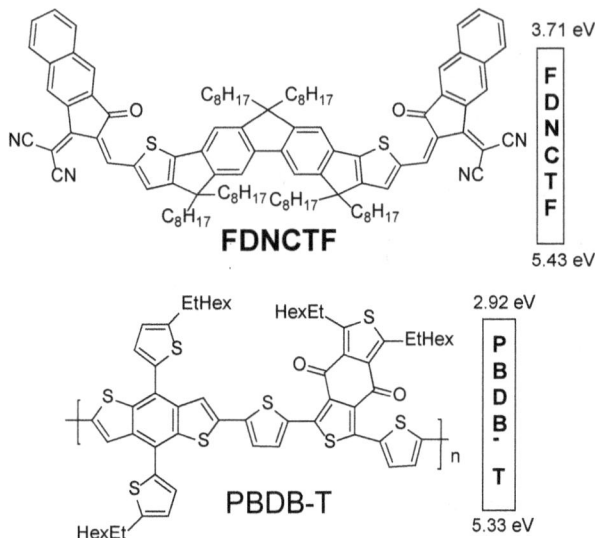

Figure 44 Fused seven membered ring – Dicyanoindenones [49].

Feng Liu et al. [50] designed and synthesized low band gap small molecule-ATT1as a non-fullerene acceptor, where dicyano-rhodanine group was attached on both sides of ATT1 (Figure 45)[50]. PTB7-Th polymer was selected as a donor in these studies. Solar cell device fabrication adopted a conventional procedure like ITO/PEDOT:PSS/PTB7-Th:ATT-1/PFN/Al adopted. Photovoltaic parameters determined were: Voc = 0.88V, Jsc = 16.18 mA, FF = 68% and PCE = 9.78% at a film thickness of 100 nm, the efficiency was found to be increased when the film thickness increased to 130 nm -- Voc = 0.87V, Jsc = 16.48 mA, FF = 70% and PCE = 10.07%. But the efficiency decreased to 9.59% at a film thickness of 160 nm. Central core structure of the molecule ATT1 was more planar, had a high molar absorption coefficient, good charge transporting quality and facilitated film forming with good morphology. Authors believe that this design has capabilities to take forward to achieve higher efficiencies.

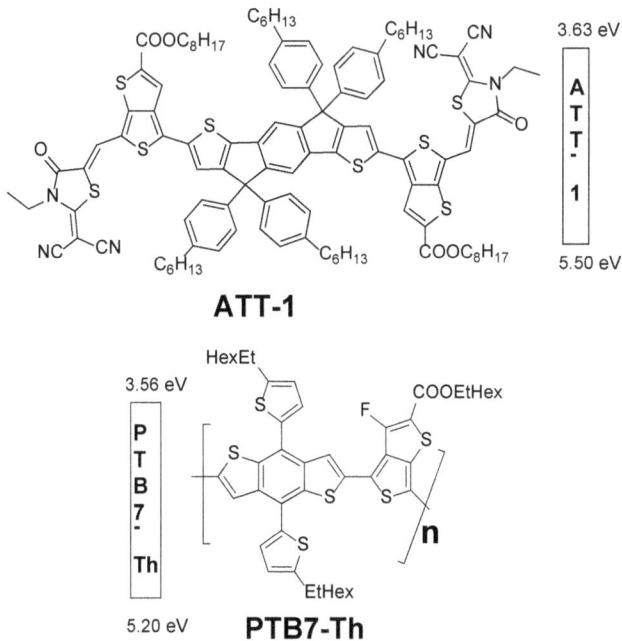

Figure 45 Fused five membered ring-thienothiophene Dicyanorhodanine [50].

Yuvraj Patil et al. [51] designed and synthesized two low band gap small non-fullerene acceptor molecules DPP7 and DPP8 (Figure 46) [51] and investigated their solar cell parameters [51]. Diketopyrrolopyrrole and tetracyano-diene fragments act as acceptor and carbazole moiety represents as donor part leading to D-A-D type architecture. Polymer was employed as donor material in the solar cell fabrications. The complimentary absorption of polymer-donor gives wide absorption for the blend. The configuration of the solar cell fabricated was: ITO/PEDOT:PSS/P: DPP7 or DPP8/PFN/Al. Photovoltaic parameters obtained for DDP7 was 4.86% efficiency and for DPP8 was 7.19% efficiency. It was informed that the superior performance of DPP8 was due to its structure, which contributed mainly to the morphology of the film as well as to the betterment of the charge mobility.

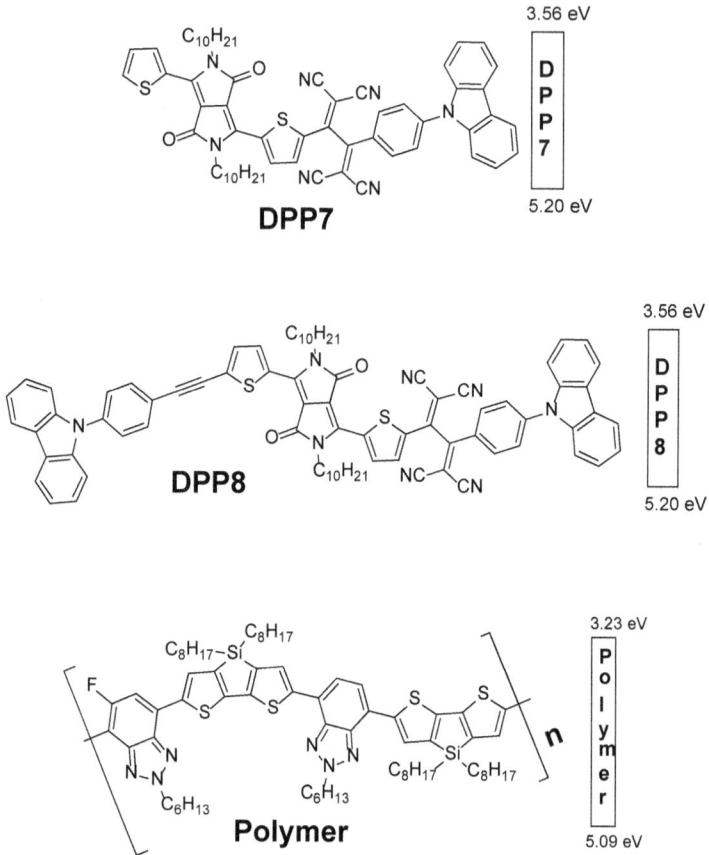

Figure 46 DPP-thiophene-tetracyanobutadiene-Carbazole hybrid [51].

Jia Sun et al. [52] synthesized two ultra low band gap small non-fullerene acceptor molecules, INPIC and INPIC-4F (Figure 47) [52]. These are interesting molecules in terms of design that there are contiguously nine rings fused together to form donor part of the molecule and flanked either side with electron withdrawing groups like dicyano-indenone (INPIC) and dicyano-difluro-indenone, making them as A-D-A type architecture. INPIC and INPIC-4F exhibit absorption 600 nm to 900 nm region there by complementing with the PBDB-T polymer acting as donor, further the blend of donor-acceptor absorption encompasses 350 nm to 900 nm. Solar cell devices were fabricated according to the given configuration: ITO/ZnO/active layer/MoO3/Ag and the photovoltaic parameters were as follows; INSPIC-4F displayed impressive parameters V_{oc} = 0.85V, J_{sc} = 21.6 mA, FF = 71.5%, PCE = 13.13%, whereas INSPIC showed only 4.31% efficiency. The difference between the two molecules INSPIC and INSPIC-4F was "fluoro" substitution and the same was reflected in solar cell efficiency improving from 4.31% to 13.13%. INSPIC-4F/PBDB-T blend morphology exhibited well defined texture, high charge mobility and improved light absorption property which were cited in favour of excellent efficiency shown by the INSPIC-4F.

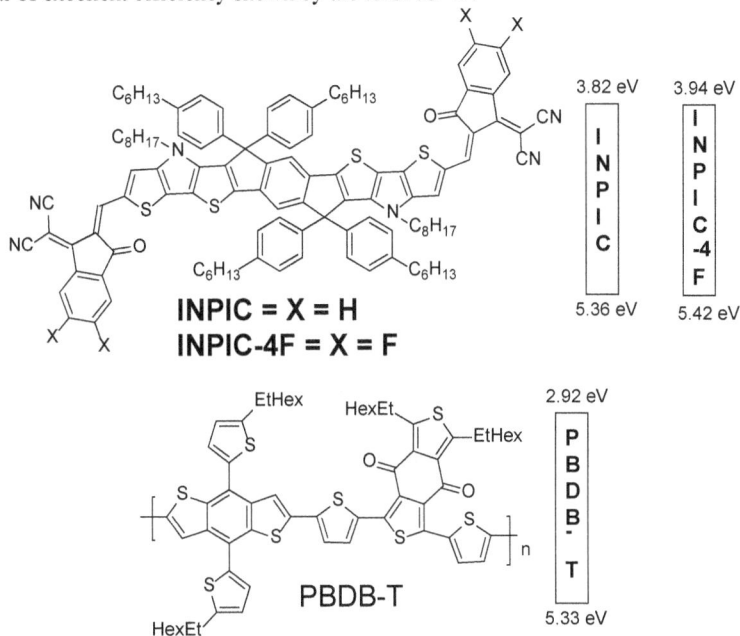

Figure 47 Fused nine membered ring -Fluorodicyanoindenone [52].

Zhuping Fei et al. [53] synthesized low band gap non-fullerene acceptor small molecule, C8-ITIC (Figure 48) [53]. Acceptor molecule has seven contiguously fused rings flanked by either side with indacenodicyano electron withdrawing group and carries four n-octyl alkylchains. Authors employed two donors like (i) PBDB-T polymer and (ii) PFBDB-T polymer. Solar cell devices were fabricated by adopting given configuration:

ITO/In$_2$O$_2$/ZnO/active layer/MoO$_3$/Ag. Impressive photovoltaic parameters were obtained with conversion efficiency of 13.2% using C8-ITIC and PFBDB-T blend. The energy loss noted in the solar cell device is less than 0.56 eV. Non fluorinated polymer PBDB-T with C8_ITIC blend recorded lower efficiency. It was informed that polymer backbone selective fluorination is another important factor to achieve higher conversion efficiencies in solar cell devices.

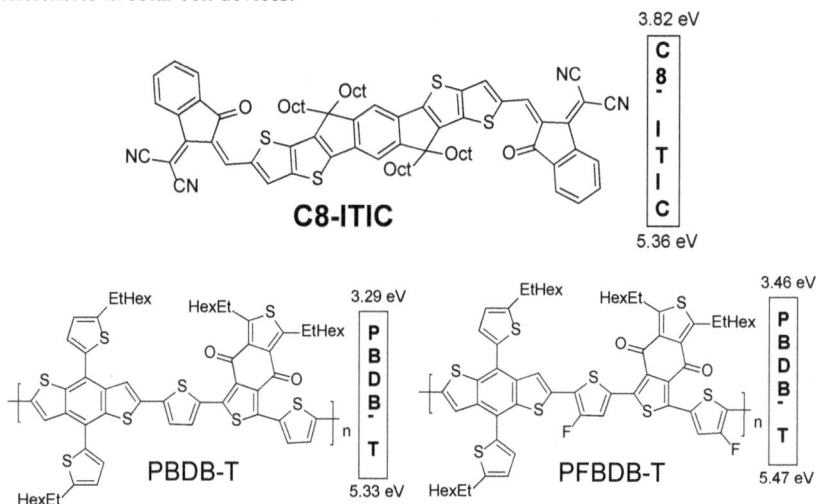

Figure 48 Fused seven membered ring – Dicyanoindenone [53].

Conclusions

Perusal of information narrated in this chapter, provides that organic solar cells (OSCs) are the promising next generation solar energy harvesting green technology sources. Organic materials required for OS cells are low cost and light weight. Fullerene and non-fullerene based OSCs are described in detail. Fullerene based OSCs have drawn great attention during the last two decades, taking the efficiency to a level of ~10%. There are some limitations with fullerene based OSCs like, cost of fullerene, light absorption property of fullerene, rigid structure of fullerene, relatively non- adjustable electrochemical property, and film forming property of fullerene which hindered further improvement in OSCs. Non-fullerene based OSCs are preferred because of several advantages like flexible nature, easy to make organic molecules in a simple way,

Materials for Solar Cell Technologies I
Materials Research Foundations **88** (2021) 176-235

Materials Research Forum LLC
https://doi.org/10.21741/9781644901090-7

modulate optical, thermal and electrochemical properties, get them in pure form, form films in an orderly nature, reducing loss of energy during functioning and many more advantages. Indeed, these non-fullerene based OSC materials have displayed over >17% PC efficiency and can be taken to higher levels, where these can be commercialized. Organic materials for OSCs must be designed with a proper insight, by invoking supra-molecular chemistry principles, such that these can lead to very orderly nature of film morphology, which may eventually provide higher PC efficiencies, satisfying commercial importance. It can be assumed organic solar cells may contribute in a big way for the global energy requirements in near future.

Acknowledgements

The present chapter was authored purely out of academic interest to familiarize the young material science researchers to appraise them about the recent work appearing in organic solar cell materials, particularly related to organic molecules. The examples covered in this review are chosen from different journals. The authors of this chapter are highly appreciative of the research articles published for their contributions in the area of organic solar cell materials. This chapter is only representative in nature and not intended to be exhaustive. The authors of this chapter further acknowledge the original contributors and publishers of the articles cited here for their potential scientific work, with a larger interest in academic excellency.

VJR thanks Dr. B. Parthasaradhy Reddy, Chairman Heterodrugs, Pvt. Ltd. and Dr K. Ratnakar Reddy, Director HR Foundation for their encouragement. VJR also thanks CSIR, New Delhi for Emeritus Scientist honour.

References

[1] Brian O'Regan, Michael Grätzel, A low-cost, high-efficiency solar cell based on dye-sensitized colloidal TiO_2 films, Nature 353 (1991) 737 – 740. https://doi.org/10.1038/353737a0

[2] Chen Yongsheng, Cao Yong, Yip Hin-Lap, Xia Ruoxi, Ding Liming, Xiao Zuo, Ke Xin, Wang Yanbo, Zhang Xin Organic and solution-processed tandem solar cells with 17.3% efficiency Science. 361 (2018) 1094–1098. https://doi.org/10.1126/science.aat2612

[3] Jianquan Zhang, Huei Shuan Tan, Xugang Guo, Antonio Facchetti and He Yan, Material insights and challenges for non-fullerene organic solar cells based on small molecular acceptors, Nature Energy, 3 (2018) 720-731. https://doi.org/10.1038/s41560-018-0181-5

[4] L. Dou, Y. Liu, Z. Hong, G. Li, and Y. Yang, Low-bandgap near-IR conjugated polymers/molecules for organic electronics, Chem. Rev. 115 (2015) 12633–12665. https://doi.org/10.1021/acs.chemrev.5b00165

[5] Akihiro Kojima, Kenjiro Teshima, Yasuo Shirai, Tsutomu Miyasaka, Organometal Halide Perovskites as Visible-Light Sensitizers for Photovoltaic Cells, J. Am. Chem. Soc., 131 (2009) 6050–6051. https://doi.org/10.1021/ja809598r

[6] Oleg V. Kozlov, Vlad G. Pavelyev, Hilde D. de Gier, Remco W. A. Havenith, Paul H.M. van Loosdrecht, Jan C. Hummelen, and Maxim S. Pshenichnikov, Ultrafast electron and hole transfer in bulkhetero junctions of low-bandgap polymers, Org. Photonics Photovolt. 4 (2016) 24–34. https://doi.org/10.1515/oph-2016-0003

[7] Qiaoshi An, Fujun Zhang, Lingliang Li, Jian Wang, Jian Zhang, Lingyu Zhou, Weihua Tang, Improved efficiency of bulk heterojunction polymer solar cells by doping low-bandgap small molecules, ACS Appl. Materials & Interfaces, 6 (2014) 6537-6544. dx.doi.org/10.1021/am500074s

[8] Yuan Fang, Ajay K. Pandey, Alexandre M. Nardes, Nikos Kopidakis, Paul L. Burn, Paul Meredith, A narrow optical gap small molecule acceptor for organic solar cells, Adv. Energy Mater., 3 (2012) 54-59. https://doi.org/.1002/aenm.201200372

[9] Jianhua Liu, Yanming Sun, PreechaMoonsin, MartijnKuik, Christopher M. Proctor, Jason Lin, Ben B. Hsu, Vinich Promarak, Alan J. Heeger, and Thuc-Quyen Nguyen, Tri -Diketopyrrolopyrrole Molecular Donor Materials for High-Performance Solution-Processed Bulk Heterojunction Solar Cells, Adv.Mater.25 (2013) 5898-5903. https://doi.org/10.1002/adma.201302007

[10] Miao Yang, Xuewen Chen, Yingping Zou, Chunyue Pan, Bo Liu and Hong Zhong; A solution-processable D–A–D small molecule based on isoindigo for organic solar cells, J Mater Sci. 48 (2013)1014–1020. https://doi.org//10.1007/s10853-012-6831-2

[11] Cira Maglione, Antonio Carella, Roberto Centore, Patricia Chavez Patrick Leveque, Sadiara Fall, Nicolas Leclerc, Novel low bandgap phenothiazine functionalized DPP derivatives prepared by direct heteroarylation: Application in bulk heterojunction organic solar cells, Dyes & Pigments, 141 (2017) 169-178. http://dx.doi.org/10.1016/j.dyepig.2017.02.012

[12] Prabhat Gautam, Rajneesh Misra, Shahbaz A. Siddiqui, and Ganesh D. Sharma, Unsymmetrical Donor−Acceptor−Acceptor−π−Donor Type Benzothiadiazole-Based

Small Molecule for a Solution Processed Bulk Heterojunction Organic Solar Cell, ACS Applied Mater., 7 (2015) 10283-10292. https://doi.org/10.1021/acsami.5b02250

[13] Mahalingavelar Paramasivam, Akhil Gupta, Aaron M. Raynor, Sheshanth V. Bhosale, K. Bhanuprakash and V. Jayathirtha Rao, Small band gap D-p-A-p-D benzothiadiazole derivatives with low-lying HOMO levels as potential donors for applications in organic photovoltaics: a combined experimental and theoretical investigation, RSC Adv., 4 (2014) 35318-35331. https://doi.org/10.1039/c4ra02700k

[14] B. Ananda Rao, K. Yesudas, G. Siva Kumar, K. Bhanuprakash, V. Jayathirtha Rao, G.D. Sharma and S. P. Singh, Application of solution processable squaraine dyes as electron donors for organic bulk-heterojunction solar cells; Photochem. Photobiol. Sci., 12 (2013) 1688–1699. https://doi.org/10.1039/c3pp50087j

[15] Mahalingavelar Paramasivam, Akhil Gupta, N. Jagadeesh Babu, K. Bhanuprakash, Sheshanath V. Bhosale and V. Jayathirtha Rao, Funnel shaped molecules containing benzo/pyrido[1,2,5]thiadiazole functionalities as peripheral acceptors for organic photovoltaic applications, RSC Adv., 6 (2016) 66978–66989. https://doi.org/10.1039/c6ra06616j

[16] Daobin Yang, Hisahiro Sasabe, Takeshi Sano, and Junji Kido, Low-Band-Gap Small Molecule for Efficient Organic Solar Cells with a Low Energy Loss below 0.6 eV and a High Open-Circuit Voltage of over 0.9 V, ACS Energy Lett., 2 (2017) 2021−2025. https://doi.org/10.1021/acsenergylett.7b00608

[17] Ming Cheng, Bo Xu, Cheng Chen, Xichuan Yang, Fuguo Zhang, Qin Tan, Yong Hua, Lars Kloo, and Licheng Sun, Phenoxazine-Based Small Molecule Material for Efficient Perovskite Solar Cells and Bulk Heterojunction Organic Solar Cells, Adv. Energy Mater. (2015)1401720. https://doi.org/10.1002/aenm.201401720

[18] Kimin Lim, Seung Yeon Lee, Kihyung Song, G. D. Sharma and Jaejung Ko, Synthesis and properties of low bandgap star molecules TPA-[DTS-PyBTTh3]3 and DMM-TPA [DTS-PyBTTh3]3 for solution-processed bulk heterojunction organic solar cells, J. Mater. Chem. C, 2 (2014) 8412-8422. https://doi.org/10.1039/c4tc01495b

[19] Bin Kan, Miaomiao Li, Qian Zhang, Feng Liu, Xiangjian Wan,Yunchuang Wang, Wang Ni, Guankui Long, Xuan Yang, Huanran Feng,Yi Zuo, Mingtao Zhang, Fei Huang, Yong Cao, Thomas P. Russell, and Yongsheng Chen, A Series of Simple Oligomer-like Small Molecules Based on Oligothiophenes for Solution-Processed Solar Cells with High Efficiency, J. Am. Chem. Soc., 137 (2015) 3886-3893. https://doi.org/10.1021/jacs.5b00305

[20] Wang Ni, Miaomiao Li, Feng Liu, Xiangjian Wan, Huanran Feng, Bin Kan, Qian Zhang, Hongtao Zhang, and Yongsheng Chen, Dithienosilole-Based Small-Molecule Organic Solar Cells with an Efficiency over 8%: Investigation of the Relationship between the Molecular Structure and Photovoltaic Performance, Chem. Mater., 27 (2015) 6077-6084. https://doi.org/10.1021/acs.chemmater.5b02616

[21] Manohar Reddy Busireddy, Venkata Niladri Raju Mantena, Narendra Reddy Chereddy, Balaiah Shanigaram, Bhanuprakash Kotamarthi, Subhayan Biswas, Ganesh Datt Sharma and Jayathirtha Rao Vaidya, A dithieno[3,2-b:20,30-d]pyrrole based, NIR absorbing, solution processable, small molecule donor for efficient bulk heterojunction solar cells, Phys. Chem. Chem. Phys., 18 (2016) 32096-32106. https://doi.org/10.1039/c6cp06304g

[22] Manohar Reddy Busireddy, Venkata Niladri Raju Mantena, Narendra Reddy Chereddy, Balaiah Shanigaram, Bhanuprakash Kotamarthi, Subhayan Biswas, Ganesh Datt Sharma, and Jayathirtha Rao Vaidya, Dithienopyrrole-benzodithiophene based donor materials for small molecular BHJSCs: Impact of side chain and annealing treatment on their photovoltaic properties, Organic Electronics 37 (2016) 312 to 325. http://dx.doi.org/10.1016/j.orgel.2016.07.003

[23] Manohar Reddy Busireddy, Narendra Reddy Chereddy, Balaiah Shanigaram, Bhanuprakash Kotamarthi, Subhayan Biswas, Ganesh Datt Sharma and Jayathirtha Rao Vaidya, Dithieno[3,2-b:20,30-d]pyrrolebenzo[c][1,2,5]thiadiazole conjugate small molecule donors: effect of fluorine content on their photovoltaic properties, Phys. Chem. Chem. Phys. 19 (2017) 20513—20522. https://doi.org/10.1039/c7cp02729j

[24] Manohar Reddy Busireddy, Chakali Madhu, Narendra Reddy Chereddy, Ejjurothu Appalanaidu, Ganesh Datt Sharma, and Jayathirtha Rao Vaidya, Optimization of the Donor Material Structure and Processing Conditions to Obtain Efficient Small-Molecule Donors for Bulk Heterojunction Solar Cells, Chem. Photo.Chem. 2 (2018) 81 – 88. https://doi.org/10.1002/cptc.201700170

[25] Manohar Reddy Busireddy, Madhu Chakali, Gontu Ramanjaneya Reddy, Narendra Reddy Chereddy, Balaiah Shanigaram, Bhanuprakash Kotamarthi, Ganesh D. Sharma, and V. Jayathirtha Rao, Influence of the backbone structure of the donor material and device processing conditions on the photovoltaic properties of small molecular BHJSCs, Solar Energy 186 (2019) 84–93. https://doi.org/10.1016/j.solener.2019.05.001

Materials Research Forum LLC
https://doi.org/10.21741/9781644901090-7

[26]   B. Ananda Rao, M. Sasi Kumar, G. Sivakumar, Surya Prakash Singh, K. Bhanuprakash, V. Jayathirtha Rao, and G. D. Sharma, Effect of Incorporation of Squaraine Dye on the Photovoltaic Response of Bulk Heterojunction Solar Cells Based on P3HT:PC70BM Blend, ACS Sustainable Chem. Eng. 2 (2014) 1743−1751. https://doi.org//10.1021/sc500276u

[27]   Jae Woong Jung, and Won Ho Jo, Low Bandgap Small Molecule as Non-Fullerene Electron Acceptor Composed of Benzothiadiazole and Diketopyrrolopyrrole for All Organic Solar Cells, Chem. Mater., 27 (2015) 6038-6044. https://doi.org/10.1021/acs.chemmater.5b02480

[28]   Christopher J. Takacs, Yanming Sun, Gregory C Welch, Louis A. Perez, Xiaofeng Liu, Wen Wen, Guillermo C. Bazan, and Alan J. Heeger, Solar cell efficiency, self-assembly and dipole-dipole interactions of isomorphic narrow bandgap molecules, J. Am. Chem. Soc., 134 (2012) 16597-16606. https://doi.org/10.1021/ja3050713

[29]   Yuze Lin, Yongfang Li, and Xiaowei Zhan, A Solution-Processable Electron Acceptor Based on Dibenzosilole and Diketopyrrolopyrrole for Organic Solar Cells, Adv. Energy Mater., 3 (2013) 724-728. https://doi.org/10.1002/aenm.201200911

[30]   Wang Ni, Miaomiao Li, Bin Kan, Feng Liu, Xiangjian Wan, Qian Zhang, Hongtao Zhang, Thomas P Russell and Yongsheng Chen, Fullerene free small molecule organic solar cells with high open circuit voltage of 1.15V, Chem. Commun., 52 (2016) 465-468. https://doi.org/10.1039/C5CC07973J

[31]   Feng Liu, Haijun Fan, Zhiguo Zhang, and Xiaozhang Zhu, Low-Bandgap Small-Molecule Donor Material Containing Thieno[3,4-b]thiophene Moiety for High-Performance Solar Cells, ACS Applied Materials & Interfaces, 8 (2015) 3661-3668. https://doi.org/10.1021/acsami.5b08121

[32]   Dongxue Liu, Ting Wang, Xin Ke, Nan Zheng, Zhitao Chang, Zengqi Xie and Yongsheng Liu, Ultra-narrow bandgap non-fullerene acceptors for organic solar cells with low energy loss, Mater. Chem. Front., 3(2019) 2157. https://doi.org/10.1039/c9qm00505f

[33]   Yong Cui, Huifeng Yao, Jianqi Zhang, Tao Zhang, Yuming Wang, Ling Hong, Kaihu Xian, Bowei Xu, Shaoqing Zhang, Jing Peng, Zhixiang Wei, Feng Gao and Jianhui Hou, Over 16% efficiency organic photovoltaic cells enabled by a chlorinated acceptor with increased open-circuit voltages, Nature Communications,15 (2019) 2515. https://doi.org/10.1038/s41467-019-10351-5

[34]   Yankang Yang, Beibei Qiu, Shanshan Chen, Qiuju Zhou, Ying Peng, Zhi-Guo Zhang, Jia Yao, Zhenghui Luo, Xiaofeng Chen, Lingwei Xue, Liuliu Feng, ChangdukYang, Yongfang Li, J. Mater. Chem. A, 6 (2018) 9613-9622. https://doi.org/10.1039/C8TA01301B

[35]   Bin Kan, Huanran Feng, Huifeng Yao, Meijia Chang, Xiangjian Wan, Chenxi Li, Jianhui Hou and Yongsheng Chen, A chlorinated low-bandgap small-molecule acceptor for organic solar cells with 14.1% efficiency and low energy loss, Science China Chemistry, (2018) 1-7. https://doi.org/10.1007/s11426-018-9334-9

[36]   Jiahui Wan, Xiaopeng Xu, Guangjun Zhang, Ying Li, Kui Feng and Qiang Peng, Highly efficient halogen-free solvent processed small-molecule organic solar cells enabled by material design and device engineering, Energy & Environmental Science, 10 (2017) 1739-1745. https://doi.org/10.1039/c7ee00805h

[37]   Huifeng Yao, Yu Chen, Yunpeng Qin, Runnan Yu, Yong Cui, Bei Yang, Sunsun Li, Kai Zhang, and Jianhui Hou, Design and Synthesis of a Low Bandgap Small Molecule Acceptor for Efficient Polymer Solar Cells, Adv. Mater., 28 (2016) 8283-8287. https://doi.org/10.1002/adma.201602642

[38]   Liyan Yang, Shaoqing Zhang, Chang He, Jianqi Zhang, Huifeng Yao, Yang Yang, Yun Zhang, Wenchao Zhao, and Jianhui Hou, A New Wide Band Gap Donor for Efficient Fullerene free All-small-molecule Organic Solar Cells,; J.Am.Chem.Soc. 139 (2017) 1958-1956. https://doi.org/10.1021/jacs.6b11612

[39]   Huifeng Yao, Yong Cui, Runnan Yu, Bowei Gao, Hao Zhang, and Jianhui Hou, Design, Synthesis, and Photovoltaic Characterization of a Small Molecular Acceptor with an Ultra-Narrow Band Gap, Angew. Chem. Int. Ed. 56 (2017)3045 –3049. https://doi.org/10.1002/anie.201610944

[40]   Oh Kyu Kwon, Jung-Hwa Park, Dong Won Kim, Sang Kyu Park, and Soo Young Park, An All-Small-Molecule Organic Solar Cell with High Efficiency Nonfullerene Acceptor, Adv.Mater.,27 ((2015)1951-1956. https://doi.org/10.1002/adma.201405429

[41]   Dan Deng, Yajie Zhang, Jianqi Zhang, Zaiyu Wang, Lingyun Zhu, Jin Fang, Benzheng Xia, Zhen Wang, Kun Lu, Wei Ma and Zhixiang Wei, Fluorination-enabled optimal morphology leads to over 11% efficiency for inverted small-molecule organic solar cells, Nature Commun. 7 (2016)13740. https://doi.org/10.1038/ncomms13740

[42]   Zhi Li, Guangrui He, Xiangjian Wan, Yongsheng Liu, Jiaoyan Zhou, Guankui Long, Yi Zuo, Mingtao Zhang, and Yongsheng Chen, Solution Processable Rhodanine-Based Small Molecule Organic Photovoltaic Cells with a Power

Conversion Efficiency of 6.1%, Adv. Energy Mater. 2 (2012)74–77.
https://doi.org/10.1002/aenm.201100572

[43]   Huifeng Yao, Long Ye, Junxian Hou, Bomee Jang, Guangchao Han, Yong Cui, Gregory M. Su, Cheng Wang, Bowei Gao, Runnan Yu, Hao Zhang, Yuanping Yi, Han Young Woo, Harald Ade, and Jianhui Hou, Achieving Highly Efficient Nonfullerene Organic Solar Cells with Improved Intermolecular Interaction and Open-Circuit Voltage, Adv.Mater. 29 (2017) 1700254.
https://doi.org/10.1002/adma.201700254

[44]   Yunlong Ma, Meiqi Zhang, Yu Yan, Jingming Xin, Tao Wang, Wei Ma, Changquan Tang, and Qingdong Zheng, Ladder-Type Dithienonaphthalene-Based Small-Molecule Acceptors for Efficient Nonfullerene Organic Solar Cells, Chem. Mater. 29 (2017) 7942-7952. https://doi.org/10.1021/acs.chemmater.7b02887

[45]   Yuze Lin, Fuwen Zhao, Shyamal K. K. Prasad, Jing-De Chen, Wanzhu Cai, Qianqian Zhang, Kai Chen, Yang Wu, Wei Ma, Feng Gao, Jian-Xin Tang, Chunru Wang, Wei You, Justin M. Hodgkiss, and Xiaowei Zhan, Balanced Partnership between Donor and Acceptor Components in Nonfullerene Organic Solar Cells with >12% Efficiency, Adv.Mater.30 (2018) 1706363.
https://doi.org/10.1002/adma.201706363

[46]   Jie Zhang, Baofeng Zhao, Yuhua Mi, Hongli Liu, Zhaoqi Guo, Guojun Bie, Wei Wei, Chao Gao, and Zhongwei An, A New Wide Band gap Small Molecular Acceptor Based on Indenofluorene Derivatives for Fullerene-Free Organic Solar Cells, Dyes & Pigments, 140 (2017) 261-268. https://doi.org/10.1016/j.dyepig.2017.01.039

[47]   Oh Kyu Kwon, Mohammad Afsar Uddin, Jung-Hwa Park, Sang Kyu Park,Thanh Luan Nguyen, Han Young Woo, and Soo Young Park, A High Efficiency Nonfullerene Organic Solar Cell with Optimized Crystalline Organizations, Adv.Mater. 28 (2015) 910-916. https://doi.org/10.1002/adma.201504091

[48]   Sunsun Li, Long Ye, Wenchao Zhao, Shaoqing Zhang, HaraldAde, and Jianhui Hou, Significant Influence of the Methoxyl Substitution Position on Optoelectronic Properties and Molecular Packing of Small-Molecule Electron Acceptors for Photovoltaic Cells, Adv. Energy Mater. 7 (2017) 1700183.
https://doi.org/10.1002/aenm.201700183

[49]   Huanran Feng, Nailiang Qiu, Xian Wang, Yunchuang Wang, Bin Kan, Xiangjian Wan, Mingtao Zhang, Andong Xia, Chenxi Li, Feng Liu, Hongtao Zhang, and Yongsheng Chen, An A-D-A Type Small-Molecule Electron Acceptor with End-

Extended Conjugation for High Performance Organic Solar Cells, Chem.Mater. 29 (2017) 7908-7917. https://doi.org/10.1021/acs.chemmater.7b02811

[50]   Feng Liu, Zichun Zhou, Cheng Zhang, Thomas Vergote, Haijun Fan, Feng Liu, and Xiaozhang Zhu, A Thieno[3,4-b]thiophene-Based Non-Fullerene Electron Acceptor for High-Performance Bulk-Heterojunction Organic Solar Cells, J. Am. Chem. Soc. 138 (2016) 15523-15526, https://doi.org/10.1021/jacs.6b08523

[51]   Yuvraj Patil, Rajneesh Misraa, M.L. Keshtov, Ganesh D. Sharma, Small molecule carbazole based diketopyrrolopyrroles with tetracyanobutadiene acceptor unit as Non-Fullerene Acceptor for Bulk Heterojunction Organic Solar Cells, J. Mater. Chem. A 5 (2017) 3311-3319. https://doi.org/10.1039/C6TA09607G

[52]   Jia Sun, Xiaoling Ma, Zhuohan Zhang, Jiangsheng Yu, Jie Zhou, Xinxing Yin, Linqiang Yang, Renyong Geng, Rihong Zhu, Fujun Zhang, and Weihua Tang, Dithieno[3,2-b:2′,3′-d]pyrrol Fused Nonfullerene Acceptors Enabling Over 13% Efficiency for Organic Solar Cells, Adv.Mater. 30 (2018) 1707150. https://doi.org/10.1002/adma.201707150

[53]   Zhuping Fei, Flurin D. Eisner, Xuechen Jiao, Mohammed Azzouzi, Jason A. Röhr, Yang Han, Munazza Shahid, Anthony S. R. Chesman, Christopher D. Easton, Christopher R. McNeill, Thomas D. Anthopoulos, Jenny Nelson, and Martin Heeney, An Alkylated Indacenodithieno[3,2-b]thiophene-Based Nonfullerene Acceptor with High Crystallinity Exhibiting Single Junction Solar Cell Efficiencies Greater than 13% with Low Voltage Losses, Adv.Mater. 30 (2018) 1705209. https://doi.org/10.1002/adma.201705209.

Materials for Solar Cell Technologies I
Materials Research Foundations 88 (2021) 236-258

Materials Research Forum LLC
https://doi.org/10.21741/9781644901090-8

Chapter 8

Absorber Materials for Solar Cells

Pallavi Jain[1], Palak Pant[2], Sapna Raghav[3], Dinesh Kumar[4*]

[1]Department of Chemistry, SRM-IST, Delhi-NCR Campus, Modinagar-210204, India

[2]Department of Computer Science, SRM-IST, Delhi-NCR Campus, Modinagar-210204, India

[3]Department of Chemistry, Banasthali Vidyapith, Banasthali, Tonk 304022, India

[4]School of Chemical Sciences, Central University of Gujarat, Gandhinagar, India

dinesh.kumar@cug.ac.in

Abstract

Solar cell production has grown rapidly in the last few decades. Essentially a solar cell (SC), known as a photovoltaic (PV) cell, is nothing more than a p-n junction, composed of a p-type and n-type semiconductor. The electric field is generated at the junction when electrons and holes pass towards the positive and negative terminals respectively. Light consists of photons, and when the light of a sufficient wavelength falls on the cells, the energy from the photon is passed to the valence band electrons, allowing electrons to move to a higher energy state called the conductive band. The entire process is carried out in the absorber layer that lies under the anti-reflective coating of the SC. Since most energy in sunlight and artificial light is within the visible range of electromagnetic radiation (EMR), a SC absorber can absorb radiation effectively at these wavelengths. Because a SC can be made using a variety of materials, its output depends solely on the properties of the material used. This chapter discusses different absorbent materials that are used for solar cells.

Keywords

Solar cell, Photovoltaic Cell, p-n Junction, Absorber Layer

Contents

Absorber Materials for Solar Cells 236

1.   Introduction..237

2. Advancement in solar cells by using up-conversion materials..........238

3. Upconversion (UC) mechanisms 239

4. Absorber materials for solar cells...239
    4.1 Indium gallium nitride (InGaN)-based SCs 239
    4.2 Copper (Cu) based absorber material...240
    4.3 Graphene based absorber material..242
    4.4 Sulphide based absorber materials 243
    4.4.1 Solar cells based on Sb_2S_3 244
    4.4.2 Tin sulphide based SCs...244
    4.4.3 Sulphide based SCs..244
    4.5 Metal nanoparticles (MNPs)..245

Conclusion...245

Acknowledgements...246

References 246

1. Introduction

The need for electronic items and gadgets is enhancing at a quick rate. However, the requirement of electric energy is fulfilled using energy resources that are limited in quantities. The by-products of these resources are also hazardous to human health as well as to the environment. Hence, water, wind, and solar energies are considered as the best alternatives to the exhausting energy resources to produce electric energy. Out of these, the most attractive energy resource is the solar energy, and owes its abundance to the sunlight. Hence, the advancement of SCs has seen significant growth since these cells convert solar energy directly to electric energy. SCs can solve the problems related to energy crisis and global warming due to carbon emissions. In general, the SCs are categorized as the first generation if these include the monocrystalline and multi-crystalline wafer-based silicon SCs (Si-SCs), the second generation include amorphous silicon (a-Si:H), gallium arsenide (GaAs), copper gallium selenium (CuGaSe), copper indium gallium sulphide (CIGS), cadmium telluride (CdTe) thin film based solar cells. The third generation involves copper zinc tin selenium (CZTSe), copper zinc tin sulphide (CZTS), dye-sensitized, quantum dot (QD), multijunction, and perovskite SCs (PSCs). Although, the recent photocell industry is mostly dominated by Si-SCs owing to their advanced technology, better stability over twenty years, and greater power conversion efficiency (PCE) [1]. Even after the advancement in technology, research on the coated SCs with solution processing at reduced temperatures has been going on to develop less priced SCs. In the last ten years, many hybrid organic-inorganic perovskites have been dominant in the industry because of its low price and facile coating process along with great carrier diffusion length, tunable bandgap, higher carrier mobility and a high coefficient of optical absorption [2-4]. Generally, organic and inorganic perovskite materials are chemically derived from CMX$_3$, where C = Inorganic cation (large, monovalent such as Rb$^+$, Cs$^+$)/organic amine cation (CH$_3$NH$_3$$^+$, NH$_2$OHNH$_2$$^+$) M= Metal cation (divalent) and X= halide anion (I$^-$, Cl$^-$, Br$^-$, or mixture). Methylammonium lead iodide (MAPbI$_3$/CH$_3$NH$_3$PbI$_3$) is predominantly the most investigated perovskite since it has a sufficient bandgap of 1.5 eV, which permits solar spectrum absorption and ambipolar attributes that are important [5]. The brilliant electrical and optical attributes of CMX$_3$ materials permitted the evolution of PSCs with an efficient power conversion, which was enhanced significantly from 3.8 % to 24.2 % over the last ten years. Thus, these became the best competitor to the traditional Si-SCs [6]. In recent years, alongside the modification, many other CMX$_3$ PSCs with super halogen anions and super alkali cations have also been described [7-10]. Also, the PSCs can be developed with the help of industrial chemicals and commonly found metals. The easy, reduced temperature and usual solution-processed, less priced spin or spray fabrication methods are very desirous for potential development of low cost PSCs. Many deposition methods like multi-step technique or one-step techniques that use anti-solvent washing, chemical vapor deposition (CVD) and thermal evaporation (TE) have been made and all these methods attained good superior polycrystalline perovskite films (pinhole free) having large grains that promote the production of PSCs with good efficiency; although, MAPbI$_3$ perovskites are moisture sensitive and also exhibit lower photo and stable thermal nature that restrains their industrial economic process [11-16].

2.   Advancement in solar cells by using up-conversion materials

The conversion of near-infrared (NIR) to visible (Vis) light by up-conversion (UC) technique is an anti-stokes process. Such materials produce high energy photon after the absorption of low energy photons. This process successively absorbs low energy NIR photons (more than one) to excite an electron to intermediary state and finally elevates the energy. The energy of emitted photon may ranges from NIR to ultraviolet visible (UV-Vis) region. The energy released in the Vis region creates the up-conversion phosphors to be helpful in luminescence implementation. In 1959, Bloembergen [17] primarily proposed this concept and in 1966, UC method was reported by Auzel [18]. Since then researchers have made several attempts to develop various up-conversion

Materials for Solar Cell Technologies I   Materials Research Forum LLC
Materials Research Foundations **88** (2021) 236-258   https://doi.org/10.21741/9781644901090-8

luminescent materials. Basically, these materials have prospective implementations in SCs, solid-state laser, and bio-imaging, hence are found to be widely investigated in the scientific field as is evident from publications in the field over the last ten years [19–21]. The three kinds of UC materials have been reported till now which are as follows:

- Transition metal-based up-conversion (TM-UC) materials
- Lanthanide (Ln)-based up-conversion (Ln-UC) materials
- Molecular up-conversion (MUC) materials

Currently, the production of different light colors has been extensively investigated in various UC materials doped with Lanthanide (Ln). These materials have also drawn interest due to potential implementations in various sectors like three dimensional (3D) displays, clean energy, fluorescence labelling, solid-state lasers, background lighting, etc. Owing to their 4f-4f electronic transitions, Ln-ions are ideal and result in durability, high quantum production and strong lines of emission [22, 23].

3.   Upconversion (UC) mechanisms

Upconversion luminescence is a process which is anti-stokes and optical in nature. In general, the UC nanostructures consisting of host matrices and dopants are important factors that determine the efficiency of luminescence. Dopants give a luminescence centre while the host matrices provide a transfer of energy platform between the dopants and direct them into positions that are optimal [24]. Regarding the UC that have a complex model, the transfer of energy dominates the UC luminescence processes [25]. Four mechanisms, namely, excited-state absorption (ESA), the energy transfer upconversion (ETU), cooperative energy transfer (CET), and photon avalanche effect (PAE) are responsible for the process of UC. ESA involves multistep excitation, whereas ETU involves successive energy transfer from a sensitizer to the activator and is more effective as compared to classical ESA. CET is different from ETU as it involves an emitting activator and a pair of sensitizer ions. In the fourth process, i.e. the PAE there is successive energy transfer but down-conversion type, while the UC step is because of ESA itself [26, 27].

4.   Absorber materials for solar cells

4.1   Indium gallium nitride (InGaN)-based SCs

Indium gallium nitride -based SCs are assumed to have poor crystalline quality [28]. The high-quality thick layers of InGaN are difficult to develop owing to the crystal defects formation and also due to the phase separation within the layer [29]. This is due to medium content of In required for high-quality thick layers of InGaN [30]. InGaN made SCs having distinct absorber designs like InGaN/GaN semi bulk [31], InGaN bulk, InGaN/GaN multiple quanta well (MQW) [32,33], InGaN/GaN superlattice demonstrates p-i-n (PIN) structures. [34]. These architectures showed a restrained In thickness and composition owing to the phase separation, strain, and 2D to 3D growth mechanisms. The excellent efficiency reported is 5.9 % for MQW structure having 46nm thickness of InGaN absorber and 19 % of indium content [35].

Hence the new concepts for designs are needed to modify completely relaxed, high crystal quality indium rich absorbers. Core-shell nanowire that contains InGaN MQW is prospective approach to remove lattice distinctions and the strain restraints and to get better In composition InGaN in contrast to planar InGaN thin films [36, 37]. Despite that, the best efficiency described is 0.3 % for 33 % indium content. In various reports, the MQW absorber is used as a substitute for thick, single-composition InGaN layers. Nano selective area growth (NSAG) technique has been initially suggested to improve and expand the thickness of a critical layer of pseudomorphic growth in distinct heterostructures [38]. NSAG has been known to be dislocation free for InGaN, even at high In content i.e., near about 40 % [39,40]. The architecture of InGaN nanopyramids (NP) for SC implementations provide rules to achieve an efficient $In_{0.3}Ga_{0.7}N$-NP made SCs. With the use of combined optical and electrical simulators, the electro-optical performance was calculated, analyzed and compared for planar and nano-designed with InGaN absorber SCs in air mass 1.5 G (AM 1.5 G) solar spectrum. The efficiency was two times higher that of the planar SCs with InGaN absorber [41].

4.2   Copper (Cu) based absorber material

Cu_2ZnSnS_4 (CZTS), is a p-type semiconductor, in contrast to other solar absorber materials (CIGS, CdTe, etc.) owing to the non-toxic nature and excess availability of the elements in it [42-46]. The semiconductors like CZTS have got a wide recognition due to non-linear optical properties, prospective implementation as solar-cell absorbers, and thermoelectric materials [47-53] and hence would cause more reduction in cost. The Cu_2ZnSnS_4 is suggested to have a bandgap in the range of 1.4-1.6 eV, which is suitable for a mono junction SC with a band 24 edge coefficient, which is more than 104 cm^{-1}. A conversion efficiency of 12.6 % has recently been attained in SCs based on $Cu_2ZnSn(S,Se)_4$. It is essential to explore the properties of some CZTS based semiconductors for improving the conversion efficiency. Later on, tin (Sn) substitutions by silicon (Si) have gained worldwide interest in their implementation as SC absorbers [54–60].

The electrical properties of two compounds of $Cu_2Zn(Sn_{0.9}Si_{0.1})S_4$, and $Cu_2Zn(Sn_{0.6}Si_{0.4})S_4$ (CZTSiS) were examined by measuring conductivity (80-300 K). The two compounds had been made from a ceramic route and are of Cu_2ZnSnS_4 and Cu_2ZnSiS_4 family. Their structural characteristics were evaluated by powder X-ray diffraction (XRD). At elevated temperatures i.e., 160-300 K, nearest-neighbour hopping (NNH) and band conduction dominated the electrical conductivity. Nevertheless, at the lower temperature (80-160 K) Mott law with the variable range hopping (VRH) process was predominated. Different characteristic parameters defining conductivity, like position length, hopping size, localization length, the density of state, characteristic temperature, and standard average hopping energy had been established and discussed. All parameters helped to understand the SCs behaviour based on CZTSiS absorber layers [61]. Due to the high degree of diffusivity, Cu^{+2} may move to the window layer, thereby weakening the system over the long term. Thus, Cu is an important doping material and with the help of controlled doping physical properties can be improved. With the goal of minimizing costs and reducing environmental Cd content, the effect of thermal air annealing on electrical, optical, and structural properties associated with electron beam surface topology formed by CdTe:Cu (5 %) thin films were reported, where the thickness of films is kept at 550 nm [62–67]. The standard thickness of the SC's absorber layer was 3-5 mm, and the controlled Cu doping was extremely potential for the development of thin films from CdTe and for concerned SC devices for achieving low ohmic contacts and resistivity. The structural study showed that annealing improvised the crystallinity as well as films had superior growth along (1 1 1) Zn blend cubic plane. The electrical current increased with voltage and decreased with annealing and showed almost ohmic behavior and decreased electrical conductivity. The surface of the films had hill and deep valley type topology, though higher ruggedness is observed for 300°C films. Transmission spectra displayed lower transmittance across the entire visible field, and maximum band distance (1.58 eV) was predicted for annealed films of 300°C. The study revealed that 300°C annealed CdTe:Cu films tend to become an effective choice for absorber layer functions [68]. To improvise the CZTS crystal surface for efficient, working p-n junctions, a perfect thermal treatment technique was developed for monograin powders, using a maintainable reactive gas process. The effect of thermal technique in sulphur (S) and tin disulphide (SnS_2) vapor was depending on the initial composition of CZTS(Se). The SC efficiency constantly improved on enhancing absorber material temperature from 823 to 973 K under 100 Torr of S vapour pressure (after annealing method). The maximum circuit current density (J_{sc}) = 18.4 mA cm^{-2} and open-circuit voltage (V_{oc}) = 720 mV were obtained in SnS_2 vapor for a device made of CZTS material [69]. Significant growth of the thin-film photovoltaic sector was possible after the development of semiconductor compounds such as CIGSe but the less abundance of

Materials for Solar Cell Technologies I   Materials Research Forum LLC
Materials Research Foundations **88** (2021) 236-258   https://doi.org/10.21741/9781644901090-8

indium alongside its high price resulted in the search for substitute materials. CZTS is one of the new potential materials having kesterite-type crystal structure, direct band-gap, and high absorptivity coefficient. In addition, all CZTS constituent elements are non-toxic and abundantly in the earth-crust, rendering it a good choice for photovoltaic thin-films. A vacuum quartz ampoule (at 1030 K) was employed to synthesize CZTS powder from its elements. To achieve the most favourable atomic stoichiometry for the compound, the raw mixture S-content in the ampoule was changed and optimized. Different techniques, like UV-Vis, raman scattering spectroscopy (RSS), energy dispersive analysis of x-ray (EDAXR), and XRD, were employed to analyze the synthesized powder. The analysis confirmed the kesterite-type structure and better atomic stoichiometry of the elements. The precise band-gap (approx. 1.45 eV) suggested CZTS powder as an absorber in SCs [70].

4.3   Graphene based absorber material

Due to the excellent absorption properties, metamaterials (MMs) may exhibit great prospective in various technical and scientific implementations. In the IR region, the absorption of gradient metasurface was 50 %, and 90 % with a single and double layer, respectively. A small distance of an absorber, which is dual-band tetrahertz, which is built on a couple of Au strip/dielectric layer, was developed. Solar absorber based on metasurface made by circular Au resonator was developed and demonstrated the absorption activities (AA) in 155-428 THz in the IR region. Further, the AA of 155-1595 THz was explored in IR, UV, and Vis regions.

A nanocomposite (plasmonic and biomimetic) was developed with a random broadband wavelength absorber. The absorption percentage of broadband achieved was 90 % from the UV to IR region. Au nanorods that are cross shaped have given a multiband resonance for absorption over the whole solar spectrum. The bulk-dirac semimetal (BDS) layer conductivity is altered by implementing a potential bias. Graphene-metal tunable frequency selective surface absorber was constructed using the algorithm random-hill climbing (RHC) to achieve the widest absorption bandwidth. The three different geometrical structures of plasmonic graphene-based absorber were characterized to achieve total unit cell absorption. Tetrahertz broadband (TB) absorber gave broad angle absorption in IR region with eight Au nanoresonators. In addition, to have insensitive polarization characteristics, TB absorber was designed with four layers of a metallic Au ring. MM light absorber blocked the solar radiation up to 80–90 %. Square-shaped MM absorber was inspected to give broadband absorption. Using the silicon-chromium-silica layer, the MM absorber was also developed for NIR and Vis spectrum having AA in 0.4-1.4 μm wavelength. MMs independent of polarization and ultra-broadband were developed with the help of ring structured resonator to gain absorption percentage of 95.5 % in every solar spectrum [71–84].

Graphene based solar absorber was developed by sandwiching the layer of graphene between resonator and dielectric layer, which resulted in the improvisation of dielectric layer absorption. The design with and without graphene layer were evaluated in terms of a uniform electric field, reflection, and absorption. Graphene layer design had given better AA from 100-1600 THz. The effects for the various physical design parameters like height of Au dielectric layer, width of Ag resonator and Au resonator as well as of total structure, were also analyzed. The increase in the width of Ag resonator and Au resonator, decreases the absorption whereas, enhancement in height of Au dielectric layer improves the absorption. The designed system was capable to absorb IR, UV, and Vis radiation more efficiently and hence was relevant for trapping light and harvesting of solar energy [85].

4.4 Sulphide based absorber materials

Binary semiconductor compounds that are composed of cheap elements have achieved worldwide recognition as potential light absorbing materials [86–88]. Several combinations of non-metal anions and metal cations give a variety of binary compounds. The most widely researched metal sulphide (MS) group have optoelectrical properties ideal for photovoltaic applications, are cheaper and obtained from easily manageable chemical sources, as well as used in versatile methods for the manufacture of thin films ranging from cheaper approaches to vacuum deposition techniques [89–94]. These binary sulphides were screened to those which consist of both excellent property of absorbing light in the film as well as significant PV effect at device level. The PV effects and optical absorption coefficient of MS consist of distinct valency like antimony sulphide (Sb_2S_3), bismuth sulphide (Bi_2S_3), copper sulphide (Cu_2S), iron sulphide (FeS_2), lead sulphide (PbS), silver sulphide (Ag_2S), and tin sulphide (SnS) were explored and attracted the attention of researches SnS and Sb_2S_3 are good prospective materials for economic usage, due to their abundance in the earth's crust, the longer stability and the non-stop improvisation in the efficiency. However, none of these layers has become a better player in the SC field due to lower device efficiency [95–106]. Hence, basic understanding and knowledge of the technical roadblocks is needed to investigate further in order to overcome these issues. In general, the following two basic steps are involved in photovoltaic energy conversion:

i. Light absorption to produce pairs of electron hole

ii. Isolation of the electron-hole pairs (EHPs) by the device structure before recombination.

The nature of the absorber film is one of the major components in establishing the success of both the measures [107–111]. Another critical factor in the interface configuration is the morphology of p-n junction. The mismatch of the lattice among junction partner materials, and absorber might result in the interfacial defect development at higher density thus causing the considerable interfacial recombination of the carriers [112].

4.4.1 Solar cells based on Sb_2S_3

The material was highly vulnerable to sulphur (S) loss in the preparation of crystalline film because of high S vapour pressure. The defects that resulted due to the deficiency of sulphur, caused considerable recombination of carrier in the device. Consequently, the S vacancy passivation was the main concern to overcome. The extrinsic selenium (Se) doping was considered to heal the defects through the incorporation of Se-atom at S vacancy in Sb_2S_3. The device made from above mentioned approach showed 4.7 % of the highest efficiency. The challenges of the unfavourable alignment of band in heterojunction Sb_2S_3/cadmium sulphide (CdS) and relatively broad band gap difference in Sb_2S_3 itself, however, are still some setbacks to be resolved.

4.4.2 Tin sulphide based SCs

Tin sulphide undergoes various challenges in preparing phase-pure absorber films, and unfavourable band alignment at the heterojunction of CdS/SnS. The matter concerning of the purity of phase issue was resolved by two procedures, namely atomic layer deposition (ALD) and thermal evaporation (TE) of SnS sources (pre-purified) had been exhibited as the best methods to coat devices with efficiencies near to 4 %. Due to the low rate of deposition and the route complexity, the development of a new kind of technique is needed in a simplistic approach. The band alignment issue was resolved by replacing CdS buffer layer by zinc (Zn) based substitutes which resulted in 4.54 % efficiency. Band alignment issue has been resolved by alternating CdS by Zn based materials to enhance the efficiency up to 4.54 % synthesized by atomic layer deposition. Zn buffer layers are also under improvement in the efficiency. This shows that there is need to study more on the formation of SnS homojunction SCs.

4.4.3 Sulphide based SCs

The heterojunction Cu_2S/CdS devices exhibited a significant competency above 10 % during the initial years of development. In addition, though, the deprivation of the operating performance of the device remained a technological challenge to be solved. While the photovoltaic effects of Ag_2S, FeS_2, and Bi_2S_3 from the photoelectrochemical

Materials for Solar Cell Technologies I   Materials Research Forum LLC
Materials Research Foundations **88** (2021) 236-258   https://doi.org/10.21741/9781644901090-8

cells have been shown, the compounds are yet the subject of study for SC applications. The increasing efficiency also expects both material understanding and its attributes and getting an optimal junction partner for ideal transfer of charge [113].

4.5   Metal nanoparticles (MNPs)

Metal nanoparticles (MNPs) disperse the solar light in an efficient manner at a wavelength of resonance where the dispersion and the absorption in the NPs is extremely affected by size, shape, volume portion of NPs in host medium, and composition of the material. The technique of transfer matrix and advanced analytical model based on the Mie theory was developed for the optimization of dispersing property of NPs as well as the simulation transfer of power from the MNPs layer i.e., Au, Ag, Cu and Aluminium (Al) to an absorber layer of n-GaAs. The performing efficiency of NPs were compared without the absorber layer and found better transfer of power of AlNPs derived structures in absorber layer than CuNPs, AuNPs, and AgNPs. The dispersion of light is a general phenomenon of interaction of light matter which has been researched widely in the laser field, biosensing and SCs etc [114–124]. The broadband dispersion and wide angle is highly advisable. The dispersion of light in the NP is an outcome of excitation and degradation of localized surface plasmons. The structure of SC device with NPs have been exhibited with improvised performance like AuNPs at the frontal side of Si-SCs and AgNPs at the frontal side of GaAs SCs decreases reflection of light which is incident on the surface and increases the device efficiency. The development in the performance of PV activated the advancement of NPs and their implementation in SCs [125, 126]. The optical property of AgNPs disperses above 97 % of the visible range of AM 1.5 solar spectrum. They have used Mie theory simulation and determined that a 30 % of the surface coverage of AgNPs of radius 105 nm, can disperses 98.9 % of photons in 380-820 nm [127, 128].

Conclusion

The chapter discussed the design of the photovoltaics and the remarkable advancement made earlier. There are many setbacks that need to be overcome before photovoltaic energy can be considered a significant part of the complete energy development.

i) Efficiencies are relatively smaller in contrast to those permitted by thermodynamics. As mentioned, photocells have higher theoretical limit of energy conversion, it is more than 33 % for mono junctions and very near to 90 % if relevant materials can be found. The efficiencies are highly affected by the material quality and are very sensitive to structural and chemical defects at lower concentrations also.

(ii) Material abundance is low, which is a major challenge and hence suitable substituent are needed to be investigated. Also, the processing cost is high which should be low. Low price can be attained using high quality materials.

(iii) Constraints of life cycle may become dominant as the developments reached their peaks where there were concerns regarding on supply chain. Several researches are directed towards these challenges and how to overcome them.

Most of the setbacks that were presented above will likely need new ways to develop. In modern era, rate of reaction and competition amongst substituent materials are enough to reach a visible penetration in the production of energy, a considerable contribution will lead to new performance levels. Performance does not only impact its cost; alongside it contributes towards the evolution of these materials. Due to above reason the chapter mostly concentrated on emerging strategies that can help to solve these problems.

Acknowledgements

Dr. Pallavi Jain is grateful to Dean and Deputy Registrar, SRM Institute of Science & Technology, Modinagar for the encouragement and support. Dr. Sapna Raghav is thankful to Department of Chemistry, Banasthali Vidyapith. Dr. Dinesh Kumar is thankful to DST, New Delhi, for the financial support extended. (Vide project sanction order F. No. DST/TM/WTI/WIC/2K17/124(C).

References

[1] M.A. Green, Y. Hishikawa, E.D. Dunlop, D.H. Levi, J. Hohl-Ebinger, M. Yoshita, A. W.Y. Ho-Baillie, Solar cell efficiency tables (Version 53), Prog. Photovolt. Res. Appl. 27 (2019) 3–12. https://doi.org/10.1002/pip.3102

[2] J.H. Noh, S.H. Im, J.H. Heo, T.N. Mandal, S.I. Seok, Chemical management for colorful, efficient, and stable inorganic–organic hybrid nanostructured solar cells, Nano Lett. 13 (2013) 1764–1769. https://doi.org/10.1021/nl400349b

[3] Q. Dong, Y. Fang, Y. Shao, P. Mulligan, J. Qiu, L. Cao, J. Huang, Electron-hole diffusion lengths>175 μm in solution-grown $CH_3NH_3PbI_3$ single crystals, Science 347 (2015) 967–970. https://doi.org/10.1126/science.aaa5760

[4] S. De Wolf, J. Holovsky, S.-J. Moon, P. L€oper, B. Niesen, M. Ledinsky, F.J. Haug, J.H. Yum, C. Ballif, Organometallic halide perovskites: sharp optical absorption edge and its relation to photovoltaic performance, J. Phys. Chem. Lett. 5 (2014) 1035–1039. https://doi.org/10.1021/jz500279b

[5] S.D. Stranks, G.E. Eperon, G. Grancini, C. Menelaou, M.J.P. Alcocer, T. Leijtens, L.M. Herz, A. Petrozza, H.J. Snaith, Electron-hole diffusion lengths exceeding 1 micrometer in an organometal trihalide perovskite absorber, Science 342 (2013) 341–344. https://doi.org/10.1126/science.1243982

[6] NREL, in: B.R.C.E. Chart (Ed.), Best Research-Cell Efficiency Chart, 2019

[7] T. Zhou, M. Wang, Z. Zang, L. Fang, Stable dynamics performance and high efficiency of ABX_3-type super-alkali perovskites first obtained by introducing H_5O_2 cation, Adv. Energy Mater. 9 (2019) 1900664. https://doi.org/10.1002/aenm.201900664

[8] M. Wang, Z. Zang, B. Yang, X. Hu, K. Sun, L. Sun, Performance improvement of perovskite solar cells through enhanced hole extraction: the role of iodide concentration gradient, Sol. Energ. Mater. Sol. Cell. 185 (2018) 117–123. https://doi.org/10.1016/j.solmat.2018.05.025

[9] T. Zhou, M. Wang, Z. Zang, X. Tang, L. Fang, Two-dimensional lead-free hybrid halide perovskite using superatom anions with tunable electronic properties, Sol. Energ. Mater. Sol. Cell. 191 (2019) 33–38. https://doi.org/10.1016/j.solmat.2018.10.021

[10] T. Zhou, Y. Zhang, M. Wang, Z. Zang, X. Tang, Tunable electronic structures and high efficiency obtained by introducing superalkali and superhalogen into AMX_3-type perovskites, J. Power. Sourc. 429 (2019) 120–126. https://doi.org/10.1016/j.jpowsour.2019.04.111

[11] E. Bi, H. Chen, F. Xie, Y. Wu, W. Chen, Y. Su, A. Islam, M. Gratzel, X. Yang, L. Han, Diffusion engineering of ions and charge carriers for stable efficient perovskite solar cells, Nat. Commun. 8 (2017) 15330. https://doi.org/10.1038/ncomms15330

[12] H. Wei, X. Zhao, Y. Wei, H. Ma, D. Li, G. Chen, H. Lin, S. Fan, K. Jiang, Flash evaporation printing methodology for perovskite thin films, NPG Asia Mater. 9 (2017) e395. https://doi.org/10.1038/am.2017.91

[13] Z. Wang, Z. Shi, T. Li, Y. Chen, W. Huang, Stability of perovskite solar cells: a prospective on the substitution of the A Cation and X Anion, Angew. Chem. Int. Ed. 56 (2017) 1190–1212. https://doi.org/10.1002/anie.201603694

[14] N.J. Jeon, J.H. Noh, Y.C. Kim, W.S. Yang, S. Ryu, S.I. Seok, Solvent engineering for high-performance inorganic–organic hybrid perovskite solar cells, Nat. Mater. 13 (2014) 897–903. https://doi.org/10.1038/nmat4014

[15] M. Becker, M. Wark, Recent progress in the solution-based sequential deposition of planar perovskite solar cells, Cryst. Growth Des. 18 (2018) 4790–4806. https://doi.org/10.1021/acs.cgd.8b00686

[16] J. Avila, C. Momblona, P.P. Boix, M. Sessolo, H.J. Bolink, Vapor-deposited perovskites: the route to high-performance solar cell production? Joule 1 (2017) 431–442. https://doi.org/10.1016/j.joule.2017.07.014

[17] N. Bloembergen, Solid state infrared quantum counters, Phys. Rev. Lett. 2 (1959) 84–85. https://doi.org/10.1103/PhysRevLett.2.84

[18] F. Auzel, Computeur quantique par transfert d'energie entre de Yb^{3+} a Tm^{3+} dans un tungstate mixte et dans verre germinate, C.R. Acad. SCi Paris 263 (1966) 819–821

[19] S.R. Bowman, L.B. Shaw, B.J. Feldman, J. Ganem, A 7-μm praseodymium based solid-state laser, IEEE J. Quan. Electr. 32 (1996) 646–649. https://doi.org/10.1109/3.488838

[20] M. Haase, H. Schafer, Upconverting nanoparticles, Angew. Chem. Int. Ed. 50 (2011) 5808–5829. https://doi.org/ 10.1002/anie.201005159

[21] H.Q. Wang, M. Batentschuk, A. Osvet, L. Pinna, C.J. Brabec, Rare earth ion doped up conversion materials for photovoltaic applications, Adv. Mater. 23 (2011) 2675–2680. https://doi.org/10.1002/adma.201100511

[22] J. Zhou, Q. Liu, W. Feng, Y. Sun, F. Li, Upconversion luminescent materials: advances and applications, Chem. Rev. 115 (2014) 395–465. https://doi.org/10.1021/cr400478f

[23] G. Liu, B. Jacquier (Eds.), Spectroscopic properties of rare earths in optical materials, Springer (2005). https://doi.org/10.1007/3-540-28209-2

[24] J.C. Boyer, L.A. Cuccia, J.A. Capobianco, Synthesis of colloidal upconverting $NaYF_4$: Er^{3+}/Yb^{3+} and Tm^{3+}/Yb^{3+} monodisperse nanocrystals, Nano Lett. 7 (2007) 847–852. https://doi.org/10.1021/nl070235+

[25] Q. Su, S. Han, X. Xie, H. Zhu, H. Chen, C.K. Chen, R.S. Liu, X. Chen, F. Wang, X. Liu, The effect of surface coating on energy migration-mediated upconversion, J. Am. Chem. Soc. 134 (2012) 20849–20857. https://doi.org/10.1021/ja3111048

[26] M. Puchalska, E. Zych, M. Sobczyk, A. Watras, P. Deren, Cooperative energy transfer in Yb^{3+}-Tb^{3+} co-doped $CaAl_4O_7$ upconverting phosphor, Mater. Chem. Phys. 156 (2015) 220–226. https://doi.org/10.1016/j.matchemphys.2015.03.004

Materials for Solar Cell Technologies I
Materials Research Foundations **88** (2021) 236-258

Materials Research Forum LLC
https://doi.org/10.21741/9781644901090-8

[27] A. Khare, A critical review on the efficiency improvement of upconversion assisted solar cells, J. Alloys Comp. 821 (2020) 153214.
https://doi.org/10.1016/j.jallcom.2019.153214

[28] A.G. Bhuiyan, K. Sugita, A. Hashimoto, A. Yamamoto, InGaN solar cells: present state of the art and important challenge, IEEE J. Photovolt. 2 (2012) 276–293.
https://doi.org/10.1109/JPHOTOV.2012.2193384

[29] Y. El Gmili, G. Orsal, K. Pantzas, T. Moudakir, S. Sundaram, G. Patriarche, J. Hester, A. Ahaitouf, J.P. Salvestrini, A. Ougazzaden, Multilayered InGaN/GaN structure vs. single InGaN layer for solar cell applications: a comparative study, Acta Mater. 61 (2013) 6587–6596. https://doi.org/10.1016/j.actamat.2013.07.041

[30] W. Zhao, L. Wang, J. Wang, Z. Hao, Y. Luo, Theoretical study on critical thicknesses of InGaN grown on (0001) GaN, J. Cryst. Growth 327 (2011) 202–204.
https://doi.org/10.1016/j.jcrysgro.2011.05.002

[31] M. Arif, W. Elhuni, J. Streque, S. Sundaram, S. Belahsene, Y. El Gmili, M. Jordan, X. Li, G. Patriarche, A. Slaoui, A. Migan, R. Abderrahim, Z. Djebbour, P.L. Voss, J.P. Salvestrini, A. Ougazzaden, Improving InGaN heterojunction solar cells efficiency using a semibulk absorber, Sol. Energ. Mater. Sol. Cell. 159 (2017) 405–411.
https://doi.org/10.1016/j.solmat.2016.09.030

[32] A. Mukhtarova, S. Valdueza-Felip, L. Redaelli, C. Durand, C. Bougerol, E. Monroy, J. Eymery, Dependence of the photovoltaic performance of pseudomorphic InGaN/GaN multiple-quantum-well solar cells on the active region thickness, Appl. Phys. Lett. 108 (2016) 161907. https://doi.org/10.1063/1.4947445

[33] C. Jiang, L. Jing, X. Huang, M. Liu, C. Du, T. Liu, X. Pu, W. Hu, Z.L. Wang, Enhanced solar cell conversion efficiency of InGaN/GaN multiple quantum wells by piezo-phototronic effect, ACS Nano 11 (2017) 9405–9412.
https://doi.org/10.1021/acsnano. 7b04935

[34] N.G. Young, R.M. Farrell, Y.L. Hu, Y. Terao, M. Iza, S. Keller, S.P. DenBaars, S. Nakamura, J.S. Speck, High performance thin quantum barrier InGaN/GaN solar cells on sapphire and bulk (0001) GaN substrates, Appl. Phys. Lett. 103 (2013) 173903.
https://doi.org/10.1063/1.4826483

[35] B.W. Liou, Design and fabrication of $In_xGa_{1-x}N$/GaN solar cells with a multiple-quantum-well structure on SiCN/Si(111) substrates, Thin Solid Films 520 (2011) 1084–1090. https://doi.org/10.1016/j.tsf.2011.01.086

[36] J.J. Wierer, Q. Li, D.D. Koleske, S.R. Lee, G.T. Wang, III-nitride core–shell nanowire arrayed solar cells, Nanotechnology 23 (2012) 194007. https://doi.org/10.1088/0957-4484/23/19/194007/meta

[37] Y. Dong, B. Tian, T.J. Kempa, C.M. Lieber, Coaxial group III−nitride nanowire photovoltaics, Nano Lett. 9 (2009) 2183–2187. https://doi.org/10.1021/nl900858v

[38] D. Zubia, S.H. Zaidi, S.R.J. Brueck, S.D. Hersee, Nanoheteroepitaxial growth of GaN on Si by organometallic vapor phase epitaxy, Appl. Phys. Lett. 76 (2000) 858–860. https://doi.org/10.1063/1.125608

[39] K. Young-Ho, K. Je-Hyung, G. Su-Hyun, K. Joosung, K. Taek, C. Yong-Hoon, Red emission of InGaN/GaN double heterostructures on GaN nanopyramid structures, ACS Photon. 2 (2015) 515–520. https://doi.org/10.1021/ph500415c

[40] S. Sundaram, Y. El Gmili, R. Puybaret, X. Li, P.L. Bonanno, K. Pantzas, G. Patriarche, P.L. Voss, J.P. Salvestrini, A. Ougazzaden, Nanoselective area growth and characterization of dislocation-free InGaN nanopyramids on AlN buffered Si(111) templates, Appl. Phys. Lett. 107 (2015) 113105. https://doi.org/10.1063/1.4931132

[41] W. El-Huni, A. Migan, Z. Djebbour, J. Salvestrini, A. Ougazzaden, High-efficiency indium gallium nitride/Si tandem photovoltaic solar cells modeling using indium gallium nitride semibulk material: monolithic integration versus 4-terminal tandem cells, Prog. Photovolt. Res. Appl. 24 (2016) 1436–1447. https://doi.org/10.1002/pip.2807

[42] M. Hamdi, B. Chrif, A. Lafond, B. Louati, C. Guillot-Deudon, F. Hle, Electrical properties of $Cu_2Zn(Sn_{1-x}Si_x)S_4$ (x=0.1, x=0.4) compounds for absorber materials in solar-cells, J. Alloys Comp. 643 (2015) 129–136. https://doi.org/10.1016/j.jallcom.2015.04.033

[43] S. Kahraman, S. Çetinkaya, H.A. Çetinkara, H.S. Guder, Effects of diethanolamine on sol-gel-processed Cu_2ZnSnS_4 photovoltaic absorber thin films, Mater. Res. Bull. 50 (2014) 165–171. https://doi.org/10.1016/j.materresbull.2013.10.043

[44] N.M. Shinde, R.J. Deokate, C.D. Lokhande, Properties of spray deposited Cu_2ZnSnS_4 (CZTS) thin films, J. Anal. Appl. Pyrolysis 100 (2013) 12–16. https://doi.org/10.1016/j.jaap.2012.10.018

[45] Y. Lin, S. Ikeda, W. Septina, Y. Kawasaki, T. Harada, M. Matsumura, Mechanistic aspects of preheating effects of electrodeposited metallic precursors on structural and photovoltaic properties of Cu_2ZnSnS_4 thin films, Sol. Energ. Mater. Sol. Cell. 120 (2014) 218–225. https://doi.org/10.1016/j.solmat.2013.09.006

[46] C.H. Ruan, C.C. Huang, Y.J. Lin, G.R. He, H.C. Chang, Y.H. Chen, Electrical properties of $Cu_xZn_ySnS_4$ films with different Cu/Zn ratios, Thin Solid Films 550 (2014) 525–529. https://doi.org/10.1016/j.tsf.2013.10.134

[47] J.W. Lekse, B.M. Leverett, C.H. Lake, J.A. Aitken, Synthesis, Physicochemical characterization and crystallographic twinning of Li_2ZnSnS_4, J. Solid State Chem. 181 (2008) 3217–3222. https://doi.org/10.1016/j.jssc.2008.08.026

[48] M.L. Liu, F.Q. Huang, L.D. Chen, I.W. Chen, A Wide-band-gap p-type thermoelectric material based on quaternary chalcogenides of Cu_2ZnSnQ_4 (Q=S, Se), Appl. Phys. Lett. 94 (2009) 202103. https://doi.org/10.1063/1.3130718

[49] J.W. Lekse, M.A. Moreau, K.L. McNerny, J. Yeon, P.S. Halasyamani, J.A. Aitken, Second-harmonic generation and crystal structure of the diamond-like semiconductors Li_2CdGeS_4 and Li_2CdSnS_4, Inorg. Chem. 48 (2009) 7516–7518. https://doi.org/10.1021/ic9010339

[50] H. Matsushita, T. Ochiai, A. Katsui, Preparation and characterization of CuZnGeSe thin films by selenization method using the Cu-Zn-Ge evaporated layer precursors, J. Cryst. Growth 275 (2005) 995–999. https://doi.org/10.1016/j.jcrysgro.2004.11.154

[51] R.A. Wibowo, E.S. Lee, B. Munir, K.H. Kim, Pulsed laser deposition of quaternary $Cu_2ZnSnSe_4$ thin films, Phys. Status Solidi A 204 (2007) 3373–3379. https://doi.org/10.1002/pssa.200723144

[52] S. Levcenco, D. Dumcenco, Y.S. Huang, E. Arushanov, V. Tezlevan, K.K. Tiong, C.H. Du, Absorption-edge anisotropy of Cu_2ZnSiQ_4 (Q=S,Se) quaternary compound semiconductors, J. Alloys Compd. 509 (2011) 4924–4928. https://doi.org/10.1016/j.jallcom.2011.01.169

[53] C. Sevik, T. Cagin, Ab initio study of thermoelectric transport properties of pure and doped quaternary compounds, Phys. Rev. B 82 (2010) 045202. https://doi.org/10.1103/PhysRevB.82.045202

[54] V. Kheraj, K.K. Patel, S.J. Patel, D.V. Shah, Synthesis and characterisation of Copper Zinc Tin Sulphide (CZTS) compound for absorber material in solar-cells, J. Cryst. Growth 362 (2013) 174–177. https://doi.org/10.1016/j.jcrysgro.2011.10.034

[55] M. Xie, D. Zhuang, M. Zhao, B. Li, M. Cao, J. Song, Fabrication of Cu_2ZnSnS_4 thin films using a ceramic quaternary target, J. Vac. 101 (2014) 146–150. https://doi.org/10.1016/j.vacuum.2013.08.001

[56] K. Ito, T. Nakazawa, Electrical and optical properties of stannite-type quaternary semiconductor thin films, J. Appl. Phys. 27 (1988) 2094–2097. https://doi.org/10.1143/JJAP.27.2094

[57] K. Tanaka, Y. Fukui, N. Moritake, H. Uchiki, Chemical composition dependence of morphological and optical properties of Cu_2ZnSnS_4 thin films deposited by sol-gel sulfurization and Cu_2ZnSnS_4 thin film solar cell efficiency, Sol. Energ. Mater Sol. Cell. 95 (2011) 838–842. https://doi.org/10.1016/j.solmat.2010.10.031

[58] J. Nelson, The physics of solar cells, Imperial College Press, London, (2003). https://doi.org/10.1142/p276

[59] M. Hamdi, A. Lafond, C. Guillot-Deudon, F. Hlel, M. Gargouri, S. Jobic, Crystal chemistry and optical investigations of the $Cu_2Zn(Sn,Si)S_4$ series for photovoltaic applications, J. Solid State Chem. 220 (2014) 232–237. https://doi.org/10.1016/j.jssc.2014.08.030

[60] W. Wang, M.T. Winkler, O. Gunawan, T. Gokmen, T.K. Todorov, Y. Zhu ,D.B. Mitzi, Device characteristics of CZTSSe thin-film solar cells with 12.6 % efficiency, Adv. Energy Mater. 4 (2014) 1301465–5. https://doi.org/10.1002/aenm.201301465

[61] W. El Hunia, S. Karrakchou, Nanopyramid-based absorber to boost the efficiency of InGaN solar cells, Sol. Energy 190 (2019) 93–103. https://doi.org/10.1016/j.solener. 2019.07.090

[62] Y. Shao, X. Li, L. Wu, D. Wang, Cu diffusion in CdTe detected by nano-metal-plasmonic enhanced resonant Raman scattering, J. Appl. Phys. 125 (2019) 013101. https://doi.org/10.1063/1.5051191

[63] J. Perrenoud, L. Kranz, C. Gretener, F. Pianezzi, S. Nishiwaki, S. Buecheler, A.N. Tiwari, A comprehensive picture of Cu doping in CdTe solar cells, J. Appl. Phys. 114 (2013) 174505. https://doi.org/10.1063/1.4828484

[64] F.D.M. Flores, J.G.Q. Galvan, A.G. Cervantes, J.S.A. Ceron, G.C. Puente, A.H. Hernandex, J.S. Salazar, M.D.L.L. Olvera, M.A.S. Aranda, M.Z. Torres, J.G.M. Alvarez, M.M. Lira, Physical properties of CdTe: Cu films grown at low temperature by pulsed laser deposition, J. Appl. Phys. 112 (2012) 113110. https://doi.org/10.1063/1.4768455

[65] M. Jhang, L. Qiu, W. Li, J. Zhang, L. Wu, L. Feng, Copper doping of MoO_x thin films for CdTe solar cells, Mater. Sci. Semicond. Process. 86 (2018) 49–57. https://doi.org/10.1016/j.mssp.2018.06.008

[66] A. Bosio, R. Ciprian, A. Lamperti, I. Rago, B. Ressel, G. Rosa, M. Stupar, E. Weschke, Interface phenomena between CdTe and ZnTe:Cu back contact, Sol. Energy 176 (2018) 186–193. https://doi.org/10.1016/j.solener.2018.10.035

[67] S. Kim, J. Jeon, J. Suh, J. Hong, T. Kim, K. Kim, S. Cho, Comparative study of Cu_2Te and Cu Back contact in CdS/CdTe solar cell, J. Korean Phys. Soc. 72 (2018) 780–785. https://doi.org/10.3938/jkps.72.780

[68] Himanshu, S.L. Patel, D. Agrawal, S. Chander, A. Thakur, M.S. Dhaka, Towards cost effective absorber layer to solar cells: Optimization of physical properties to Cu doped thin CdTe films, Mater. Lett. 254 (2019) 141–144. https://doi.org/10.1016/j.matlet.2019.07.037

[69] M. Kauk, K. Muska, M. Altosaar, J. Raudoja, M. Pilvet, T. Varema, K.Timmo, O. Volobujeva, Effects of sulphur and tin disulphide vapour treatments of $Cu_2ZnSnS(Se)_4$ absorber materials for monograin solar cells, Energy Procedia 10 (2011) 197–202. https://doi.org/10.1016/j.egypro.2011.10.177

[70] V. Kheraj, K.K. Patel, S.J. Patel, D.V. Shah, Synthesis and characterisation of Copper Zinc Tin Sulphide (CZTS) compound for absorber material in solar-cells, J. Cryst. Growth 362 (2013) 174–177. https://doi.org/10.1016/j.jcrysgro.2011.10.034

[71] M. Li, U. Guler, Y. Li, A. Rea, E.K. Tanyi, Y. Kim, N.A. Kotov, Plasmonic biomimetic nanocomposite with spontaneous subwavelength structuring as broadband Absorbers, ACS Energy Lett. 3 (2018) 1578–1583. https://doi.org/10.1021/acsenergy lett.8b00583

[72] B.X. Wang, C. Tang, Q. Niu, Y. He, T. Chen, Design of narrow discrete distances of dual-/triple-band terahertz metamaterial absorbers, Nanoscale Res. Lett. 14 (2019) 1–7. https://doi.org/10.1186/s11671-019-2876-3

[73] W. Guo, Y. Liu, T. Han, Ultra-broadband infrared metasurface absorber, Opt. Express 24 (2016) 20586–20592. https://doi.org/10.1364/OE.24.020586

[74] F. Xiong, J. Zhang, Z. Zhu, X. Yuan, S. Qin, Ultrabroadband, more than one order absorption enhancement in graphene with plasmonic light trapping, Sci. Rep. 5 (2015) 16998. https://doi.org/10.1038/srep16998

[75] D. Katrodiya, C. Jani, V. Sorathiya, S.K. Patel, Metasurface based broadband solar absorber, Opt. Mater. 89 (2019) 34–41. https://doi.org/10.1038/srep20347.

[76] Y. Jiang, W. Xinguo, J. Wang, J. Wang, Tunable terahertz absorber based on bulk-Dirac-semimetal metasurface, IEEE Photonics J. 10 (5) (2018) 1–7. https://doi.org/10.1109/JPHOT.2018.2866281

[77] A.D. Khan, A.D. Khan, S.D. Khan, M. Noman, Light absorption enhancement in trilayered composite metasurface absorber for solar cell applications, Opt. Mater. 84 (2018) 195–198. https://doi.org/10.1016/j.optmat.2018.07.009

[78] J. Li, Y. Chen, Y. Liu, Mathematical simulation of metamaterial solar cells, Adv. Appl. Math. Mech. 3 (6) (2011) 702–715. https://doi.org/10.4208/aamm.11-m1109

[79] A.K. Azad, W.J. Kort-Kamp, M. Sykora, N.R. Weisse-Bernstein, T.S. Luk, A.J. Taylor, H.T. Chen, Metasurface broadband solar absorber, Sci. Rep. 6 (2016) 20347. https://doi.org/10.1038/srep20347

[80] B.X. Wang, X. Zhai, G.Z. Wang, W.Q. Huang, L.L. Wang, Design of a four-band and polarization-insensitive terahertz metamaterial absorber, IEEE Photonics J. 7 (2015) 1–8. https://doi.org/10.1109/JPHOT.2014.2381633

[81] E.S. Torabi, A. Fallahi, A. Yahaghi, Evolutionary optimization of graphene-metal metasurfaces for tunable broadband terahertz absorption, IEEE Trans. Antennas Propag. 65 (2017) 1464–1467. https://doi.org/10.1109/TAP.2016.2647580

[82] B.X. Wang, L.L. Wang, G.Z. Wang, W.Q. Huang, X.F. Li, X. Zhai, Theoretical investigation of broadband and wide-angle terahertz metamaterial absorber, IEEE Photonics Technol. Lett. 26 (2014) 111–114. https://doi.org/10.1109/LPT.2013.2289299

[83] D. Hu, H.Y. Wang, Q.F. Zhu, Design of an ultra-broadband and polarization-insensitive solar absorber using a circular-shaped ring resonator, J. Nanophotonics 10 (2016) 026021. https://doi.org/10.1117/1.JNP.10.026021

[84] H. Deng, Z. Li, L. Stan, D. Rosenmann, D. Czaplewski, J. Gao, X. Yang, Broadband perfect absorber based on one ultrathin layer of refractory metal, Opt. Lett. 40 (2015) 2592–2595. https://doi.org/10.1364/OL.40.002592

[85] S.K. Patel, S. Charola, C. Jani, M. Ladumor, J. Parmar, T. Guo, Graphene-based highly efficient and broadband solar absorber, Opt. Mater. 96 (2019) 109330. https://doi.org/10.1016/j.optmat.2019.109330

[86] H.E. Suess, H.C. Urey, Abundances of the elements, Rev. Mod. Phys. 28 (1956) 53. https://doi.org/10.1103/RevModPhys.28.53

[87] P.D. Matthews, P.D. McNaughter, D.J. Lewis, P. O'Brien, Shining a light on transition metal chalcogenides for sustainable photovoltaics, Chem. Sci. 8 (2017) 4177–4187. https://doi.org/10.1039/C7SC00642J

[88] C. Wadia, A.P. Alivisatos, D.M. Kammen, Materials availability expands the opportunity for large-scale photovoltaics deployment, Environ. Sci. Technol. 43 (2009) 2072–2077. https://doi.org/10.1021/es8019534

[89] L. Yu, Y. Lv, G. Chen, X. Zhang, Y. Zeng, H. Huang, Y. Feng, A generally synthetic route to semiconducting metal sulfide nanocrystals by using corresponding metal powder and cysteine as metallic and sulfuric sources, respectively, Inorg. Chim. Acta 376 (2011) 659–663. https://doi.org/10.1016/j.ica.2011.06.046

[90] Z. Zhuang, X. Lu, Q. Peng, Y. Li, A facile "dispersion–decomposition" route to metal sulfide nanocrystals, Chem. Eur J. 17 (2011) 10445–10452. https://doi.org/10.1002/chem.201101145

[91] R. Scheer, H.W. Schock, Chalcogenide photovoltaics: physics, technologies, and thin film devices, John Wiley & Sons, 2011. https://doi.org/10.1002/9783527633708

[92] D. Abou-Ras, T. Kirchartz, U. Rau, Advanced characterization techniques for thin film solar cells, John Wiley & Sons, 2016. https://doi.org/10.1002/9783527636280

[93] N.P. Dasgupta, X. Meng, J.W. Elam, A.B. Martinson, Atomic layer deposition of metal sulfide materials, Acc. Chem. Res. 48 (2015) 341–348. https://doi.org/10.1021/ar500360d

[94] S.M. Ho, T. Anand, A review of chalcogenide thin films for solar cell applications, Indian J. Sci. Technol. 8 (2015) 67499. https://doi.org/10.17485/ijst/2F2015/2Fv8i12/2F67499

[95] H. Noguchi, A. Setiyadi, H. Tanamura, T. Nagatomo, O. Omoto, Characterization of vacuum-evaporated tin sulfide film for solar cell materials, Sol. Energ. Mater. Sol. Cell. 35 (1994) 325–331. https://doi.org/10.1016/0927-0248(94)90158-9

[96] H. Pathan, P. Salunkhe, B. Sankapal, C. Lokhande, Photoelectrochemical investigation of Ag_2S thin films deposited by SILAR method, Mater. Chem. Phys. 72 (2001) 105–108. https://doi.org/10.1016/S0254-0584(01)00319-4

[97] Y. Guo, H. Lei, B. Li, Z. Chen, J. Wen, G. Yang, G. Fang, Improved performance in Ag_2S/P_3HT hybrid solar cells with a solution processed SnO_2 electron transport layer, RSC Adv. 81 (2016) 77701-77708. https://doi.org/10.1039/C6RA19590C

[98] D.H. Yeon, B.C. Mohanty, C.Y. Lee, S.M. Lee, Y.S. Cho, High-efficiency double absorber PbS/CdS heterojunction solar cells by enhanced charge collection using a ZnO nanorod array, ACS Omega 2 (2017) 4894–4899. https://doi.org/10.1021/acsomega.7b00999

[99] O. Agnihotri, B. Gupta, R. Thangaraj, $Cd_{1-x}Zn_x$/PbS heterojunctions prepared by spray pyrolysis, Solid State Electron. 22 (1979) 218–220. https://doi.org/10.1016/0038-1101(79)90118-7

[100] P. Sinsermsuksakul, L. Sun, S.W. Lee, H.H. Park, S.B. Kim, C. Yang, R.G. Gordon, Overcoming efficiency limitations of SnS-based solar cells, Adv. Energy Mater. 4 (2014) 1400496. https://doi.org/10.1002/aenm.201400496

[101] O. Savadogo, K. Mandal, Characterizations of antimony tri-sulfide chemically deposited with silicotungstic acid, J. Electrochem. Soc. 139 (1992) L16–L18. https://doi.org/10.1149/1.2069211

[102] R.R. Ahire, R.P. Sharma, Photoelectrochemical characterization of Bi_2S_3 thin films deposited by modified chemical bath deposition, Indian J. Eng. Mater. Sci. 13 (2006) 140–144

[103] A. Kirkeminde, R. Scott, S. Ren, All inorganic iron pyrite nano-heterojunction solar cells, Nanoscale 24 (2012) 7649–7654. https://doi.org/10.1039/C2NR32097E

[104] S. Yuan, H. Deng, X. Yang, C. Hu, J. Khan, W. Ye, J. Tang, H. Song, Postsurface selenization for high performance Sb_2S_3 planar thin film solar cells, ACS Photonics 4 (2017) 2862–2870. https://doi.org/10.1021/acsphotonics.7b00858

[105] A. Ennaoui, S. Fiechter, C. Pettenkofer, N. Alonso-Vante, K. Buker, M. Bronold, C. Hpfner, H. Tributsch, Iron disulfide for solar energy conversion, Sol. Energ. Mater. Sol. Cell. 29 (1993) 289–370. https://doi.org/10.1016/0927-0248(93)90095-K

[106] R. Mane, B. Sankapal, C. Lokhande, Photoelectrochemical (PEC) characterization of chemically deposited Bi_2S_3 thin films from non-aqueous medium, Mater. Chem. Phys. 60 (1999) 158–162. https://doi.org/10.1016/S0254-0584(99)00099-1

[107] A. Collord, H. Xin, H. Hillhouse, Combinatorial exploration of the effects of intrinsic and extrinsic defects in $Cu_2ZnSn(S,Se)_4$, IEEE J. Photovolt. 5 (2015) 288–298. https://doi.org/10.1109/JPHOTOV.2014.2361053

[108] M. Bohm, R. Kern, H. Wagemann, The Influence of Grain-Boundary Recombination and Grain Size on the I (V)-characteristics of polycrystalline silicon

solar cells. Fourth EC Photovoltaic Solar Energy Conference, Springer, (1982) 516–521. https://doi.org/10.1007/978-94-009-7898-0_84

[109] J.S. Park, S. Kim, Z. Xie, A. Walsh, Point defect engineering in thin-film solar cells, Nat. Rev. Mater. 3 (2018) 194–210. https://doi.org/10.1038/s41578-018-0026-7

[110] L.L. Kazmerski, The effects of grain boundary and interface recombination on the performance of thin-film solar cells, Solid State Electron. 21 (1978) 1545–1550. https://doi.org/10.1016/0038-1101(78)90239-3

[111] J. Just, C.M. Sutter-Fella, D. Lutzenkirchen-Hecht, R. Frahm, S. Schorr, T. Unold, Secondary phases and their influence on the composition of the kesterite phase in CZTS and CZTSe thin films, Phys. Chem. Chem. Phys. 18 (2016) 15988–15994. https://doi.org/10.1039/C6CP00178E

[112] D.G. Moon, S. Rehan, D.H. Yeona, S.M. Lee, S.J. Parka, S.J. Ahn, Y.S. Choa, A review on binary metal sulfide heterojunction solar cells, Sol. Energ. Mater Sol. Cell. 200 (2019) 109963. https://doi.org/10.1016/j.solmat.2019.109963

[113] O. Popov, A. Zilbershtein, D. Davidov, Random lasing from dye-gold nanoparticles in polymer films: enhanced gain at the surface-plasmon-resonance wavelength, Appl. Phys. Lett. 89 (2006) 191116. https://doi.org/10.1063/1.2364857

[114] E. Heydari, R. Flehr, J. Stumpe, Influence of spacer layer on enhancement of nanoplasmon-assisted random lasing, Appl. Phys. Lett. 102 (2013) 133110. https://doi.org/10.1063/1.4800776

[115] J. Ziegler, M. Djiango, C. Vidal, C. Hrelescu, T.A. Klar, Gold nanostars for random lasing enhancement, Opt. Exp. 23 (2015) 15152–15159. https://doi.org/10.1364/OE.23.015152

[116] T.L. Temple, G.D.K. Mahanama, H.S. Reehal, D.M. Bagnall, Influence of localized surface plasmon excitation in silver nanoparticles on the performance of silicon solar cells. Sol. Energ. Mater Sol. Cell. 93 (2009) 1978–1985. https://doi.org/10.1016/j.solmat.2009.07.014

[117] H.A. Atwater, A. Polman, Plasmonics for improved photovoltaic devices, Nat. Mater. 9 (2010) 205–213. https://doi.org/ 10.1038/nmat2629

[118] Z. Ouyang, X. Zhao, S. Varlamov, Y. Tao, J. Wong, S. Pillai, Nanoparticle-enhanced light trapping in thin-film solar cells, Prog. Photovolt. 19 (2011) 917–926. https://doi.org/10.1002/pip.1135

Materials for Solar Cell Technologies I
Materials Research Foundations **88** (2021) 236-258

Materials Research Forum LLC
https://doi.org/10.21741/9781644901090-8

[119] W. Liu, X. Wang, Y. Li, Z. Geng, F. Yang, J. Li, Surface plasmon enhanced GaAs thin film solar cells, Sol. Energ. Mater Sol. Cell. 95 (2011) 693-698. https://doi.org/10.1016/j.solmat.2010.10.004

[120] L. Hong, Rusli, X. Wang, H. Zheng, L. He, X. Xu, H. Wang, H. Yu, Design principles for plasmonic thin film GaAs solar cells with high absorption enhancement, J. Appl. Phys. 112 (2012) 054326. https://doi.org/10.1063/1.4749800

[121] A.A. Miskevich, V.A. Loiko, Light absorption by a layered structure of silicon particles as applied to the solar cells: theoretical study, J. Quant. Spectrosc. Radiat. Transf. 146 (2014) 355–364. https://doi.org/10.1016/j.jqsrt.2013.12.008.

[122] A.J. Haes, R.P. Duyne, A unified view of propagating and localized surface plasmon resonance biosensors, Anal. Bioanal. Chem. 379 (2004) 920–930. https://doi.org/10.1007/s00216-004-2708-9

[123] K.A Willets, R.P. Van Duyne, Localized surface plasmon resonance spectroscopy and sensing, Annu. Rev. Phys. Chem. 58 (2007) 267–297. https://doi.org/10.1146/annurev.physchem.58.032806.104607

[124] C.S. Kealley, M.D. Arnold, A. Porkovich, M.B. Cortie, Sensors based on monochromatic interrogation of a localized surface plasmon resonance, Sens. Actuators B Chem. 148 (2010) 34–40. https://doi.org/10.1016/j.snb.2010.05.023

[125] F.J. Beck, A. Polman, K.R. Catchpole, Tunable light trapping for solar cells using localized surface plasmons, J. Appl. Phys. 105 (2009) 114310. https://doi.org/10.1063/1.3140609

[126] D.M. Schaadt, B. Feng, E.T. Yu, Enhanced semiconductor optical absorption via surface plasmon excitation in metal nanoparticles, Appl. Phys. Lett. 86 (2005) 063106. https://doi.org/10.1063/1.1855423g

[127] K. Nakayama, K Tanabe, H.A. Atwater, Plasmonic nanoparticle enhanced light absorption in GaAs solar cell, Appl. Phys. Lett. 93 (2008) 121904. https://doi.org/10.1063/1.2988288

[128] T.L. Temple, D.M. Bagnal, Broadband scattering of the solar spectrum by spherical metal nanoparticles, Prog. Photovolt. 21 (2013) 600–611. https://doi.org/10.1002/pip.1237

Keyword Index

Absorber Layer 236

Carbon Nanomaterials 86

Carbon 62

Carbonaceous Hollow Nanostructures 129

Device 1

Dye-Sensitized Solar Cells (DSSCs) ... 29

Efficiency 1

Energy Materials 176

Energy Transfer Process 176

Environmental Monitoring 62

Food Safety 62

Fullerene Acceptors 176

Graphene 29, 62

Hollow Nanostructures 129

Indium Tin Oxide 86

Low Band Gap 176

Metal Mesh 86

Metal Nanowires 86

Metallic Electrodes 129

Monocrystalline 148

Multijunction 1

Nanocrystalline 1

Nanomaterials 62

Organic Solar Cells (OPV) 29, 176

PCE 148

Perovskite Solar Cells (PSCs) 29

Perovskite Solar Cells 176

Perovskites 1

Photo Conversion Efficiency 176

Photovoltaic Cell 236

Photovoltaic Parameters 176

Photovoltaic 1, 148, 129

p-n Junction 236

Poly (3 4-ethylene dioxythiophene): poly (styrenesulfonate) PEDOT: PSS 86

Power Conversion Efficiency (PCE) 1

Quantum Dot Sensitized Solar Cell (QDSSC) 1

Quantum Dots 1

Reduced Graphene Oxide 62

Schottky Junction 29

Shockely-Queisser 148

Shockley-Queisser (SQ) Limit 1

Transparent Conducting Electrodes 86

About the Editors

Dr. Inamuddin is working as Assistant Professor at the Department of Applied Chemistry, Aligarh Muslim University, Aligarh, India. He obtained Master of Science degree in Organic Chemistry from Chaudhary Charan Singh (CCS) University, Meerut, India, in 2002. He received his Master of Philosophy and Doctor of Philosophy degrees in Applied Chemistry from Aligarh Muslim University (AMU), India, in 2004 and 2007, respectively. He has extensive research experience in multidisciplinary fields of Analytical Chemistry, Materials Chemistry, and Electrochemistry and, more specifically, Renewable Energy and Environment. He has worked on different research projects as project fellow and senior research fellow funded by University Grants Commission (UGC), Government of India, and Council of Scientific and Industrial Research (CSIR), Government of India. He has received Fast Track Young Scientist Award from the Department of Science and Technology, India, to work in the area of bending actuators and artificial muscles. He has completed four major research projects sanctioned by University Grant Commission, Department of Science and Technology, Council of Scientific and Industrial Research, and Council of Science and Technology, India. He has published 176 research articles in international journals of repute and nineteen book chapters in knowledge-based book editions published by renowned international publishers. He has published 115 edited books with Springer (U.K.), Elsevier, Nova Science Publishers, Inc. (U.S.A.), CRC Press Taylor & Francis Asia Pacific, Trans Tech Publications Ltd. (Switzerland), IntechOpen Limited (U.K.), Wiley-Scrivener, (U.S.A.) and Materials Research Forum LLC (U.S.A). He is a member of various journals' editorial boards. He is also serving as Associate Editor for journals (Environmental Chemistry Letter, Applied Water Science and Euro-Mediterranean Journal for Environmental Integration, Springer-Nature), Frontiers Section Editor (Current Analytical Chemistry, Bentham Science Publishers), Editorial Board Member (Scientific Reports-Nature), Editor (Eurasian Journal of Analytical Chemistry), and Review Editor (Frontiers in Chemistry, Frontiers, U.K.) He is also guest-editing various special thematic special issues to the journals of Elsevier, Bentham Science Publishers, and John Wiley & Sons, Inc. He has attended as well as chaired sessions in various international and national conferences. He has worked as a Postdoctoral Fellow, leading a research team at the Creative Research Initiative Center for Bio-Artificial Muscle, Hanyang University, South Korea, in the field of renewable energy, especially biofuel cells. He has also worked as a Postdoctoral Fellow at the Center of Research Excellence in Renewable Energy, King Fahd University of Petroleum and Minerals, Saudi Arabia, in the field of polymer electrolyte membrane fuel cells and computational fluid dynamics of polymer electrolyte membrane fuel cells. He is a life member of the Journal of the Indian

Chemical Society. His research interest includes ion exchange materials, a sensor for heavy metal ions, biofuel cells, supercapacitors and bending actuators.

Dr. Tauseef Ahmad Rangreez is working as a postdoctoral fellow at National Institute of Technology, Srinagar, India. He completed his Ph.D in Applied Chemistry, from Aligarh Muslim University, Aligarh, India on the topic "Development of Nanostructure Organic-Inorganic Composite Materials based Sensors for Inorganic Pollutants". He worked as a Project Fellow under the UGC Funded Research Project entitled "Development of Nanostructured Conductive Organic Inorganic Composite Materials based sensors Functionalities for Organic and Inorganic Pollutants". He completed his Masters in Chemistry from Jamia Hamdard, New Delhi. He has published several research articles of international repute. He has edited books with Springer and Materials Science Forum LLC, U.S.A. His research interest includes ion exchange chromatography, development of nanocomposite sensors for heavy metals and biosensors.

Dr. Mohd Imran Ahamed received his Ph.D degree on the topic "Synthesis and characterization of inorganic-organic composite heavy metals selective cation-exchangers and their analytical applications", from Aligarh Muslim University, Aligarh, India in 2019. He has published several research and review articles in the journals of international recognition. Springer (U.K.), Elsevier, CRC Press Taylor & Francis Asia Pacific and Materials Research Forum LLC (U.S.A). He has completed his B.Sc. (Hons) Chemistry from Aligarh Muslim University, Aligarh, India, and M.Sc. (Organic Chemistry) from Dr. Bhimrao Ambedkar University, Agra, India. He has co-edited more than 20 books with Springer (U.K.), Elsevier, CRC Press Taylor & Francis Asia Pacific and Materials Research Forum LLC (U.S.A) and Wiley-Scrivener, (U.S.A.). His research work includes ion-exchange chromatography, wastewater treatment, and analysis, bending actuator and electrospinning.

Dr. Rajender Boddula is currently working with Chinese Academy of Sciences-President's International Fellowship Initiative (CAS-PIFI) at National Center for Nanoscience and Technology (NCNST, Beijing). He obtained Master of Science in Organic Chemistry from Kakatiya University, Warangal, India, in 2008. He received his Doctor of Philosophy in Chemistry with the highest honours in 2014 for the work entitled "Synthesis and Characterization of Polyanilines for Supercapacitor and Catalytic Applications" at the CSIR-Indian Institute of Chemical Technology (CSIR-IICT) and Kakatiya University (India). Before joining National Center for Nanoscience and Technology (NCNST) as CAS-PIFI research fellow, China, worked as senior research associate and Postdoc at National Tsing-Hua University (NTHU, Taiwan) respectively in the fields of bio-fuel and CO_2 reduction applications. His academic honors include University Grants Commission National Fellowship and many merit scholarships, study-abroad fellowships from Australian Endeavour Research Fellowship, and CAS-PIFI. He has published many scientific articles in international peer-reviewed journals and has authored around twenty book chapters, and he is also serving as an editorial board member and a referee for reputed international peer-reviewed journals. He has published edited books with Springer (UK), Elsevier, Materials Science Forum LLC (USA), Wiley-Scrivener, (U.S.A.) and CRC Press Taylor & Francis group. His specialized areas of research are energy conversion and storage, which include sustainable nanomaterials, graphene, polymer composites, heterogeneous catalysis for organic transformations, environmental remediation technologies, photoelectrochemical water-splitting devices, biofuel cells, batteries and supercapacitors.